Studies in Industry and Society
Glenn Porter, General Editor

*Published with the Assistance
of the Hagley Museum and Library*

Control through
Communication

Control through Communication

The Rise of System in American Management

JoAnne Yates

The Johns Hopkins University Press
Baltimore and London

Johns Hopkins Paperbacks edition, 1993

The Johns Hopkins University Press
2715 North Charles Street
Baltimore, Maryland 21218-4319
The Johns Hopkins Press Ltd., London

Library of Congress Cataloging-in-Publication Data
Yates, JoAnne, 1951–
Control through communication : the rise of system in American
management / JoAnne Yates.
   p. cm. — (Studies in industry and society)
   Bibliography: p.
   Includes index.
   ISBN 0-8018-3757-X (alk. paper) ISBN 0-8018-4613-7 (pbk.)
   1. Communication in management. 2. Communication in management—
United States—History—Case studies. I. Title. II. Series.
HD30.3.Y38   1989
658.4'5—dc19   88-13745

A catalog record for this book is available from the British Library.

Contents

Illustrations

Foreword

JoAnne Yates's *Control through Communication* is an important addition to the Industry and Society series of the Johns Hopkins University Press and the Hagley Museum and Library. By now, informed Americans are well aware that business has undergone a revolution in the last century. The modern corporation emerged first in the railroads, the great transportation enterprises of the nineteenth century, and then spread into manufacturing, retailing, finance, and the entire range of industries that contribute to the economy of the United States. These large, complex, hierarchical organizations expanded across the face of the nation and then around the globe. Far more than any previous bureaucracies within business, the military, government, and religion, they required communication on an unprecedented scale and at unparalleled speeds. Only through such communication could managers have any hope of coordinating the many physically separated individuals and activities required to make the modern corporation work.

Anyone who has worked in even a small organization knows how hard it is to make all those who need to know aware of policies, data, and changes. A set of formal and informal networks arises to facilitate such information flows, but they always work imperfectly. As business struggled to acquire, record, and disperse data and decisions, a rising tide of technical and organizational innovations appeared, all designed to improve communication and control in complex organizations.

In this book JoAnne Yates has documented and analyzed the introduction and adoption of these new technologies and organizational responses. Her technique of comparative cases studies is a fruitful one. By examining in detail the history of three individual businesses—The Illinois Central Railroad, Scovill Manufacturing Company, and E. I. Du Pont de Nemours & Company—she is able to understand the human and organizational contexts in which change came to the modern economy. By simultaneously analyzing the coming of the new office technologies and the rise of systematic management, she is able to place her case studies in the larger setting of the business system as a whole.

The study's findings are fascinating, both as history and as a lens through which to view the more recent evolution of business. This is not a story of technological determinism, in which advances appeared and were quickly and widely adopted because they obviously represented progress. Neither is it a tale of the inevitable triumph of large-scale, impersonal forces in history, in which accumulating chaos called forth timely and logical responses. This story is somewhat less tidy, but still compelling. Except in extreme circumstances, it appears that firms could function for long periods in relatively inefficient ways. Protected in one way or another from harsh, immediately pressing competitive forces, businesses (and other organizations) perhaps have been governed largely by inertia. Only the introduction of new and vigorous leadership, the intervention of outside elements such as government, or life-threatening competition seem to have been sufficient to cause a change in direction. These are case studies in which active, creative managers made real differences for their firms. It is very good indeed to have the first thoroughly researched, well-documented analysis of the process through which the complex machinery of modern business communication and control came into being. The result is a significant contribution to the history of business, of technology, of communication, and of management. It is a pleasure to add such a book to this series.

This volume is one, like David Hounshell's contribution to the series, *From the American System to Mass Production*, that addresses changes that affected virtually the whole of the American economy. We continue to seek such studies as well as those with a focus on the Mid-Atlantic states. This dual emphasis mirrors the nature of the research collections at Hagley and the interests and programs of our Center for the History of Business, Technology, and Society.

The Hagley Museum and Library is especially pleased to be associated with the publication of *Control through Communication* because Professor Yates did some of her research with the help of a grant-in-aid from our institution. She spent much time at Hagley, and our staff enjoyed the opportunity to work with her and to assist her. As many of the notes and illustrations in this volume attest, Hagley's research collections played an important part in Professor Yates's research.

In my own work with her as general editor for this series, I was much impressed with Professor Yates's scholarship. She has incorporated prodigious research into this study, and her analysis and interpretations have always been keen. In addition, she has the ability, rare among authors, to edit her own prose mercilessly and skillfully. The successive versions of this manuscript yielded increasingly cogent narrative and more forceful interpretation. The result of JoAnne Yates's labors is, I am confident, a major work that does much to explain the means by which modern society has learned to control more and more disparate undertakings through improved systems of communication.

The world could not have developed in anything like its present form without the human and technical controls made possible by the history Yates analyzes. As has so often been the case with organizational and technological changes since the rise of modern science and the beginnings of the Industrial Revolution, our collective challenge is always to see that our awesome powers are used as positively as they can be.

GLENN PORTER, *Director*
Hagley Museum and Library

Acknowledgments

This book is the culmination of a long period of research, during which I have incurred many debts. The encouragement of Alfred D. Chandler, Jr., in the early stages of the project meant a great deal to me. Daniel Nelson and Glenn Porter both read drafts of the book and provided invaluable comments, suggestions, and encouragement. My colleague and friend Harriet Ritvo, a source of support throughout the project, read parts of a late draft of the book and offered useful suggestions for my final revisions. Loretta Caira bore with me in my seemingly endless revisions of the manuscript, entering set after set of changes on the computer at an amazing speed. I thank all of these people.

The archival community has provided welcome enthusiasm and interest as well as invaluable research assistance. Helen Samuels, archivist at the Massachusetts Institute of Technology, put me in touch with this community in the first place. A summer spent among archivists at the University of Michigan's Bentley Historical Library in a Mellon seminar led by Francis X. Blouin, Jr., proved stimulating to my research. I wish to thank the staffs at the three archives I used extensively in writing my case studies: the Newberry Library, the Manuscripts and Archives Department at the Baker Library of Harvard Business School, and the Hagley Museum and Library. Florence Lathrop at the Baker Library was supportive and helpful throughout my use of the Scovill Manufacturing Company records. My debts at the Hagley Library, where I spent the longest time, are enormous. The entire staff helped make my stays productive and enjoyable. In particular, Richard Williams and the late Betty-Bright Low were supportive and helpful in my early stay there. At a later stage, Michael Nash not only facilitated my work on the Du Pont materials but also encouraged me to approach Glenn Porter, director of the Hagley Museum and Library, about submitting the manuscript to his series at the Johns Hopkins University Press. It has indeed been a privilege to work with Glenn.

Everyone associated with the Johns Hopkins University Press has been very helpful. I have especially appreciated the support of Henry Tom, executive editor of the press, throughout the review and publication process.

My work on the book was facilitated by research funds from several sources. An Old Dominion Fellowship from MIT's Department of Humanities allowed me a term off for research early in the project. A grant from the Hagley Foundation helped support me during my longest stay at the Hagley Library. A Mellon Fellowship supported my participation in the archival seminar at the Bentley Historical Library. Summer research funds from MIT's Sloan School of Management provided aid in later stages of the project.

Last, but certainly far from least, is my family. My parents cheered me on throughout the project. My husband, Craig Murphy, provided intellectual and emotional support at every stage of my study. Our discussions helped me refine my thinking, and the knowledge that he was working in the next room helped me through many 5:00 A.M. writing sessions.

Introduction

The Transformation of Internal Communication

In early nineteenth-century America, business enterprises were generally small, family affairs. The internal operations of a firm were controlled and coordinated through informal communication, principally by word of mouth except when letters were needed to span distance. Today, massive corporations are held together by networks of communication up, down, and across hierarchies. Employees at all levels read and write countless memoranda, letters, and reports, and spend untold hours in meetings.

When and how did this transformation occur? It began with the railroads in the mid-nineteenth century and spread to manufacturing firms beginning around 1880. During the years from 1850 to 1920, a new philosophy of management based on system and efficiency arose, and under its impetus internal communication came to serve as a mechanism for managerial coordination and control of organizations. The informal and primarily oral mode of interaction gave way to a complex and extensive formal communication system depending heavily on written documents of various sorts. Quill pens and bound volumes gave way to typewriters, stencil duplicators, and vertical files that aided in creating, copying, storing, and retrieving documents. These technologies also affected the function and form of communication within the firm. New types or genres of communication developed, with orders, reports, memoranda, and meetings evolving to suit their managerial and technological contexts.

The transformation in the nature, extent, and role of communication within American firms was substantially complete by the end of World War I. Since then, of course, communication—and especially its technology—has continued to evolve. Nevertheless, by 1920 the major elements of the modern communication system and its role as a tool of managerial control had been established.

In spite of the critical role communication plays within organizations, its evolution has received little attention. Business historians have studied the development of various specific business functions, such as accounting, and the communication flows related to those functions.[1] The internal communication system as a whole, however, links all the functions. The historian who has had the most to say about developments in internal information flows is Alfred D. Chandler, Jr. In his pathbreaking works on the evolution of the modern corporation and modern management, *Strategy and Structure: Chapters in the History of the American Industrial Enterprise* and *The Visible Hand: The Managerial Revolution in American Business,*[2] he frequently refers to increasing and changing needs for information and analysis, and consequently for record keeping and reporting, in growing businesses of the period.[3] His study of the evolution of American management has provided me with a framework; my debt to his work is evident throughout this book. Nevertheless, his focus is not on the communication system itself, and thus he leaves many aspects of it unexplored.

Historians of technology have examined the invention and dissemination of some key communication technologies, such as the typewriter or telephone, but have not pursued the evolving interactions between communication technology and managerial uses and forms of communication within firms.[4] Recently, James R. Beniger has looked more generally at the development of information technologies as a mechanism of control during the same period.[5] His scope is extremely broad, however, including everything from gyroscopic regulators for controlling machinery to advertising for controlling consumer buying. As a result of the broad focus, he neglects specific interactions between communication or information technology and managerial needs. Thus, the role played by technology in the evolution of the communication system within firms still needs scholarly attention.

Finally, those who study the history of business communication have generally studied earlier periods, ignoring the genres of communication that emerged in the late nineteenth and early twentieth centuries.[6] In addition, they have tended to focus more on published materials than on the actual documents used by businesses, and more on changes in style than on changes in function of documents as managerial tools.

This book focuses on the evolution of the communication system as an integrated whole—its functions, technologies, and genres—in conjunction with the rise of system in American management. How and why did formal internal communication become the principal tool for managerial control, displacing the traditional, ad hoc methods of management?

The word *control* in a managerial context deserves some attention here, since it tends to carry strong associations. Richard Edwards, for example, defines it narrowly as "the ability of capitalists and/or managers to obtain desired work behavior from workers."[7] For him, control is exercised only in relation to low-level workers and is essentially coercive in nature. Beniger, on the other hand, uses control in a very broad sense, as "purposive influence toward a predetermined goal."[8] I intend the term in a broader and less negative sense than that employed by Edwards, but in a narrower sense than that used by Beniger. Managerial control—over employees (both workers and other managers), processes, and flows of materials—is the mechanism through which the operations of an organization are coordinated to achieve desired results. Managerial control is essentially management as we now think of it.

During the period from 1850 to 1920, formal internal communication emerged as a major tool of management, exerted toward the goal of achieving system and, thus, efficiency. By the end of that period, control through communication was a fact of life in the workplace.

Control through Communication

The development of internal communication systems in firms was not simply an incidental by-product of their growth. Rather, firm growth precipitated a search for new theories and methods of management that would help achieve efficient coordination of large, multifunctional firms. The managerial philosophy that emerged, first in the railroads and later in manufacturing firms, sought to achieve better control of business processes and outcomes by imposing *system*, in great part through formal communication. According to this philosophy or theory, which has been designated by Joseph Litterer as "systematic management," efficiency was to be gained by substituting managerially mandated systems for ad hoc decisions by individuals, whether owners, foremen, or workers.[9] These systems were established, operated, evaluated, and adjusted—that is to say, managed or controlled—all on the basis of flows of information and orders.

Control through communication as it unfolded during this period operated on at least three different levels. First, flows of downward communication from all levels of management conveyed information, procedures, rules, and instructions to control and coordinate processes and individuals at lower levels. Second, upward flows of communication drew data and analyses up the hierarchy to serve as the basis for managerial control of finances, facilities, materials, and processes. Information was progressively summarized and analyzed as it proceeded up the hierarchy, serving on each level as the basis for monitoring and evaluating individuals and operations at lower levels. As a by-product of these vertical flows, lateral flows of communication developed at middle levels to coordinate and document interactions.

In the early twentieth century a third, less direct form of control through communication emerged. Systematic management was built on the assumption that individuals were less important than the systems they functioned within. Negative reactions to this depersonalization led to the development of such communication tools as the in-house magazine, which attempted to humanize the workplace for workers and managers, and managerial meetings, which were intended to gain the goodwill and harness the efforts of lower- and middle-level managers. Yet, even these channels of communication reinforced the values of systematic management at the same time that they repersonalized certain aspects of the workplace.

Although the emergence of systematic management to replace ad hoc management was the major force driving the development of formal internal communication, a series of changes in the technology of written communication played an enabling role. Communication technology circumscribed the universe of possible and practical methods of communication at any given time, but over time that technology itself responded to changing managerial needs. The typewriter, for example, greatly increased the speed and reduced the cost with which a written document could be produced, thus facilitating increased use of written communication. On the other hand, business demand for efficient production of documents created demand for the typewriter and spurred improvements in it. Thus, while technological developments supported the growth of internal communication, that growth in turn encouraged further technological evolution by providing an expanding market for innovations.

Finally, new communication genres developed as a product of organizational needs and available technologies. Circular letters, reports, and manuals were shaped by the demands of their production and use. Older customs of form and style gave way in the face of a new desire to make documents more efficient to create and to use. Thus, developments in managerial methods, communication technologies, and communication genres fed on one another in the evolution of the communication system.

Approach and Organization

This study uses both printed and archival sources from the period 1850 to 1920. To keep it within feasible proportions, I limited it in several ways. First, it concentrates on railroads and manufacturing firms. As Chandler has shown, the railroad industry pioneered in creating big business, and during the second half of the nineteenth century these companies developed many of the management techniques and communication practices later adopted by manufacturing companies. In addition, highly specialized uses of information such as accounting systems and production control systems receive only superficial attention. Similarly, the representative workers' associations or "company unions" that emerged at the very end of the period studied and are precursors of modern labor relations are given only brief consideration.

The study falls into two parts, the first laying out the argument in general terms and the second consisting of specific case studies. The first three chapters draw primarily on printed sources to outline developments in the functions, technologies, and genres of internal communication. Chapter 1 traces the evolution of managerial theory and the new demands it placed on internal communication. Chapter 2 looks at developments in the technologies of communication, starting with the telegraph and then proceeding to devices for producing, reproducing, and storing written documents. Chapter 3 examines the developing genres of internal communication—that is, classes of communication (such as the report or memo) that share characteristics of form and function. Nineteenth- and early twentieth-century periodical and textbook treatments of these issues, along with relevant secondary research, provide the sources for these chapters. The growing professionalization of management and the emergence of formal management education during this period assured that many of the issues were discussed and debated in published sources.

The second part presents three case studies of evolving internal communication systems based on archival materials. Studying the emerging communication system in the context of real organizations provides a better understanding of the interactions among managerial method, communication technology, and genre. These chapters describe and analyze the development of the communication systems in three companies: the Illinois Central Railroad, Scovill Manufacturing Company, and E. I. du Pont de Nemours & Company.[10]

The bound volumes and files are, of course, only the skeletal remains of the communication systems that once controlled and coordinated these companies; nevertheless, they provide both explicit and implicit evidence of changes in these systems. Some documents contain explicit statements about the evolution of communication as a managerial tool. Given the volume of material in question (literally thousands of cubic feet of records) and the even greater volume that did not survive, finding such explicit references was often a matter of luck and persistence. In addition, all documents serve as *examples* of communication genres. As such, they provide implicit clues about their functions and physical evidence of the technologies used to create and handle them. From the skeletal remains of the communication system, the muscle and flesh can be deduced.

All three case studies illustrate the interactions between managerial method, communication technology, and genre, but they highlight different aspects of this relationship. The Illinois Central Railroad established a limited formal communication system quite early in its corporate life to provide the coordination and control necessary for safety, consistency, and honesty in a relatively large and geographically dispersed enterprise. Unlike the most progressive railroads, however, the Illinois Central did not develop the managerial tools necessary for improving efficiency until the late 1880s, driven by regulation under the Interstate Commerce Act and a change in upper management. The case studies of the two manufacturing companies, Scovill and Du Pont, illustrate contrasting patterns in the adoption of the philosophy of systematic management and the communication system necessary to that philosophy's realization. Both companies trace their development back to 1802, when Scovill began as a manufacturer of brass buttons and Du Pont of gunpowder. During the late nineteenth century, Scovill began a gradual systematizing of both management and communication, which accelerated greatly in the years immediately before and during World War

I. In contrast, Du Pont's adoption of the communication system necessary for systematic control and coordination came suddenly and rapidly
after a change in the company's management in 1902. Thus, each of the
three companies illustrates different forces and patterns in the development of control through communication.

Control through Communication

Chapter 1
Managerial Methods and the Functions of Internal Communication

The period from 1850 to 1920 was one of great change in the size and structure of American firms. Before 1850, the economy was dominated by small firms owned and managed by a single individual or a partnership and operating in a local or regional market. The spread of the telegraph and of railroads around the middle of the century encouraged firms to serve larger, regional and national markets, while improvements in manufacturing technology created potential economies of scale. Beginning with the railroads themselves, many firms grew and evolved structurally in response to these opportunities. As Alfred Chandler demonstrated in *Strategy and Structure*, first the functionally departmentalized enterprise, then the decentralized, multidivisional corporation developed.[1]

During this period of change, experience soon showed that the ad hoc managerial methods that had worked satisfactorily for small, owner-managed firms in a less competitive environment were inadequate for larger firms run by managerial hierarchies and competing in expanded markets. The philosophy of management that evolved in response to new needs, later to be labeled *systematic management*, promoted rational and impersonal systems in preference to personal and idiosyncratic leadership for maintaining efficiency in a firm's operations. This general philosophy spawned many specific techniques and movements, including its most famous offspring, the scientific management movement. Systematic management attempted to improve control over—and thus the efficiency of—managers, workers, materials, and production processes. In the early years of the twentieth century, it became clear that reliance on impersonal systems contributed to morale problems among workers and managers. Attempts to repersonalize certain aspects of work life, such as the paternalistic corporate welfare movement, arose to supplement systematic management.

These two lines of development in managerial philosophy demanded the evolution of extensive formal communication within growing firms.[2] As this chapter demonstrates, the informal, incidental, and primarily oral communication within the traditional small firm could satisfy neither the requirements of systematic management nor those of later attempts to humanize the workplace. More formal and systematic modes of communication, primarily written but also including documented oral communication, were essential. Regular flows of upward, downward, and lateral communication as well as detailed record-keeping procedures played a critical role in the new "systems." The early twentieth-century humanizing efforts demanded the evolution of further modes of communication.

Hence, it was not growth per se that required the development of the internal communication system, but the managerial philosophy that evolved in response to growth. Formal internal communication became a managerial tool for coordination and control.

Management and Communication in Early Manufacturing Firms

Up until the late nineteenth century, manufacturing firms were generally managed by ad hoc methods that required little formal internal communication. Most firms were managed by their owner(s), sometimes with the help of skilled artisans or foremen.[3] These individuals could gather information, give orders, and coordinate operations orally. Written communication was limited to formal documentation of monetary transactions and informal correspondence across physical distances.

A firm's accounts recorded financial transactions, primarily with the external world. The accounts were simple and descriptive, kept in the traditional double-column form. In the various bound journals, ledgers, waste books, and day books, all debits and credits were recorded first by strict chronology, then chronologically within individual accounts.[4] Agents who sold the firm's goods also provided such accounts, generally on a quarterly or semiannual basis.[5] These transaction-based accounts allowed the owner to figure profits (i.e., sales minus expenses) and to check the state of financial relations with individual suppliers, agents, or buyers, but they gave little other information. Neither the nature of the accounting system nor the size of the enterprises encouraged any detailed consideration of operating expenses.

Internal correspondence served primarily as a way to bridge distances, not as a managerial tool. When an owner was on the road visiting suppliers or agents, he might write to a partner or a responsible employee (such as a foreman or a bookkeeper) to exchange information, orders, and opinions.[6] If an owner's residence was not located right next to his plant, when urgent business arose he might also communicate across the intervening distance by means of notes carried by messengers. In both cases, the correspondence responded to a specific need to bridge distance; it was not part of a regular flow of information. Correspondence with agents, who were generally independent commission agents rather than salaried employees, was only marginally more formalized. Agents responded to requests and instructions, provided information on markets and competitors, and transmitted orders. Except for the quarterly or semiannual accounts, this correspondence responded to specific needs rather than forming part of a systematic flow of communication. In all of these cases, letters might be frequent, but they were not regular in content or timing, nor were they differentiated in form from external correspondence.

Even in the factories that began appearing in the early and mid-nineteenth century, neither managerial methods nor internal communication evolved substantially. Early American factories, starting with the New England textile mills, brought together more workers and sometimes more processes (e.g., spinning and weaving) in a single location than previous manufacturing operations had. Manufacturing in these facilities, unlike in most small firms, was broken down into discrete processes that were organized into rudimentary departments run by foremen or skilled workers.[7] Moreover, ownership was often separated from management, with an "agent" or factory manager running the entire operation for owners located in distant cities.

But in spite of these differences from the traditional owner-managed firm, before 1880 managerial methods in factories did not change substantially. As Daniel Nelson has observed, "In many industries internal management techniques, particularly those involving relations between the factory managers and workers, were not fundamentally different from what they had been in the craftsman's shop."[8] Because the factories were still relatively small, and because the owners were more concerned with financial problems than managerial ones, he continued, "The operation of the plant was generally left to the foremen and the skilled workers," who ran their departments almost as independent shops. In describing the management of these factories, Daniel Wren has noted that "problems were met and solved on an *ad hoc* basis, and only a few managers could learn from the experiences of others in solving factory problems or handling people. The general view of leadership was that success or failure to produce results depended upon the 'character' of the leader, upon his personal traits and idiosyncrasies, and not upon any generalized concepts of leadership."[9] Individuals, not systems, were still primary.

As in traditional small firms, the foremen managed the workers and the factory manager coordinated the foremen primarily by word of mouth. No changes in either managerial methods or internal communication occurred at this level. At the highest level, the separation of ownership from management (a separation that was usually physical as well as functional) dictated that the factory manager communicate the accounts and other information to the owners in writing. In some cases, this extra step led to the development of slightly more sophisticated accounting and reporting practices. In the Slater textile mills, for example, the Slaters introduced a very rudimentary form of cost accounting, a set of accounts giving them the cost of labor and incidentals per yard of cloth, to help them retain a measure of control over their mills.[10] The plant managers reported the financial and operating results each month, enabling the Slaters to compare the results over time and between mills. Other textile mills made similar small advances in using accounting as a managerial tool. Yet in general, factories, like smaller manufacturing establishments, kept traditional accounts and were managed primarily by oral methods.[11] It was the railroads, not the factories, that led the way in both managerial methods and formal internal communication.

Railroads: Innovators in Management and Communication

The railroads were important early innovators in managerial theory and practice, pioneering in the use of formal internal communication for control. The railroads and the telegraph gave other types of firms ready access to larger markets, thus opening the way for these firms to expand at the end of the nineteenth century. These transportation and communication industries themselves, however, experienced that growth first, making them, of necessity, pioneers in the management of large firms.[12] In the midcentury period they arrived at some of the major principles of management that were later to become part of the philosophy of systematic management, principles that required the development of internal communication systems.

From the beginning of the American railroad industry in the 1830s, its financial and physical attributes had implications for managerial structure and for internal communication. The large capital requirements of even a small railroad dictated that ownership be distributed among a number of stockholders, and thus be separated from management. As in the early factories, salaried managers ran the day-to-day affairs of the companies and reported the results to the owners (or to the directors as representatives of the owners). This relationship demanded a regular, though not necessarily frequent, flow of financial and sometimes operational information from management to the directors.

In addition, the physical characteristics of railroads had inherent implications for management and communication at lower levels. The danger of derailments and collisions, especially on single-track lines, demanded that equipment be well maintained and that train movement be carefully coordinated by some central authority. These tasks were made even more complicated by the fact that a railroad was necessarily geographically dispersed. Traditional oral methods of communicating orders and gathering operating information were clearly inadequate in this industry. From the beginning, written timetables and basic rules were necessary for coordinating traffic flow, and correspondence was needed to bridge the distance between stations in order to communicate nonroutine exceptions to standard rules. Moreover, any operating information needed by the central management had to be sent from the stations in written form.

Although these characteristics necessitated heightened awareness of the roles of management and communication, initially both were rather haphazard. The demands, first of safety, then of efficiency, prompted certain innovative railroad managers to articulate new principles of management. These principles in turn expanded the role of formal internal communication.

Safety and Early Innovations in Communication Systems

As early as 1841, a series of collisions on the Western Railroad (running between Worcester and West Stockbridge, Massachusetts) prompted its managers and directors to investigate and to tighten managerial control. The initial investigation revealed that written orders were generally used to communicate changes in standard timetables, but that such matters were being handled quite loosely.[13] The final report from management to the directors decreed that, in the future, "the time & manner of running the trains shall be established and published by the Engineer," and that "no alteration in the times of running or mode of meeting & passing of trains shall take effect, until after positive knowledge shall have been received at the office of the superintendent, that [written] orders for such change have been received & are understood by all concerned."[14] Thus, to assure safety, the company established clear channels of authority and of communication. Downward communication, including the published schedule and rules as well as specific written orders, was formalized as a managerial tool.

Furthermore, the final report also established rudimentary records and upward reporting procedures by which management could create a body of knowledge about railroad operations to aid both middle and top management in learning from past mistakes. Certain managers, such as the road master and the master mechanic, were required to keep records about operations and to report to their superiors once a month. The rationale and method for the road master's reports were stated as follows: "The Road Master shall keep a journal of his operations stating the several points at which labor has been performed—the nature and extent of the same and results produced in order that the experience thus acquired may be rendered serviceable in subsequent operations. . . . He will at the End of Each month make a report to the Engineer of his proceedings—with such suggestions as he may deem necessary."[15]

An 1842 follow-up report on procedures for avoiding collisions added the stipulation that "each agent, under the chief engineer, in all grades of service, is to report once a month, to the engineer, his operations, for the preceding month, so that the whole management of the road shall, at short, stated intervals, come under the review of the engineer's department."[16] This procedure thus required records to be maintained and information from those records to be summarized and passed up to higher levels at regular intervals.

These reports contained, in embryonic form, some of the goals and communication mechanisms later to be articulated more fully by proponents of systematic management. The published rules, the journal of operations, and the monthly reports all reflected a desire to rise above the individual memory and to establish an organizational memory tied to job positions and functions, rather than to specific individuals. Downward communication via written rules and orders provided a mechanism for embodying the acquired wisdom at any given moment. The regular, though still slight, flow of upward communication provided a mechanism for updating that wisdom in light of experience. The chief engineer was to "review" the management of the road by means of these reports, a process that hints at the later movement's use of upward reports as the basis for comparing and evaluating managerial performance. Moreover, top management of the Western Railroad revealed a new trust in procedures and systems, most of them involving written documents, rather than solely in the good judgment of individuals. Finally, in instituting these changes, the Western Railroad showed a self-consciousness, a desire to step back from day-to-day management in order to examine and improve managerial methods. Previous managers had rarely taken time for such self-scrutiny. These attitudes and techniques, originally intended to improve safety, would continue to evolve as managers tackled the problem of efficiency in large companies.

*Efficiency and the Further Development of
Communication Systems*

As railroads grew, so did the need to improve efficiency. By the mid-1850s many managers were beginning to understand that growth by itself did not bring increased profits. In fact, inefficiencies that were relatively harmless in small companies seemed to multiply in larger ones, creating diseconomies rather than economies of scale. In an 1885 memorandum on the "Organization of Railroads," Charles E. Perkins of the Burlington Railroad noted the differences in operation between small and large railroads:

> In deciding the question of organization it will be necessary to consider two stages, so to speak, of railroad development. The first stage where the volume of traffic is not sufficient to make necessary or to warrant the highest degree of physical efficiency; and the second stage where the volume of traffic is so great as not only to warrant the expenditure, but also to make it economical to maintain the physical efficiency at the highest point.[17]

Beginning in the mid-1850s, the demands of managing second-stage railroads led to further evolution of managerial principles and practices, and in particular to changes in the extent and uses of upward communication.

In an 1856 superintendent's report made by Daniel C. McCallum to the president of the New York and Erie Railroad Company, McCallum argued that the greater cost per mile generally experienced on longer roads at this time resulted from "the want of a system perfect in its details, properly adapted and vigilantly enforced."[18] Thus, he was extending the dependence on systems initiated years earlier by the Western Railroad. He went on to state the principles underlying such a system, which may be summarized as follows: (1) set up clear responsibilities for each position, making each individual responsible directly to his immediate superior; and (2) set up a system of reports and checks through which upper management can evaluate the efficiency of all subordinates and of all aspects of operations.

In this second principle McCallum went even farther than the Western Railroad officials (and prefigured the systematic management philosophy even more closely) in establishing routine upward communication as a managerial tool. Monthly, weekly, daily, and even hourly reports were to draw data on railroad operations up the hierarchy for use as the basis for evaluation and decision making at the top. This upward flow of information was central to his managerial philosophy: "In my opinion a system of operations, to be efficient and successful, should be such as to give to the principal and responsible head of the running department a complete daily history of details in all their minutiae."[19]

As Chandler has pointed out, McCallum's innovations in managerial principles and techniques, as well as innovations by a few other railroad managers, were widely publicized through such railroad periodicals as Henry Varnum Poor's *American Railroad Journal*.[20] Poor, who extolled the innovations, generalized and extrapolated from them in ways that emphasized the connection between managerial theory and systematic communication. In his own theorizing, Poor proposed three general principles of the "science of management": organization, communication, and information.[21] *Organization* referred to the clear designation of duties and responsibilities and to a clear chain of command. Such careful designation would almost certainly require rule books and job descriptions, a form of downward communication. *Communication*, for Poor, "meant primarily a method of reporting throughout the organization which would give the top management an accurate and continuous account of the progress of operations, and which in so doing would assure the necessary accountability all along the line." Thus, the regular flow of formal upward communication was in itself one of his managerial principles. Finally, Poor used the term *information* to designate "recorded communication—that is, a record of the operational reports systematically compiled and analyzed." These analyzed records and re-ports formed an organizational memory used as a basis for understand-ing, maintaining, and improving operational efficiency and financial performance.

Poor's final principle signals the fact that railroad managers of the mid- to late nineteenth century did more than simply increase the amount of upward communication: they changed its basic nature. As Francis X. Blouin, Jr., has commented, "Record keeping thus shifted from serving a descriptive function to serving as an analytical tool."[22] The simple records of financial transactions and the descriptions of op-erations such as those the Western Railroad managers recorded were supplemented by much more diverse types of records and reports provid-ing the data and analysis needed to evaluate financial and operational performance. In Chandler's words, "For the middle and top managers, control through statistics quickly became both a science and an art. This need for accurate information led to the devising of improved methods for collecting, collating, and analyzing a wide variety of data generated by the day-to-day operations of the enterprise."[23] The analysis was par-ticularly important, since simply sending vastly increased amounts of raw data up the hierarchy would have created massive information over-load at higher levels. The data needed to be summarized and analyzed in ways that would enable upper management to evaluate performance quickly and efficiently.

One important area in which the upward flow of data both increased and changed radically in nature was accounting. "To meet the needs of managing the first modern business enterprise," Chandler has pointed out, "managers of large American railroads during the 1850s and 1860s invented nearly all of the basic techniques of modern accounting," refin-ing financial accounting and inventing capital and cost accounting.[24] The first two types of accounting are, for the most part, outside the scope of this book, though one development in financial accounting is worth mentioning. Railroads went beyond using simple balance sheet, profit-and-loss statements and began using the ratio of operating ex-penses and gross revenues, called the "operating ratio." This financial measure, unlike the balance sheet figures, could be used to compare the performance of one railroad to that of another.

The third type of accounting, cost accounting, was developed as a method for achieving McCallum's second goal—helping upper manage-ment compare and evaluate internal operations on the basis of upward flows of information—and thus deserves a little more consideration. Just as new techniques in financial accounting allowed financiers to judge a railroad's performance over time or in comparison to other rail-roads', cost accounting, developed in the late 1860s, allowed upper man-agement to judge the performance of one section or division of the rail-road over time or in comparison to others'. Doing this required creating a standard measure that could legitimately be compared. For the railroads, the principal measures used were cost per ton mile for freight and cost per passenger mile for passenger traffic. They became important managerial tools in monitoring and evaluating operating efficiency, not just financial results.

In their search first for safety, then for efficiency, railroad managers were pioneers in managerial theory and practice. They anticipated the systematic management philosophy in arguing for the need to systematize procedures independent of the individuals involved and to use systematically gathered operational information as the basis for evaluation and decision making at higher levels. The major tools they established to achieve these goals were regular internal flows of communication.

Systematic Management: Control through Communication

In manufacturing as in railroads, managerial principles and practices were only scrutinized and reformulated when growth failed to produce expected economies of scale. In the second half of the nineteenth century, influenced both by improvements in the technology of production and by larger markets made accessible by telegraph and railroads, companies grew and departmentalized. By the final decades of the century, however, growth was failing to produce the expected profits. Instead, large manufacturing firms were plagued by confusion and disorder. As long as what one writer of the 1890s called "the old slipshod way of our forefathers" prevailed,[25] the promise of improved technology and wider markets could not be fully realized.

The problems faced by firms were many. At the production level, departments or subunits were generally run by foremen or job contractors who were accustomed to working autonomously and exercising their prerogatives in matters of personnel and schedule.[26] Both horizontal and vertical coordination broke down.[27] The middle and upper managers, many of them engineers by training, found themselves farther and farther from the production line, but with no tools to control what went on there or to coordinate their own actions effectively.[28] The horizontal flow of orders and production processes was impeded by the lack of coordination among separate units and subunits. The consequent inefficiencies drove the manufacturing managers to follow the lead of the railroad managers in developing more effective managerial theory and practice.

Starting in the 1870s and continuing into the early decades of the twentieth century, a new literature of management theory and technique began to appear, initially in the engineering publications and later in the newly evolving management organs.[29] This body of literature and the underlying philosophy that it expounded, which Joseph Litterer has aptly designated as *systematic management*, represented an attempt to "eliminate confusion, oversight and neglect; coordinate efforts, return firm control to the top people in the organization; accomplish these things through the use of standardized procedures on routing managerial work through 'Method' or 'System.' "[30] Only thus, the managers felt, could they achieve the efficiency they were striving for. Like the theories of railroad management discussed above, which were harbingers of this broader-based movement, systematic management based its reassertion of control and coordination on record keeping and flows of written information up, down, and across the hierarchy. The spread of systematic management theory and methods reshaped the communication system within manufacturing enterprises.

Systematic management, a broad and somewhat amorphous philosophy, has received far less attention than the scientific management theories of Frederick W. Taylor and his followers.[31] It is important, then, to clarify the relationship between these two philosophies of management. Although the advocates of systematic management were concerned with improving managerial methods at all levels of the organization, Taylor and his circle focused primarily on the factory floor, attempting to create some scientific basis for efficient production. Scientific management was widely associated with three specific innovations: time and motion studies of factory work methods, functional foremen, and differential piece rate payments. For all its specialized techniques, however, scientific management was essentially a part of the broader movement. Daniel Nelson has argued convincingly that Taylor's work was "a refinement and extension of systematic management."[32] Taylor insisted on instituting managerial systems of the sort advocated by many other systematizers before adding the specific innovations that became associated with scientific management. Moreover, scientific management as it was proposed by Taylor was written and talked about far more than it was implemented. The broader "search for order and integration" (to borrow one of Litterer's titles) of the systematic management proponents had a more widespread influence on managerial methods in American firms than Taylor's much-publicized but narrower efforts.

Many different themes were intertwined in the writings of systematizers, but central to the idea of systematic management were the two principles first articulated by the Western Railroad and by McCallum of the New York and Erie, respectively: (1) a reliance on systems mandated by top management rather than on individuals, and (2) the need for each level of management to monitor and evaluate performance at lower levels. Mariann Jelinek has stated these two trends at the most abstract level: (1) "a continuing attempt to transcend dependence upon the skills, memory, or capacity of any single individual," and (2) "an attempt to rise above the concrete details of the task to think about what is being done, rather than merely to do it." Using a more concrete definition, Litterer has described the two common factors in the proposals of systematizers in terms of the specific processes involved: "One is a careful definition of duties and responsibilities coupled with standardized ways of performing these duties. The second is a specific way of gathering, handling, analyzing, and transmitting information."[33] However these principles are stated, they both contributed to the decline of ad hoc, word-of-mouth management and to the rise of formal internal communication.

Transcending the Individual

The need to transcend reliance on the individual in favor of dependence on system became a central tenet of systematic management. It enabled managers in manufacturing firms to regain the control over production then held by the semi-autonomous foremen, job contractors, and skilled workers. The skills and memory of the individuals were to be put into the service of the organization to improve efficiency. This required, as Frederick Taylor himself put it, "the deliberate gathering in on the part of those on management's side of all the great mass of traditional knowledge, which in the past has been in the heads of the workmen, and in the physical skill and knack of the workman, which he has acquired through years of experience."[34]

David Noble has described this desire to transcend the individual as a purposeful act of aggression by the managerial class against the working class: "As managers in industry, engineers now undertook to expropriate and systematize the intelligence of production, to place it in the hands and handbooks of management, and to use it to reorganize the production process for maximum output and profit."[35] Yet, a closer look at the first principle of systematic management reveals inadequacies in Noble's viewpoint. Shifting power from workers and foremen to management was not the only motive for transcending the individual; proponents of the new managerial theories also advocated the transcendence of individual managers in favor of systems and procedures at higher levels. In fact, Litterer, the scholar who has studied the literature of systematic management most thoroughly, defines its main focus in terms of systematizing the jobs of managers: "The common idea [of those writing about systematic management] appeared to involve an over-all approach to operating a business which had specific jobs for all members of management, those jobs linked by a set of procedures which would ensure all necessary work being done."[36]

While Taylor and some systematizers focused for the most part on regaining control over and improving the efficiency of workers and foremen on the factory floor, others such as Alexander Hamilton Church, a major figure in the systematic management movement, concentrated on making management itself more efficient.[37] Church pointed out the problem with depending on individual knowledge and memory even at the top of the hierarchy: "How many concerns languish when the care of their founder is withdrawn and why? Simply because he cannot transfer the multitudinous details of organization from his memory to that of a successor. It is these details that are essential and it is their absence that must be fatal unless their place is supplied anew."[38] Only by replacing individual idiosyncrasy with system, individual memory with organizational memory, and personal skills with firm-specified skills at all levels did the systematizers feel that they could achieve the current and future efficiency they sought. In organizations, as in machines, interchangeability of parts should promote efficiency. Thus, the "expropriation" of skills that Noble saw as a weapon in the class struggle between the managerial and working classes actually occurred within the managerial class as well.

Noble's description of systematization, although marred by its focus solely on workers, suggests the nature of one of the major mechanisms used to transcend the individual: handbooks and other written records and communications. Processes once retained in workers' or managers' heads and hands were now recorded in handbooks and passed back down the hierarchy to govern work methods. In Jelinek's explanation of the need to transcend the individual, she also suggests the importance of the internal communication system to achieving that transcendence: "By recording the specifics of a task, a given outcome could be replicated: built into a formal system, institutionalized and independent of the individual."[39] Paperwork flowing across the hierarchy also aided in coordinating operations. In the 1880s, systematizers such as Henry R. Towne, Henry Metcalfe, and John Tregoing developed systems of shop orders to control the flow of orders through factories.[40] The system was thus embodied in various written forms that existed independently of the individuals involved.

The proponents of systematic management were conscious of their reliance on written records and communications. One of the systematizers writing around the turn of the century, Horace Lucian Arnold, asserted the importance of shop books and factory accounts, noting, "Even if entire honesty and sincerity prevailed at all times in all business transactions, the mere differences due to variations in individual understandings of orders would render it impossible to conduct any business of magnitude on verbal specifications."[41]

In the past, only transactions with the external world had been documented; now the systematic management movement suggested that internal practices and procedures also be documented. Henry Metcalfe argued: "Now, administration without records is like music without notes—by ear. Good as far as it goes—which is but a little way—it bequethes [sic] nothing to the future. Except in the very rudest industries, carried on as if from hand to mouth, all recognize that the present must prepare for the demands of the future, and hence records, more or less elaborate, are kept."[42] Church also emphasized the importance of establishing an organizational memory: "Under rational management the accumulation of experience, and its systematic use and application, form the first fighting line."[43]

The literature of systematic management proposed many different administrative systems. At the bottom of all of them, however, lay the need to transcend the individual worker and manager in favor of systems institutionalized through records and flows of written communication.

Monitoring and Evaluating Performance

If the desire to transcend the individual produced changes in record keeping and downward communication, the second trend in systematic management—the need for upper levels of management to analyze, evaluate, and adjust performance at lower levels—produced even greater changes in upward communication. It was not enough simply to systematize existing processes; managers wanted "to rise above the concrete details of the task to think about what is being done." To do so, they needed to establish "a specific way of gathering, handling, analyzing, and transmitting information."[44] Records and reports documented actual operations and transmitted this information up to progressively higher levels to provide the basis upon which upper management could analyze and evaluate what was being done at the lower levels. As Alexander Hamilton Church stated it, "The object of the commercial, or, as it might also be termed, the administrative organization scheme, should be to collect knowledge of what is going forward, not merely qualitatively, but quantitatively: It should also provide the means of regulating as well as the means of recording."[45] This function required a change in kind as well as amount of internal communication. As had happened earlier in the railroads, records and reports changed from descriptive to analytic.

While the handbooks, systematized procedures, and horizontal flows of shop orders helped solve the problem of horizontal coordination, upward flows of reports helped solve the problem of vertical coordination. Such systems for drawing information and analysis up the hierarchy were instituted in a variety of areas. For controlling and assessing production, the systematizers attached upward flows of information onto the horizontal flow of materials and shop orders. Frequent routine reports communicated to higher levels the amounts and types of work done on various orders. This "automatic flow of information to higher management," according to Litterer, "permitted it to know the status of work and frequently provided the data for the analysis of problems and the proposal for corrective action."[46]

Similar developments in cost accounting constituted a major technique of systematic management.[47] Just as railroads developed ways of figuring cost per ton mile and per passenger mile for comparing operational performance, manufacturing firms developed tools that allowed the manager at each level to compare, or, in the words of one innovator in cost accounting, to place "artificially in juxtaposition," the performances of units at lower levels.[48]

At the simplest level, Metcalfe saw the (relatively primitive) system of shop accounts he proposed in 1885 in terms of flows of information into and out of a central point:

> The proposed system of shop accounts is based on two compensating principles.
>   1. The radiating from a central source, let us say the office, of all authority for expenditure of labor or material. These being, however they may be disguised, the elemental forms of all internal expenditure.
>   2. The conveying toward the office from all circumferential points, of independent records of work done and expenses made by virtue of that authority.[49]

As cost accounting systems grew more complex, such records were sent up the hierarchy for progressive analysis. Managers used the analyses in comparing performance and judging efficiency at lower levels. Hence, higher level managers were no longer reduced to depending on the foreman's word and a gut feeling. Cost accounting provided them with a more objective tool.

In addition to cost accounting and production control systems, the literature of systematic management proposed many other administrative systems.[50] Systems were proposed for aid in controlling inventory and labor, for example. Even office work, itself created in great part by the paperwork necessary for the other systems to function, was systematized.[51] All of these systems depended on documentation and upward flows of information to allow middle and upper management to monitor and control what went on at lower levels. As in the railroads, the establishment of regular and frequent upward flows of information was even more important than the institution of downward flows of orders in enabling upper management to control, evaluate, and improve the efficiency of large organizations.

Extensions and Influences of Systematic Management

Over time, the principles of systematic management spread to new areas, were extended by analogy to new applications, and were popularized (and consequently diluted) in the popular press. As growing firms vertically integrated by internalizing functions such as marketing and sales, which had previously been handled by agents, the principles of systematic management were extended into areas other than production. Such functional integration was only profitable, in the long run, if the functions could be coordinated more efficiently within a company than they could be by market transactions between companies. Systematic management provided the tools, in the form of internal communication flows, needed for efficient coordination. Moreover, since the sales function was usually geographically dispersed, its need for extensive written communication was even greater than that of functions contained in a single location. Other functions, such as research and development and purchasing, soon came to be handled by central staff departments that were also run on principles of systematic management.

As the techniques of systematic management spread, some were extended by analogy. The documentary impulse evidenced in downward and upward communication, for example, extended to lateral communication. One text on handling and filing correspondence noted:

> Of considerable importance in every large organization is the inter-department correspondence—the notes from one department head to another. Every department head finds it necessary at times to request information from other departments. Even with an inter-communicating telephone system with which every large office and plant should be equipped, many of these requests are of a nature that, to guard against misunderstandings, demand written communications.[52]

In other words, the individual memory needed to be transcended in lateral, as well as vertical, communication. This extension of the concept of establishing an organizational memory via written documentation may have served in part as a rationalization for politically motivated communication. As companies grew, allegiances to and rivalries between departments created friction, with each department wanting to document its side. Nevertheless, the notion of transcending individual memory came to be seen as applying to all interactions: "It is necessary in business that orders should be definitely given, questions specifically and intelligently asked and answered, and that some record of all communications be kept."[53]

The principles of systematic management entered public discourse early in the twentieth century. When the Eastern Rate Case of 1910–11 popularized the term *scientific management* for Taylor's particular brand of systematic management, it also popularized the whole concept of efficiency and system.[54] *System, efficiency,* and *scientific* became catchwords in the business world and beyond. Although this period of wild popularity perhaps diluted the meaning of these concepts, it also gave them enormous publicity and consequently helped spread them. While the notion of system became broader, it still retained its association with flows of communication, particularly sets of forms and reports.[55] By the beginning of World War I, efficiency was a widely accepted goal and business systems were accepted methods of achieving that goal.

Repersonalizing Management: Indirect Control through Communication

Systematic management principles required depersonalizing the workplace in the interests of control and efficiency. Both workers and managers, however, frequently resented and resisted the substitution of impersonal systems for personal relations. In response to this resistance, attempts to repersonalize certain aspects of the workplace emerged. The corporate welfare movement—with its clubhouses, libraries, health care, and beautification programs—contributed to a systematic humanizing of certain aspects of worker life.[56] Comparable efforts attempted to involve low and middle level managers in a more personal way. The welfare movement and other attempts to elicit cooperation by humanizing organizational life adopted new modes of internal communication as tools. Although these forms of communication were designed to promote a more humane and cooperative atmosphere, they also served indirectly to reinforce control in the interests of efficiency.

The welfare movement had roots both in earlier experiments in industrial paternalism and in its contemporary society. Stuart D. Brandes has traced the origins of what he calls "welfare capitalism" in America to the paternalism of early textile mills, which sometimes provided housing, protection, and even instruction to their employees.[57] These late eighteenth- and early nineteenth-century precedents, relatively limited in scope and frequency, were established primarily to guarantee a supply of good labor. The welfare movement also drew from the broader progressive reform movement of the turn of the twentieth century, with its concern about a whole range of social problems affecting the poor and working classes. Many industrial welfare advocates were simply a subset of such reformers who concerned themselves particularly with improving and "elevating" the lives of workers and their families.[58] For these individuals, as one of them said, " 'betterment work must be considered an end in itself; it must be prompted largely by humanitarian considerations.' "[59]

Business's most pressing motive for adopting corporate welfare programs, however, was immediate and practical: the need to reduce widespread labor unrest. As firms had grown in the late nineteenth century, the loyalty of workers to their owners or managers, often described in terms of a family relationship, faded. Moreover, in its reliance on systems rather than individuals, systematic management had further depersonalized firms and further lessened the already weakened loyalty of workers to their firms. Workers increasingly opposed, either directly or indirectly, actions taken by management. The labor confrontations and strikes that started in the 1870s and rose in intensity in the last decade of the nineteenth century and the first two decades of the twentieth were effective checks to the efficiency that growth and systematic management had promised. Taylor had presented scientific management as a "partial solution to the labor problem," though in 1895 he saw that problem more as one of systematic "soldiering" or intentional inefficiency than of labor strife.[60] He thought that his rate system, though impersonal, would be perceived by workers as "scientific" and therefore "fair"; consequently, it would defuse hostility and encourage hard work to earn more money. The 1911 strike at the Watertown Arsenal, a factory using the Taylor system, seriously eroded that hope.[61]

Meanwhile, managers in the most systematically managed firms were beginning to search for answers to their labor problems. They were coming to realize that "while it was important to develop procedures which would be independent of the individuals involved, no system would work without leadership on the one hand and voluntary co-operation on the other."[62] Many recognized that cooperation, so necessary to the efficiency they sought, could be gained by showing concern for worker welfare and by humanizing at least some aspects of the workplace. They looked at the welfare programs in terms of how they might improve efficiency. As Noble puts it, "They remade the innovations of welfare workers into elaborate industrial-relations programs, to try to foster a spirit of voluntary cooperation among workers, to transform the energy of potential conflict into a constructive, profitable force within a larger corporate framework."[63] By systematically showing concern for human needs and by partially repersonalizing their relations with workers, they hoped to regain the cooperation and loyalty they had lost, but without giving up the impersonal systems that gave them efficiency.

One of the tools that they used in humanizing the workplace was a new form of internal communication, the in-house magazine or shop paper. Its function was described in a 1918 article: "Many shops have outgrown the one-man stage. No longer can the head of the organization interpret his policies personally to the workmen. But a factory house organ, whether it be a single typewritten sheet or a 24–page magazine, offers an opportunity to bind personal interest closer in the small shop and keep management from becoming mechanical in the large plant."[64]

This goal of personalizing large plants was often stated in terms of recreating a family feeling, as in a 1919 article entitled "Fostering Plant Spirit through a Plant Paper":

> One of the words often used today to express the spirit desired in industrial organization is "family." The ideal sought for is that all workers from the president to the office boy shall feel that they belong to one big family and have the loyalty which that relationship implies. The employer wants his men to work not for him but with him. The aim in all industrial service work is hearty cooperation by each because of the recognition of common interests.[65]

These shop papers were a form of downward communication, ultimately aimed at improving cooperation, control, and efficiency, but they did so indirectly rather than directly.

Around the end of the war, yet another communication tool grew in part out of the efforts of the welfare movement: representative shop committees, otherwise known as company unions. Where welfare work was not enough to quell worker unrest and the threat of strikes or organization by national unions, firm management sometimes turned to an employee representation plan.[66] The shop committees, members of which were elected by labor, provided a forum for two-way communication and negotiation between workers and management on a variety of issues. While sometimes simply short-lived alternatives to trade unions, these committees were seen by some managers, according to Daniel Nelson, as a way to "build on the welfare program to restore 'personal relationships' through the 'principle of representation.' "[67] Like the shop paper, representative shop committees introduced in this spirit indirectly reinforced firm control over a restless workforce by providing a mechanism for communication aimed at eliciting cooperation:

> The shop committee is not merely a means for bargaining collec-
> tively with employees; that has been and is now being done by
> trade unions where these are recognized. The shop committee is
> and does far more than this. It is a device of management, an in-
> strument of efficiency, a method of integrating men and manage-
> ment in such fashion that the unused resources lying dormant in
> the minds of the workers will be brought into active use. The shop
> committee is more than a scheme for allaying unrest by bringing
> labor into partnership with capital. It is a plan, in the last analysis,
> for securing efficiency in production, though the fact is by no
> means unimportant that efficiency of production calls for the es-
> tablishment of the same kind of friendly human relations between
> employer and employee that prevailed when industry was smaller
> and less exacting.[68]

Because they only emerged in great numbers after the period covered by this book, these committees are only of passing interest here. Nevertheless, they illustrate another communication mechanism introduced to allow limited humanizing of the workplace in the interests of efficiency.

Better developed during the first two decades of the twentieth century was another use of committee meetings as a tool for control. Managerial resistance to systematic management methods was less public than that of workers, but it was perhaps even more potentially damaging to the goals of the movement. As firms became systematized, lower and middle level managers lost much of the autonomy they had once held, and their jobs were depersonalized, just as those of the workers had been. Moreover, they were just as capable of "soldiering" as workers were, though they did so in very different ways. In an 1898 article discussing the need for better managerial systems, H. L. Arnold described a scenario in which "independent subordinate managers of either real or recognized authority are secretly or openly jealous of each other, and where each one sees that the less anyone else knows about the actual facts of his department, the better the chances are for his continuance in his position."[69] While the system of job tickets Arnold proposed in the article was intended to solve this problem, clearly any system dependent on flows of information could be sabotaged by a lower level manager through delays or misinformation. Cooperation from managers was as necessary as from workers.

By the second decade of the twentieth century, committee meetings or "shop conferences" of managers gained currency as a way to elicit necessary cooperation at the lower levels of management.[70] As one executive put it, "The old way of going out into the factory and shouting at the [fore]men, or calling them down . . . has passed, I believe[,] in all shops that are properly managed."[71] When the systems by themselves were inadequate to elicit cooperation, they were supplemented by shop conferences. These meetings, which often involved several levels of management, provided a forum for multidirectional communication. Higher level managers solicited suggested improvements in processes and systems from lower level managers (especially foremen and department heads), at the same time encouraging cooperation with existing systems. The meetings allowed lower level managers to learn from each other while providing a forum at which managerial "soldiering" was likely to be exposed through comparison of performance.[72]

Thus, the meetings fulfilled several functions simultaneously. One systematizer described their purpose with characteristic ambiguity:

> The primary idea is to enlist the cooperation of the [fore]men in the shop in forming plans and offering suggestions for the good of the company. By frequent meetings and a thorough airing of opinions[,] an esprit de corps and a feeling of responsibility for the success of the business as a whole is established. In its method this system is the opposite of the military method of management. The committee system is especially well adapted to furnishing a means by which the discontented can give expression to their feelings, and affords a valuable aid to the management in locating the cause of any disaffection. Furthermore, it is claimed for this system that it provides a method of overseeing whereby an executive totally ignorant of shop and sales processes is provided with reliable data concerning any weak spot in the production, buying or selling departments.[73]

In the same paragraph, the advocate has claimed both a democratizing and a controlling role for such committee meetings. And indeed, they fulfilled both functions. Ultimately, however, the committees supported rather than undercut systematic management. First, the same writer maintained, "The purpose of all committees is to act as advisory bodies only." Furthermore, in response to arguments that meetings took participants away from their work, he noted "that any loss due to slackened production will be more than compensated through increased harmony and the dependence which can be placed upon the [job] bosses." Although this medium of communication humanized the new managerial systems by introducing a more personal (though still formal) mode of communication, its ultimate justification was also efficiency.

The in-house magazine, the representative workers' committees, and shop management committees all evolved in part as reactions to the depersonalization of jobs brought about by firm growth and systematic management techniques. These reactions were ultimately minor correctives, for they did not seriously challenge the underlying premises of systematic management.

Conclusion: Management and Communication

Systematic management as it evolved in the late nineteenth and early twentieth centuries was built on an infrastructure of formal communication flows: impersonal policies, procedures, processes, and orders flowed down the hierarchy; information to serve as the basis for analysis and evaluation flowed up the hierarchy; and documentation to coordinate processes crossed the hierarchy. These flows of documents were primary mechanisms of managerial control. It is perhaps indicative of the movement's dependence on written communication that Frederic A. Parkhurst, a follower of Taylor, noted, "The factory mail system is one of the first features to be installed in connection with the new form of management."[74]

Internal communication also played a role in reactions to the depersonalization of systematic management. When opposition to the policies of upper management led to a corrective humanizing of certain aspects of work life, it was achieved in part through new types of internal communication. The in-house magazine and worker and management committees were designed ultimately, however, to reinforce the same goal toward which systematic management strove: efficiency.

While the systematic management philosophy drove the rapid evolution of internal communication, simultaneous developments in communication technology played a role in supporting that emergence. The next chapter traces the evolution of communication technology and its relationship to internal communication.

Chapter 2
Communication Technology and the Growth of Internal Communication

While formal internal communication was emerging as a control mechanism in response to firm growth and the new managerial philosophy, communication technologies (defined broadly here to include both physical devices and related conceptual schemes) were also evolving. The telephone, one of the most dramatic innovations, functioned primarily to extend the range of existing, informal communication. Other technologies were more important to the newly emerging formal communication system. The telegraph opened up new possibilities for rapid but systematic exchanges across distance, though it was more often used for ad hoc communication. Most important to the new forms and uses of internal communication, however, were the technologies of written communication: the typewriter, duplicating methods, and filing systems. They shaped the way businesses could produce, reproduce, and store documents.

These innovations did not cause the internal communication system to develop as it did. In fact, in some cases a technology was available long before changing managerial techniques brought it into widespread use. By reducing the time and expense involved in creating and using documents, however, the technological innovations discussed in this chapter made feasible and economic—and thus indirectly encouraged—the rapid growth and effective use of systematic communication in management. Without them, the growth of formal internal communication might have been severely limited, and systematic managers might have adopted very different coordination and control mechanisms.

The failure of the telephone to displace internal written communication illustrates the limits of a technology's influence on the communication system. In 1876 Alexander Graham Bell demonstrated the telephone's potential as a management tool by using the instrument to call his assistant Thomas A. Watson to him.[1] Moreover, early installations of point-to-point telephones (before the telephone exchange was invented in 1877–78) often linked the office and factory of a single firm. Beginning in the 1890s, private branch exchanges were widely adopted to link many locations within large facilities. Just as companies expanded to the point at which word-of-mouth management posed physical problems, the invention of the telephone made possible informal oral exchanges between individuals located some distance apart.

Nevertheless, this same period witnessed an explosion of written documents in large firms. The telephone bridged the widening distance between individuals in growing companies, and its universal adoption testifies to its utility in informal exchanges. Yet it could not and did not displace the written communication that was growing up. Oral exchanges, whether face-to-face or by telephone, were idiosyncratic, often inexact, and undocumented. The ideology of systematic management demanded increasing written communication to provide consistency, exactness, and documentation.

Although managerial needs thus imposed limits on the role of technologies in shaping intrafirm communication, several technological innovations were important as facilitators, enablers, and encouragers of formal information flow within the firm. The telegraph played an important role in the systematic management of geographically dispersed firms such as the railroads. The typewriter sped the production of documents while separating it from their creation. Carbon paper and duplicators provided cheap and convenient methods for making the copies necessary for mass downward communication. Vertical filing allowed written documents to be organized accessibly and retrieved efficiently, and thus to serve as a usable organizational memory. Moreover, new duplicating and filing technologies in combination permitted decentralized files, which both encouraged and were encouraged by the growth of internal documentation. Together, these late nineteenth-century innovations revolutionized the handling of written communication in organizations and removed potential impediments to its rapid growth.

The Telegraph and Internal Communication across Distance

The telegraph, introduced to America in 1844, had a dramatic impact on many aspects of public, private, and especially business life.[2] Before its introduction, news and information traveled only as fast as did people, using horses, stagecoaches, ships, or the new (and still short) railroads. The telegraph reduced transmission time to almost nothing, bringing individuals and the public in general much closer together. In the business world, it accelerated market transactions and improved market functioning.[3] Moreover, along with the railroad it helped open wider markets to companies and thus encouraged their growth.

Although the telegraph was considerably more expensive than the mails, its speed made it more valuable for urgent matters. Up to 1845, postal rates varied by distance from six cents per sheet for under thirty miles to twenty-five cents per sheet for over four hundred miles.[4]

Spurred by private competition and by Britain's demonstration that low-ering postal rates could actually increase postal revenues, the U.S. Postal Service lowered its rates twice, first in 1845 and again in 1851.[5] By 1851 the cost of most prepaid letters traveling up to three thousand miles was only three cents. In contrast, the 1850 telegraph rate for the first ten words between New York and Chicago was $1.55.[6] Yet the higher per-word cost was relatively unimportant for time-sensitive exchanges in-volving an immediate decision or transaction. After all, the mail took days or weeks, while the telegraph took minutes.

The telegraph played an important role in communication within ge-ographically dispersed firms. Initially, it was used primarily for urgent but ad hoc communication. Only gradually did railroads and some other geographically dispersed firms begin to use it to speed the routine flows of information necessary to systematic management.

The Railroads and the Telegraph

Because telegraph lines were frequently strung along a railroad right-of-way, the railroads enjoyed a special relationship with the telegraph from the very beginning. Although railroads could obtain service at spe-cial low rates (or even free), however, they did not immediately take advantage of their position.

American railroads were slow (much slower, in fact, than their British counterparts)[7] to adopt the telegraph for controlling the movement of trains. To assure safety, railway companies strictly followed the lessons about standard printed rules and written special orders learned at such cost by the Western Railroad. Initially it was almost unthinkable to substitute telegraphic communication, which seemed to occur by some mysterious and not necessarily reliable process, for strict adherence to written orders. In describing how the first, government-funded tele-graph line between Washington and Baltimore had been employed, Sam-uel Morse briefly noted its one-time use to hold a train in Washington while a special train traveled south from Baltimore.[8] Yet neither Morse nor the railroad apparently saw the potential importance of this inci-dent, since it was not repeated.

The New York and Erie Railroad was the first line to initiate regular telegraphic control of train movement, called dispatching. In 1851 Charles Minot, then superintendent of the Erie, telegraphed ahead a special order to clear the track for his train. His action was viewed as so radical and dangerous that his engineer refused to accept Minot's order to continue, and Minot had to run the train himself.[9] This incident opened the way for regular use of the telegraph in running the Erie.

The Erie's operating procedures soon incorporated routinely monitor-ing the whereabouts of trains and moving them by telegraphic order whenever abnormalities in the schedule arose.[10] The telegraph carried a regular flow of information from stationmasters to the central dispatching point, where the dispatcher entered it on tabular forms that revealed the location of all trains. When the normal schedule had been disrupted, the telegraph carried special orders from the dispatcher to conductors and engineers, and acknowledgments back from them. While the operators transmitted the messages in Morse code, the orders were written out (rather than spoken) at both ends to emphasize their similarity to the written orders they replaced. This system of telegraphic monitoring and intervention allowed the flow of trains to be managed for safety and efficiency. In spite of the Erie's leadership in this area, however, telegraphic dispatching was not routinely used in most American railroads until the 1860s, when they entered a period of growth and systematization.[11]

Dispatching was a specialized railroad function, but the telegraph could also be used for more generic managerial functions. Once again, the Erie Railroad led the way, this time under the general superintendency of Daniel McCallum. As his 1856 Superintendent's Report explained, hourly telegraphic reports were a key element of the system he designed to control and evaluate performance.[12] While these reports carried the information on train positions necessary to dispatching, they also included data useful in distributing cars and engines more efficiently. This information was recorded on special forms and filed for later use by management in analyzing and evaluating the use of engines, the distribution and movement of loaded and unloaded cars, and many other aspects of operations. Daily reports from the conductors and station agents conveyed additional kinds of data to be analyzed. In regular reporting, as in dispatching, the Erie was well in advance of other railroads.

The telegraph made possible (and the special relationship between telegraph and railroad companies made economical) regular and rapid flows of information to improve railroad efficiency. Without the telegraph, reports could not have been delivered to a central point every hour. Yet until McCallum developed his principles of reporting, the telegraph was not used systematically. Even after McCallum's innovations, other railroads were slow to imitate the Erie.

The Telegraph in Other Dispersed Firms

Although other types of businesses did not enjoy the railroads' radically reduced telegraph costs, some of them also found the telegraph useful in making dispersed operations economical. Chandler tells us, for example, that the integrated meat packing and marketing companies that emerged beginning in the 1880s were dependent on both refrigerated railroad cars and regular telegraphic communication to control the flows of perishable meat.[13] Before that time, cattle had been shipped east for slaughter at many small, independent processing plants that served local markets. With the advent of refrigerator cars, companies like Swift and Armour established a few large and efficient meat-packing plants in the

Midwest. Then they shipped only edible beef east to be sold through networks of distribution houses.

This system, which reduced both production (slaughtering) and distribution (shipping) costs, depended on telegraphic communication. Every day each distribution house wired its orders to the central offices in Chicago, which monitored and controlled the flows of orders to the packing plants and of meat from them. The close coordination allowed by the telegraph was essential to minimize spoilage of the meat, since even in refrigerator cars the meat had a very limited life. Swift and Armour paid about $200,000 a year in telegraph costs.[14] Without virtually instantaneous communication, however, much greater losses might have been incurred, when cars full of beef were stranded on side lines or when too much meat arrived at a single distribution point, and cattle might have continued to be shipped east on the hoof.

Although the high telegraph costs were justified by the need to minimize spoilage, the companies soon saw that the data used for coordination could be passed up the hierarchy for further analysis to improve efficiency: "As in the case of railroads a generation earlier, the managers at headquarters were soon employing the data used in coordinating flows to evaluate managerial performance."[15] The telegraph played a crucial role in enabling such companies to systematize their operations for maximum efficiency.

The development of the telegraph, then, opened the way for regular upward flows of data used initially to coordinate work flow and later to provide data for analysis and evaluation, as well. Because of high costs, however, routine flows of data by telegraph were only likely to be established in a few industries: the railroads, which had a special relationship with the telegraph industry, and industries where work flow was both geographically dispersed and highly time-sensitive, such as the meat-packing industry.

The Early Technology of Written Communication

Much more important to the establishment of systematic flows of communication in most manufacturing industries were the technologies of written communication. Changes in production, reproduction, and storage of documents gradually reduced various deterrents to extensive use of internal communication.

For manufacturing companies of the early and middle nineteenth century, written communication consisted almost exclusively of external correspondence, both outgoing and incoming. The total volume of correspondence was generally small enough to be handled by the owner(s) and frequently one or more salaried clerks. Methods for generating, copying, and storing letters were primitive but adequate for that level of correspondence. These methods underwent minor alterations in the decades following the midcentury mark, but several problems remained.

Outgoing Correspondence

Early in the century outgoing letters were written by hand, initially with a quill but by midcentury with a steel pen.[16] Before they were sent out, they were hand-copied onto the blank pages of a bound copy book. These chronological records of all outgoing correspondence were saved indefinitely, since documentation of external communication and transactions had long been standard business practice.

The first mechanical method of copying to gain widespread use in American business was press copying, first patented by James Watt in 1780 but not widely adopted in business until much later.[17] As the technology came into common use, a screw-powered letter press was used in conjunction with a press book, a bound volume of blank, tissue paper pages. A letter freshly written in a special copying ink was placed under a dampened page while the rest of the pages were protected by oilcloths. The book was then closed and the mechanical press screwed down tightly (see fig. 2.1). The pressure and moisture caused an impression of

2.1   A letter press in action. *(Catalogue for Yawman and Erbe, 1905. Hagley Museum and Library.)*

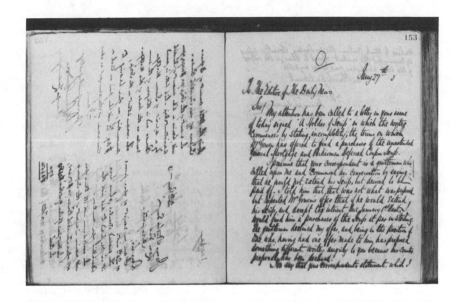

2.2   An open letter press book. *(Pennsylvania Railroad Collection, Hagley Museum and Library.)*

the letter to be retained on the underside of the tissue sheet. This impression could then be read through the top of the thin paper. (See fig. 2.2. The right-hand page shows a copy as it was meant to be read through the tissue, while the left-hand page shows the side on which the previous letter was actually copied. Previous and subsequent letters can be seen through the thin tissue paper, as well.) Press copying was clearly more rapid than hand copying, especially when several different letters were copied at once.

These letter presses were used by some individuals and businesses in the first half of the nineteenth century, but they only came into general use in the second half of the century. Their delayed acceptance, in spite of the obvious savings they offered in time and money, stemmed both from early inadequacies of the technology itself and from the absence of a compelling need in most companies of the earlier period. Initially, even the so-called copying inks produced dim copies that faded rapidly. In 1856 the first aniline dye was invented. These dyes produced clearer and more permanent copies and even allowed an occasional second copy to be made from the original.[18] (Both copies had to be made while the ink was still fresh, however; after a day or so, the ink set and even a first copy was unsatisfactory.) Aniline dyes undoubtedly made press copying a more attractive technology in the second half of the nineteenth century.

In addition, the initial speeding up of business activity that accompanied the spread of the telegraph and the railroad increased the need for rapid and accurate copying. At low volumes of correspondence, a general clerk could copy letters by hand as well as keep accounts and perform other tasks.[19] Thus, the labor involved in hand copying was probably not perceived as a separate cost. As business activity and thus correspondence picked up, this labor increased and became more visible. A letter press reduced the labor cost, both by decreasing copying time and by allowing an office boy to do the copying once performed by a more expensive clerk. At the same time, it eliminated the danger of miscopying. Copies were now facsimiles of the letter sent, down to the signature. Thus, the technology fulfilled the growing need for faster, cheaper, and more accurate copying. In fact, the technique was so satisfactory for moderate volumes of correspondence that it continued to be used by some organizations well into the twentieth century.[20]

Although the process of press copying was clearly more efficient than that of hand copying, the resulting press books did not differ much from hand copy books: both served as centralized and chronological storage systems. Generally a business used a single copy book at any one time, starting a new volume whenever one was filled; occasionally a businessman would keep an extra copy book at home for correspondence he sent from there. Because making a readable extra copy of a document was difficult even after the invention of aniline dyes, second copies were made only under special circumstances.

Usually one central set of books, kept at the main office, contained all the copies of outgoing correspondence used in maintaining the business. Both the volumes themselves and the letters copied within the volumes were chronologically organized. To aid in locating specific letters, alphabetical indexes were provided in the front or back of press books. In well-run businesses, the name of each person or company with which the firm corresponded was entered in the index, and each time a letter to the recipient was copied, its page number was listed after the name (see fig. 2.3). Each volume was separately indexed, and the index usually listed only the page numbers, not the subjects, of all letters to a given recipient.

Incoming Correspondence

Although outgoing correspondence was copied and stored in a bound hand copy or press copy book, loose incoming correspondence was handled quite differently. Through the middle of the century, the pigeonhole was the primary storage device for such correspondence, sometimes supplemented by a desk spindle on which papers could be impaled.[21] Generally, a businessman folded his incoming letters and put them in the pigeonholes above his desk. He might sort them by sender or just put them into the handiest pigeonhole. Letters were frequently "abstracted" on the outside—that is, the recipient would write the sender's

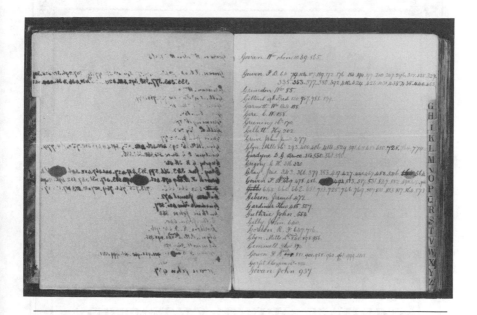

2.3   The index of a letter press book. *(Pennsylvania Railroad Collection, Hagley Museum and Library.)*

name, the date of the letter, and sometimes an indication of its subject on the outside of the folded letter to help in locating it later. When the pigeonholes filled up, letters were often tied up in bundles and stored in a safe place.

At low levels of correspondence, such a storage system was satisfactory. The letters of the last year or two could be kept in the pigeonholes, so that by the time letters were retired they were not being actively used. As business—and correspondence—increased in the midcentury period, however, pigeonhole desks and cabinets became less satisfactory storage systems. In spite of the burdensome folding and abstracting, locating a given document became increasingly difficult. If abstracts contained only the sender's name and the date, someone seeking one letter might have to unfold several to locate it. Capacity problems were even more pressing than efficiency issues. By the late 1860s and early 1870s, the yearly volume of incoming correspondence in many firms already far exceeded the space in a typical pigeonhole desk. Several preliminary responses to these problems preceded the advent of vertical filing in the 1890s.

The first such development, which addressed only the capacity issues, was more colorful than significant. The pigeonhole desk was expanded to its utmost (though still inadequate) extent in the Wooton Patent

THE DESK OF THE AGE!

—THE ⚜ WOOTON—

—PATENT—

CABINET·OFFICE·SECRETARY.

THE BEST OFFICE APPLIANCE IN THE WORLD.

A REMARKABLE SUCCESS.

THIS celebrated appliance was patented and introduced to the public in October, 1874, since which it has found its way into all portions of the civilized world.

It has had a large sale in Great Britain, France and Germany, and orders have reached us from the remotest countries on the globe. South America, Mexico, China, Japan, India, Egypt, Turkey and Australia have all paid tribute to its superior merits.

Wherever presented, at home or abroad, it has at once commanded the admiration of the best business classes, and has spontaneously elicited from these the appellation of THE DESK OF THE AGE. Nor has this title been accorded it without most excellent foundation. Considering its wonderful conveniences and its admirable adaptation to all classes of business, any less enconium would fall short of the praise to which it is fairly entitled.

THE WOOTON DESK MFG. CO.,

Indianapolis, Ind., U.S.A.

FRANK H. SMITH, PRINT., INDIANAPOLIS

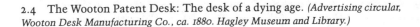

2.4   The Wooton Patent Desk: The desk of a dying age. *(Advertising circular, Wooton Desk Manufacturing Co., ca. 1880. Hagley Museum and Library.)*

Desks.[22] Patented in 1874, the Wooton Desks were elaborate cabinet desks with locking, swing-out cases containing pigeonholes and drawers of various sizes and shapes (see fig. 2.4). An advertising circular described one of the models as follows:

> Everything that ingenuity can suggest or devise to facilitate desk labor, has been introduced in our Secretary. . . . Its comprehensive character is such that ample accommodations are afforded for the requirements of the *most voluminous business*. Every facility is furnished for a thorough and systematic classification of books, papers, memorandums, etc. Through its aid the usual fret and worry of office work is converted into a positive pleasure. It is a MINIATURE COUNTING-HOUSE, with a combination of such conveniences as are found best adapted for the manipulation of office work, and these *all under one lock and key*.[23]

The desk catered to the one-man business, the ad copy continued, in which "The operator having arranged and classified his books, papers, etc., seats himself for business at the writing table [of the Wooton Desk], and realizes at once that he is *'master of the situation.'* "[24]

While the desks may have been "miniature counting-houses" for the one-man business, this concept of business was already giving way to the development of the modern office. In fact, in the 1880s the company's advertising rhetoric sometimes took forms more likely to appeal to progressive businessmen interested in systematic management: "Awkward desks which served the requirements of past decades are no longer adequate for the busy *to-day*, and progressive businessmen throughout the country are rapidly adopting our labor-saving appliances. They possess merits of so positive a character that an appreciative public has readily accorded them the title of " 'THE DESKS OF THE AGE.' "[25]

Even the transformation of the "miniature counting-house" into a "labor-saving appliance" failed to save the pigeonhole concept. The inefficiency of retrieving folded items remained. And because the Wooton Desk, like a normal pigeonhole desk, was not infinitely expandable, it was only a temporary expedient in this expansionary period. While it might handle the personal papers of such Wooton owners as John D. Rockefeller and Jay Gould,[26] or the firm records of very small businesses, it could not begin to handle the external and internal correspondence and records of growing, systematically managed firms. In the 1890s, the Wooton Desk ceased production.[27]

Flat filing, which emerged in the second half of the nineteenth century, was a more satisfactory interim storage technology that increased efficiency of retrieval as well as capacity.[28] The various types of flat files, in which documents were stored flat rather than folded, eliminated abstracting, reduced retrieval problems, and allowed unlimited expansion. Forms of flat files ranged from bound volumes, simple imitations of the bound press book used for copies of outgoing correspondence, to rear-

SHIPMAN'S PATENT GUMMED LETTER AND INVOICE FILE

By this method papers are inserted on the gummed stubs, thus making them in book form.

2.5   Equipment for flat filing in bound form. *(Catalogue and price list, Lucas Brothers, Inc., ca. 1912. Hagley Museum and Library.)*

rangeable cabinet flat files, harbingers of the vertical filing systems that were to revolutionize storage and retrieval of documents.

An early method of filing incoming correspondence flat, used by some of the railroads, involved creating a bound, chronological record comparable to, although not merged with, the copy books of outgoing correspondence.[29] Bound volumes of narrow paper stubs with indexes at the front or back usually served as the base of this system (see fig. 2.5). As they were received, incoming letters and other documents were numbered, indexed as in press books, and glued onto the paper stubs. Occasionally the incoming correspondence was sewn directly into bindings or glued onto full blank pages. Like the copy books, these bound volumes prevented the loss of documents but fixed them in purely chronological order. Moreover, while indexing required less work than abstracting, and searching for a given letter did not require unfolding and refolding each letter checked, the chronologically based system was inflexible and retrieval still slow.

More common forms of flat filing—in letter boxes and horizontal cabinet files—stored documents in unbound form, so they could be arranged and rearranged as desired and removed when needed. In these files the papers themselves, rather than index entries, could be arranged alphabetically by correspondent or by some other appropriate scheme.

The box file or letter box, forms of which are still available today for home use, was the most common flat filing device (see fig. 2.6). Typically, it "consisted of a box, its cover opening like a book, with twenty-five or twenty-six pages or pieces of manila paper, tabbed with the letters of the alphabet and fastened into the box at one side, the papers being filed between these sheets."[30] When not in use, the letter boxes were stored on edge on a shelf, as if they were bound books. Each box generally held all the correspondence received during a given time period, arranged alphabetically by first letter and chronologically within a given alphabetical division. When one box was full, another was started. Letter boxes without alphabetical dividers were also used to hold all the letters from a frequent correspondent. Because they were unbound, such letters could be moved from the alphabetical sequence into a separate box when they became too bulky.

2.6   A box file for incoming correspondence. *(Catalogue and price list, Lucas Brothers, Inc., ca. 1912. Hagley Museum and Library.)*

Some railroads, which had extensive internal and external correspondence, organized it in letter boxes using a numerical system adopted from British government registry systems for documents.[31] In this system, every time a new major correspondent or subject appeared, a new letter box was assigned the next consecutive number. A separate alphabetical index of correspondents and subjects with the corresponding box numbers was maintained to allow access. This system allowed ready expansion at the end of the series but required the extra indexing step for storage and retrieval.

Letter boxes were common in the midcentury period, but around 1868, cabinet flat files were introduced, probably by Amberg File and Index Company.[32] These predecessors of vertical filing cabinets were simply wooden file sections or cabinets with drawers of the same dimensions as letter boxes, but inserted horizontally rather than shelved on edge (see fig. 2.7). The drawers, like letter boxes, could be organized and indexed in the ways just discussed. A variant form of flat file introduced around this time was the arch file, often called the Shannon file after one trade name. These devices held correspondence in place by piercing it with arched metal clamps (see fig. 2.8). The papers were fixed to prevent loss (as in bound storage methods), but the clamps could be removed to rearrange them. Shannon files came in box or in cabinet form and could be organized and indexed in various ways. By 1881 Cameron, Amberg, and Co., Amberg's successor and a major source of filing apparatus during this period and into the twentieth century, carried full lines of box files, cabinet flat files, and Shannon files.[33] Other companies followed their lead.

These flat files improved on the pigeonhole desk by being expandable and by making documents more readily retrievable; they improved on the bound volumes of incoming correspondence by being rearrangeable according to the user's needs. The advantages of such flat filing systems over other available systems for handling incoming correspondence quickly made them popular. They were the "modern" way of storing correspondence in the 1870s.

By 1870, the quill pen, hand copy book, and pigeonhole had given way to the steel pen, letter press, and flat file. Technology both responded to and facilitated the increase in correspondence caused by the midcentury growth and speeding up of business. The press copier, for example, was available by the end of the eighteenth century but did not gain broad use until after the middle of the nineteenth century, with the invention of aniline dyes and the general speeding up of business. By creating a need for faster and less expensive methods of copying, business growth spurred the diffusion of the technology. On the other hand, without the technology of press copying, handling increasing correspondence would have been much more difficult as well as more expensive. The evolving technology clearly facilitated growth in a company's external transac-

2.7   Cabinet flat files. *(Catalogue, Yawman and Erbe Manufacturing Co., 1910. Hagley Museum and Library.)*

Genuine Shannon Sectional Cabinets

THE Shannon Arch method has been for twenty years, and is today, the safest way to file letter or cap size papers.

Prices quoted include stock index (please specify in your order the *style* of indexes desired).

Price *"with lock"* means the *"I and A"* combination locking device described on page 6.

Compressor on cover keeps papers compactly together; printed form shows record of transfers for locating back correspondence.

No. 1020. 3-Dr. Letter-size Shannon Half Section with Document or Utility Vertical File, $10.40.
No. 1018. Cap-size, $10.65.

No. 19. 3-Dr. Letter-size Section, $6.50.
No. 23. 3-Dr. Cap-size Section, $6.75.

No. 20. 9-Dr. Letter-size Section, $16.25.
No. 18. 9-Dr. Cap-size Section, $16.75.

2.8   Shannon arch files in cabinet sections. *(Catalogue, Yawman and Erbe Manufacturing Co., 1910. Hagley Museum and Library.)*

tions. Far greater growth was on its way, however, and these technologies would soon be inadequate.

Problems in the Office

By the 1870s and 1880s, the railroad and telegraph, along with new mass production technologies, had spurred even greater growth in the pace and size of business. Moreover, the nature of the firm was changing rapidly as departments proliferated, the vertical hierarchy stretched, and the principles of systematic management gained acceptance. With this growth and change came both greater external correspondence and the development of internal correspondence. The weaknesses of the pen, press book, and flat file system of handling written correspondence soon became evident.

The speed with which documents could be handwritten was limited.

In 1853 the speed record for handwriting was thirty words per minute, while normal writing was much slower.[34] One method of reducing the correspondent's time was available but rarely used. Since the time of the Roman Empire, various systems of shorthand had allowed one person to record the words of another, uttered at a normal speaking rate.[35] At this time, however, shorthand systems were used primarily to allow court reporters to record court proceedings and newspaper reporters to make rapid notes on political speeches. Stenography, or phonography (as it was often called)—taking dictation by shorthand and later transcribing it—seems to have played only a limited role in business before the typewriter. Private commercial schools taught regular penmanship, along with bookkeeping and commercial arithmetic, as their normal curriculum; stenography was taught only occasionally.[36] Had it become more widespread in business at this time, stenography might have reduced the amount of time the correspondent spent on routine letters. Stenography alone, however, did not completely solve the problem of speed; it only shifted the time-consuming task of handwriting from an owner or manager whose time had a high value to a stenographer whose time had a lower value.

Copying technology also left much to be desired. While the letter press now allowed the reproduction of one or sometimes two copies of a freshly written outgoing letter much faster than by hand, no method existed for copying incoming documents or any document more than one day after its creation. Moreover, press copying still took time, and it was messy and inconvenient. In addition, it could not produce large numbers of copies. When the railroads, driven by the need for safety and efficiency in geographically dispersed firms, wanted to issue notices and orders to large numbers of employees, they were forced to have them printed, an expensive and slow process. More firms would face this problem as they grew and turned to the principles of systematic management.

Finally, the storage system had several weaknesses that created more and more problems as external correspondence increased and as internal correspondence grew up. With both box and cabinet files, retrieval of documents was still slow and laborious (though faster than with folded documents in pigeonholes), and rearrangement, while possible, was not easy.[37] To locate correspondence in an opened box file or a horizontal cabinet file, all the correspondence on top of the item sought had to be lifted up. Since the alphabetically or numerically designated drawers in horizontal cabinet files filled up at different rates, correspondence was transferred out of active files into back-up storage at different rates as well. And the drawers could not be allowed to get too full, since then papers would catch and tear as the drawers were opened. Letter boxes had to be taken down from a shelf and opened up, a time-consuming operation when large amounts of filing were done. Moreover, when each letter box had its own set of alphabetical dividers, expansion required starting a new box with a new alphabetical series. As the physical basis for filing, box and cabinet flat files were an improvement on pigeon-holes, but still far from perfect.

Press books, which stored outgoing documents in bound chronological form, caused even more problems. The press books at a firm's main location generally formed a single, centralized series, requiring everyone to go to one place to find copies of outgoing correspondence. Moreover, outgoing and incoming documents were stored in two completely different systems organized in different ways. As figure 2.9 illustrates, locat-

2.9   An illustration of how difficult it was to locate correspondence in press books and letter boxes. *(Catalogue for Yawman and Erbe, "Rapid Roller Letter Copier," 1905. Hagley Museum and Library.)*

ing all of the correspondence with a given party or on a given issue was an increasingly laborious and unsure process.

The evolution of internal written communication created more pressures on the storage system. It was not clear, for example, whether internal correspondence should be handled as incoming, outgoing, or both. Moreover, centralized storage of internal correspondence posed problems. In the case of downward communication, for example, employees needed easy access to general orders and policies as well as to special orders addressed specifically to them. Finally, press book indexes, which generally included only the page numbers of all letters to a given recipient, became less useful as the number of communications to a single internal correspondent increased.

Thus, the technologies of written communication created barriers to the growth of external and internal correspondence. As other factors encouraged its growth, however, the technological barriers began to be surmounted. A veritable revolution in communication technology took place. Although innovations in the various parts of the system for handling written communication progressed more or less simultaneously and affected each other, they can be broken up into three categories: in production, the typewriter; in reproduction, carbon paper and duplicators; and in storage, vertical filing.

Revolution in Production: The Typewriter

The physical creation of correspondence had always been a laborious and time-consuming process. In the second half of the nineteenth century, when so much else was speeding up, it was natural that people should look for ways of speeding up the production of documents. The typewriter, which appeared in the final quarter of the century during the period of most rapid firm growth and transformation, provided the necessary speed.

Attempts to mechanize the production of documents had begun as early as 1714, but for a century and a half none of them progressed beyond the experimental stages to a saleable product.[38] Some succeeded in making machines that produced printlike writing, but because none of the machines worked faster than handwriting, they were not attractive to those who might have bought them.[39] Consequently, these machines never even approached commercial viability. Finally, in 1868, Christopher Latham Sholes, along with Carlos Glidden and Samuel W. Soulé, registered the first of a series of U.S. patents that would at last produce a commercially viable product (see fig. 2.10).[40]

That first patent was only a beginning, and the typewriter progressed slowly toward commercial success. The machine underwent four more years of development under the direction of James Densmore before it first came to market.[41] During this period Sholes settled on the current keyboard configuration, beginning with an alphabetical arrangement

2.10   A typewriter model built on the Sholes patent. *(From William Henry Leffing-well, ed.,* The Office Appliance Manual, *1926.)*

and rearranging keys to minimize their clashing.[42] The first twenty-five nonexperimental machines were manufactured in a makeshift private factory in 1871–72. Although these machines were eventually sold, it soon became clear that production expertise was needed to produce a typewriter of acceptable quality in large quantities. At this point the developers approached the Remingtons, whose factory, one of the largest of the time, manufactured firearms and sewing machines. An agreement was reached, and the first machines from the Remington factory reached the market in 1874.[43]

Ironically, Sholes and Densmore, unlike the developers of the telegraph and telephone, did not immediately target the business world as their major market. They originally looked to a much smaller and more specialized user: the court reporter.[44] The first typewriter Sholes sold, an early experimental model, was bought by a court reporter named Charles E. Weller for use in his work.[45] Indeed, most of the first small batch of pre-Remington typewriters went to court reporters and telegraphers, not to "ordinary business firms."[46] An advertising brochure issued for the first Remington-made machines put court reporters at the top of the list of prospective users, followed by lawyers, editors, authors, clergymen, and others. Only in its last sentence does it suggest that "the merchant, the banker, ALL men of business can perform the labor of letter writing with much saving of valuable time."[47]

Sales were slow in the late 1870s but increased rapidly in the 1880s. In the first five years on the market, 1874–78, only four thousand of the Remington typewriters were sold.[48] Meanwhile, a few improvements, including the shift key to allow upper and lower case, made them more appealing to buyers. By 1880, a second company entered the market.

The business world's discovery of the typewriter in the 1880s undoubtedly accounts for the rapid sales growth, competition, and further technological innovation of that decade. From 1881 to 1884, the two companies sold a total of almost eighteen thousand machines, over four times as many as were sold in the first five years.[49] As the market grew, more competition entered it and soon spurred more improvements and more sales. By 1886, according to the *Scientific American*'s estimate at that time, a total of fifty thousand typewriters of all makes had been manufactured. " 'Five years ago the type writer was simply a mechanical curiosity,' " noted one observer in 1887. " 'Today its monotonous click can be heard in almost every well regulated business establishment in the country. A great revolution is taking place, and the type writer is at the bottom of it.' "[50] Annual capacity of all companies in 1886 was about fifteen thousand, but by 1888 the Remington Standard Typewriter Company (now a separate company bought from the Remingtons in 1886) made more than that in one year. In the 1890s, Underwood introduced the first successful "visible typewriter," which allowed the typist to see what was being typed (see fig. 2.11).[51] Soon all typewriters had this feature, and more were being sold than ever. In 1900, according to the 12th Census, 144,873 typewriters were sold.[52] After a slow start, perhaps partly attributable to the developers' initial misreading of the market, the typewriter rapidly became a familiar part of office life.

Sholes's typewriter undoubtedly owed its success in business in part to the fact that it, unlike its predecessors, was faster than the pen. In the late 1870s the *Typewriter Magazine*, started by a Remington sales agent, pointed out that the typewriter could type seventy-five words per minute, while the pen typically produced only about twenty-four words per minute.[53] After the development of touch typing in the 1880s, typing

VISIBLE WRITING ILLUSTRATED

Gentlemen.
    We are in receipt of your favor of the 2 nd..instant
and beg to assure you that the matter referred to will be treated
as strictly confidential.
    The several other matters treated in your communication will
receive our careful attention.
                    Very truly yours.

The Underwood Typewriter, Model No. 5, as it appears after
writing the last word of a business communication.

2.11 An Underwood visible typewriter. *(Catalogue, Underwood Typewriter Company, ca. 1905. Hagley Museum and Library.)*

became even faster: typing speed contests in the early twentieth century produced speeds of around one hundred twenty net words per minute.[54] Although the average typist did not approach this speed, a normal typing rate of eighty words per minute was more than three times faster than the normal handwriting rate. This improvement in efficiency appealed to the expanding businesses of the late 1870s and the 1880s.

The typewriter also opened the way for another change in the procedure by which documents came into being in business establishments: creation was completely separated from final production. Before the typewriter, the writer sometimes drafted a document (or even, in a few cases, dictated it to a stenographer) and turned it over to a clerk to copy it out in final form. But such separation was by no means universal. An owner or manager might compose and produce his own correspondence some or all of the time. Conversely a clerk, who frequently was in training for advancement in the firm, might compose as well as produce many routine letters. But the almost universal separation of those functions only occurred in conjunction with the typewriter. A whole new class of clerical workers arose to operate the new machines, as well as to take dictation and perform other clerical functions related to the handling of written documents.[55]

The statistics on typists, stenographers (who typed up their shorthand notes), and secretaries during the period following business adoption of the typewriter suggest how rapidly the severance of composing from producing took place. In 1890, there were already 33,000 people employed as stenographers and typists.[56] By 1900, the U.S. Census showed 134,000 in the broader occupational category including stenographers, typists, and secretaries.[57] Such workers continued to flood into business and government in increasing numbers. By 1910 that occupational category had almost tripled to 387,000, and by 1920 it had doubled again to 786,000.

The almost universal separation of composing from typing had both technological and social roots. It reflected the skill needed for rapid typing, a technological factor, as well as the systematizers' conscious search for efficiency at all levels in business. Increasing subdivision of tasks and specialization of jobs, with the techniques of systematic management to coordinate the various specialized elements, helped achieve this efficiency. The typewriter provided an obvious opportunity to extend this principle to office work.

Although the earliest typists (often referred to as "typewriters") were trained primarily by the manufacturers, the private commercial schools that taught penmanship and bookkeeping soon began training typists.[58] As early as the mid-1870s, before the typewriter became popular, one major commercial school magnate began to hold special classes in shorthand and typing.[59] Soon it became clear that stenography and typing could be linked to achieve rapid production of documents, and the two skills began to feed on each other.[60] In 1888 the U.S. Commissioner of Education noted the spread of departments and even whole schools of stenography and typing.[61] By 1900 the amanuensis course, which included both skills, accounted for 36 percent of the enrollment in commercial schools. By the end of the century, public high schools had also begun to offer typing and other business skills.[62] The commercial and public schools together trained an ever-growing workforce to operate typewriters.

Of those learning to type, an increasing proportion were women. The commercial schools had welcomed women long before the coming of the typewriter; nevertheless, in 1871 only 4 percent of total commercial school enrollment was female.[63] The popularization of the typewriter, however, brought women pouring into the commercial schools and business offices. The proportion of women in commercial schools increased to 10 percent by 1880, 28 percent in 1890, and 36 percent in 1900. The employment statistics are even more revealing: as early as 1890 64 percent of all stenographers and typists were women, and by 1900 the figure stood at 77 percent. The composition of the office workforce had changed radically.[64]

The typewriter and those who used it also transformed the production and use of documents. By taking physical production out of the hands of those who composed the messages, stenographers and typists reduced the amount of time the highly paid executive spent on correspondence. The typewriter itself decreased the amount of less expensive clerical time taken to transcribe correspondence from dictation or from a handwritten draft. Moreover, by replacing illegible scrawls with neat, print-like text, the typewriter reduced both the time needed to read the document initially and the time needed to locate a given document in the files later. Finally, the large workforce of trained typists and secretaries helped standardize the formats and conventions for the new genres of internal written communication discussed in the next chapter.

Early in the twentieth century another office machine appeared to help reduce even further the time and money involved in producing written communication: the dictating machine (see fig. 2.12). Within

The Edison Commercial System
Conducted with the Business Phonograph

FROM BRAIN
TRADE MARK
Thomas A. Edison.
TO TYPE

Appliances Manufactured and installed by the
National Phonograph Company, Commercial Dep't
Main Office Orange, N.J.U.S.A.

2.12   The Edison phonograph as adapted to business uses in the early twentieth century. *(Advertising circular, National Phonograph Company, 1907. Hagley Museum and Library.)*

ten years of Edison's invention of the phonograph in 1877, a dictating machine known as the Graphophone was being marketed as an office machine for recording dictation.[65] Other brands followed. By 1911 a government study found that many large companies, such as the Pennsylvania Railroad, were using dictating machines in conjunction with typewriters to eliminate or reduce the more expensive use of stenographers.[66] In fact, the study reported, in the Pennsylvania Railroad's correspondence department, "the installation of these machines enabled the typewriters [i.e., typists] to produce from 60 to 80 letters per day, whereas under the old system the average output of each typewriter was only 30 to 40 letters daily. It also reduced the cost per letter from 5.2 cents to 2.7 cents." Although dictaphones never became as universal as typewriters, where they were used they further increased the speed and decreased the cost of document production.

Typewriters and related innovations accelerating the production of documents both fed on and facilitated enormous growth in external and internal written communication. The expanding market for these devices encouraged competition and continued technological evolution, while the reductions in time and cost associated with the typewriter increased the attractiveness of written communication as a managerial tool. Finally, the typewriter opened up the way for new methods of duplicating and filing documents that made certain types and uses of internal communication practical for the first time.

Evolution in Reproduction: Carbon Paper and Duplicators

As the revolution in *producing* written communication was progressing, a quieter but equally important evolution in *reproducing* documents was also occurring. Carbon paper soon replaced the letter press for making routine file copies, while various sorts of duplicators provided, for the first time, rapid and inexpensive methods of mass reproduction. Around the turn of the century, a method of photo-reproducing copies of documents after their creation was developed. These innovations were closely linked with other technological developments (the typewriter and vertical filing), as well as with the growth of organizations and systematization of management. The increasing need for internal written communication spurred the creation and diffusion of these new copying methods, while they, in turn, spurred the evolution of the internal written communication system so integral to the systematic management movement.

Beyond the Press Book

With the press book, the copying of correspondence at its point of origin was directly linked to its storage. The typewriter did not necessarily displace this system. Typewriter ribbons could be impregnated with the aniline copying inks used in pens, and the typed originals could be used to make press copies in the press book. Initially, in fact, many companies did just that.

Other factors, however, encouraged change. As long as copies of outgoing correspondence were made in bound volumes, outgoing and incoming correspondence on the same subject had to be stored in different systems, creating the retrieval problems mentioned earlier. The development of internal communication further complicated that problem. Flexibility in storage and retrieval required a way of reproducing unbound file copies. In the late nineteenth century, two such copying methods emerged: the rolling press copier and carbon paper. Eventually carbon paper became the method of choice.

The rolling copier, a simple adaptation of press copying, involved technology that was already available but had been ignored until the need arose. It used a pair of rollers, rather than a screw press, to produce press copies on a continuous roll of tissue paper (see fig. 2.13).[67] The paper was then cut up to provide loose copies. One of Watt's original eighteenth-century press copiers had used rollers rather than a screw-press to produce loose copies—on individual sheets of tissue paper rather than a continuous roll—but it had eventually disappeared for lack of demand.[68] (At that time, bound volumes were preferred, presumably because they more closely resembled hand copy books; amounts of correspondence were not yet great enough to make retrieval a problem.) This older form of press copying technology reemerged as a popular alternative when increased external correspondence and the emergence of internal written communication created a need for more flexible storage. The loose press copies could now be stored in flat files with originals of incoming or internal correspondence on the same subject.

The rolling copier had some drawbacks, however. Press copying was still a messy operation that took extra time before letters were sent out and often produced blurred copies. Also, since copies were cut from a continuous roll, they varied in length and thus produced disorderly files. After the introduction of vertical filing in the 1890s, the problem of disorderly files was exacerbated by the fact that the flimsy tissue paper copies would not stand on edge but collapsed to the bottom of the file folders. Finally, because the copier was generally kept in a central mail room, making copies of internal correspondence that would not otherwise need to go to the mail room was inconvenient. The rolling press copier still left room for improvement.

Carbon copying, a more significant advance in reproducing documents, was also based on an older technology. Ralph Wedgewood invented carbon paper (which he called "carbonic" or "manifold" paper) in the first decade of the nineteenth century, and by 1823 it was available in the United States.[69] But carbon paper could not be used successfully with the pens of the time. Carbon copying required pressure that could not be applied to a quill, steel, or gold pen without ruining the tip or

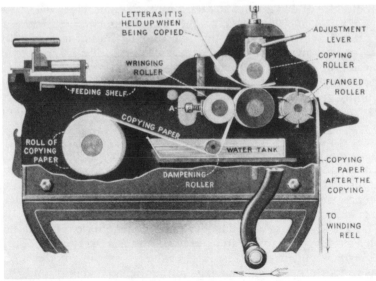

2.13 Rolling press copier, external and cut-away view. *(Catalogue, Yawman and Erbe, "Rapid Roller Letter Copier," 1905. Hagley Museum and Library.)*

tearing the paper. Moreover, the carbon paper of the early nineteenth century was soaked through with carbonic ink, and thus was double-sided. It was intended to be used between a top sheet of thin tissue paper and a bottom sheet of normal letter paper, with a stylus as the writing instrument. The "original," under this system, was the carbon copy on the letter paper, while the "copy" was the tissue paper sheet through which the writing could be seen. With the later development of single-sided carbon paper, a pencil and regular letter paper could be used for the original. Since neither pencil nor carbon copy was considered appropriate for business letters, however, such "manifold" systems had only limited uses such as for receipts, orders, and telegrams.

Only with the advent of the typewriter did carbon paper challenge the supremacy of press copying. The link between the typewriter and carbon paper began in the prehistory of the typewriter. The Pratt machine of 1866, one of the failed writing machines preceding the ultimately successful Sholes machine, used carbon paper instead of an inked ribbon to produce the typed document.[70] Sholes's earliest experimental model also used carbon paper, which he borrowed from a friend at Western Union.[71] Although the inked ribbon quickly replaced carbon paper for making the primary typed document, Sholes and his associates understood the potential for using carbon paper to make copies. When the typewriter was first demonstrated to the Remingtons in 1872, a manufacturer of carbon paper attended and demonstrated how the two could be used together.[72]

This pairing of new and old technology revolutionized small-scale reproduction of documents at their point of origin. Here was a way to copy, not just one, but several copies of a document at the same time the original was produced. Yet carbon copying did not immediately replace press copying. Many used the press book or a rolling copier in conjunction with the typewriter. Gradually, however, the advantages of carbon copying became clear. From 1910 to 1913 the Taft Commission on Economy and Efficiency studied the office practices of large firms in order to recommend more efficient office methods in government. In 1912 it noted that "by the almost universal practice of business concerns, the carbon copy has supplanted the press copy as a record of outgoing correspondence."[73] This statement was based primarily on large businesses: many smaller companies continued to use the rolling copier and even press books for some years.[74] The commission's report provides evidence, however, that by 1912 the larger companies had embraced carbon copying.

The Taft Commission study, especially when viewed in conjunction with a similar government study of 1906,[75] reveals the reasons for carbon copying's ascendancy. Both studies evaluated copying methods on the basis of four criteria: permanency, authenticity, economy, and adaptability. The first, permanency (which was of less importance to business than to government), favored carbon copying. The studies found that press copies faded significantly in fifty years, while carbon copies did not. Authenticity, that is, the legal standing of the copy as evidence of the original, was widely debated in government and business,[76] though it was relevant only for external, not internal, business correspondence. Some legal experts argued that press copies were better evidence, because they were made immediately before mailing and thus showed the final form of the document with any corrections and the signature. Carbon copies did not show the signature; also, changes could be made to the original after the carbon was made. On the other hand, some authorities argued that carbon copies were better evidence, since they were made at the same time as the original. The 1906 commission did not resolve this debate. The Taft Commission requested a definitive legal opinion, which supported the claims of carbon copies.[77]

While these first two issues may have played some role in business's switch to carbon copying, economy and adaptability were probably the central factors. By 1912 carbon copying held a distinct advantage in economy or cost, perhaps the most important criterion for business. In the 1906 study, carbon copying was considered more expensive than press copying, though the basis for that determination was not entirely clear. By the 1912 study, there was no doubt that carbon copying was much less expensive than press copying. The report cited a materials cost of $.56 per thousand for carbon copying and $2.80 per thousand for press copying into bound books (materials for press copying with a rolling copier would be less, since no binding would be required, but probably not less than for carbon copying). Moreover, it noted the substantial labor cost for those employed partially or solely to make press copies.

In both studies, carbon copying clearly won out in adaptability, which involved indirect as well as direct costs. First, by eliminating an extra step it helped get the mail out more rapidly and efficiently. Second, the copies were neater and easier to read. Third, the onionskin paper on which carbon copies were usually made had more body than the soft and flimsy tissue of press copying, making carbon copies easier to handle and less likely to fall to the bottom of vertical files.

Finally, another issue not treated in either government report may have influenced business adoption of the technology: carbon copying was better suited to the newly developing internal communication. Many internal documents needed to reach not just one person, but several. Because press copying could produce at most two (not very satisfactory) copies, sending the same notice to several department heads required either retyping the notice multiple times or circulating a single copy from person to person, according to a list. In the latter case, none of the department heads got to retain a copy; to refer to it later, they had to send to the central files. Carbon copying, on the other hand, could produce up to ten readable copies at a single typing, thus allowing each department head to retain a copy. Moreover, carbon copying could occur anywhere in a company where there was a typewriter, a characteristic particularly useful in internal communication. Carbon paper's suitability to the developing internal communication system undoubtedly spurred business's adoption of the technology.

With the help of the typewriter, carbon paper had revolutionized the reproduction of small numbers of copies at the point of origin. Such reproduction was now simple, inexpensive, and convenient for almost any written communication.

Mass Duplication of Internal Communication

As companies grew and as systematic management techniques took hold in the last quarter of the century, owners and managers increasingly wanted to distribute written rules, instructions, and announcements to large numbers of employees. Notices could be circulated or posted, but such systems were often slow and unreliable. Days and weeks might pass before the communication reached all relevant employees; and without copies of their own, employees might forget a new instruction after they read it. Sending notices out to be printed was a slow and expensive alternative. Earlier, the railroads had been forced to adopt it anyway, since size, geographical dispersion, and safety demanded mass distribution of written notices and orders, but manufacturing companies were not eager to incur the expense of following their example.

Firms clearly needed a fast, inexpensive, and convenient method of making large numbers of copies on their own premises. The final quarter of the century witnessed the emergence of several devices for mass duplication, both satisfying the growing companies' needs to communicate in writing with their employees and facilitating further growth and systematization. The technologies fell into two general types: those using aniline dyes, and those using stencils.

The aniline dye methods of this era used a gelatin bed or roll to transfer dye from an original to up to one hundred—thus the common name *hectograph*—copies (see fig. 2.14).[78] An original written or typed in aniline copying ink was placed face down on the gel, transferring the ink to the bed. The original was then removed, and sheets of blank paper were placed one after another on the gelatin bed, from which they picked up the (usually purple) dye. One popular version of the process was developed in 1880 in Germany by Alexander Schapiro, whose name was attached to one brand of gelatin apparatus, the Schapirograph (see fig. 2.14).[79] In the 1920s the gelatin process of transferring aniline dye was to be replaced by the more convenient spirit process.[80]

In stencil duplicating, copies were made by passing ink through holes in an otherwise impervious stencil master.[81] The first stencil process available in America was invented by Thomas Edison.[82] In 1876 he patented his electric pen and duplicating press to be used for what he called "autographic" printing (see fig. 2.15). This pen had a vibrating needle, powered by an electric motor, for its point. As the user wrote with it, the needle point made a series of tiny holes in the paper. The perforated

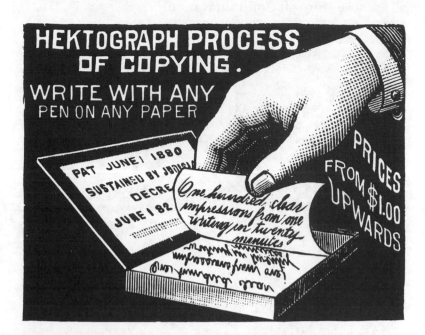

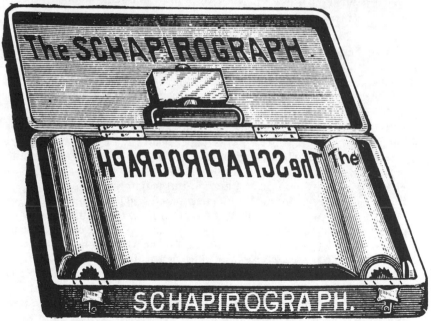

2.14   The Hektograph and the Schapirograph, two devices for dye and gelatin duplicating. *(Catalogue and price list, Lucas Brothers, ca. 1912. Hagley Museum and Library.)*

This Duplicating Apparatus is especially adapted to the needs of Railway and Telegraph Companies.

It is unequaled for the speed, economy and convenience with which circulars, notices, diagrams, etc., can be prepared and issued.

2.15   Edison's Electric Pen, used to make the stencil master, with duplicating press, used to make the copies. *("Catalogue of Telegraph Instruments and Supplies," Western Electric Company, 1983. Hagley Museum and Library.)*

stencil master was then placed in a frame, and ink, applied by means of a roller, passed through the holes to paper below. A single master could be used to make up to five thousand copies, according to one advertising circular.[83]

Edison put his system on the market immediately after patenting it. Initially, advertising emphasized its use in external communication. An 1876 pamphlet advertised it as an eye-catching and economical way of reproducing everything from price lists and advertising circulars to musical scores.[84] Far down the list, in small print, the brochure also suggested that "for insurance reports, notices to sub-agents from general agents, and the small private printing of the company generally, it is invaluable." Within a few years, the importance of duplicating to internal communication was better understood, and this duplicating system was advertised as "especially adapted to the needs of Railway and Telegraph Companies," particularly for circulars and notices.[85] The electrical pen was, however, clearly technological overkill at the time: an electrical device was neither a necessary nor a reliable method of perforating paper to make a stencil.

Mechanical methods of producing stencil masters of handwriting soon emerged to replace it. The most common of these in America was the file plate method, which involved placing waxed paper on a file-like grooved surface, then writing on it with a metal stylus to create perforations in the wax. This method was devised in England in the late 1870s but patented in America by Edison. Starting in 1887, the A. B. Dick Company marketed it here as the Edison Mimeograph.[86] Rid of the complications of the electric pen, stencil reproduction of handwriting gained considerable popularity in business. A stencil could produce far more copies than the hectograph, and, according to the advertisements, made much less mess. With each improvement, the process became more accepted in the expanding office of the 1880s.[87]

Yet just when stencil copying of handwriting was gaining acceptance, handwriting was giving away to the typewriter in business. With the spread of the typewriter, stencil technology was adapted to make it suitable for typewriting and further developed to make it faster and less messy.[88] Because typewriter keys could not make an adequate impression on the existing stencil papers, special porous paper was imported from Japan beginning in 1888. With this new paper, the typewriter could join forces with stencil duplicating to make large numbers of clear, legible copies. In the 1890s, spurred by growing demand for duplicating devices, manufacturers developed new methods of speeding up the production of copies after the stencil had been created. First "automatic" flat bed stencil copiers, then rotary stencil copiers were devised, the latter very similar to some still in use today (see fig. 2.16). These rapid copiers made the technology even more convenient.

By the end of the nineteenth century, aniline dye and stencil duplicating methods made it quick, inexpensive, and convenient to disseminate

Edison's Rotary Mimeograph No. 76

LATEST MODEL

PRINTING CAPACITY UP TO 7¼x14 INCHES,
ACCOMMODATING PAPER 8½x16 INCHES OR SMALLER

A duplicating machine is a practical and absolute necessity in every progressive business office.

2.16   A 1910 model of Edison's Rotary Mimeograph. *(Catalogue and price list, Lucas Brothers, ca. 1912. Hagley Museum and Library.)*

instructions, orders, and information to large groups of employees. A little later, some firms came to use the technologies for reproducing in-house magazines, as well. They could even duplicate forms, though the most frequently used forms were still printed. Together with the type-writer, duplicating technologies opened up many possibilities for down-ward communication that would otherwise have been difficult. Indeed, without such technologies, the uniform employee communication de-sirable to systematic managers could only have been attained at great and perhaps prohibitive cost.

Unplanned Copying after the Point of Creation

The duplicating technologies developed during the last three decades of the nineteenth century allowed businesses to make few or many cop-ies of documents as long as they planned ahead. Carbon copies could be made at the time of the document's creation, and press copies could be made up to twenty-four hours afterwards if copying ink had been used originally. Mass duplication required a special dye or stencil master made at the time of creation. No nineteenth-century technology, how-ever, allowed the duplication of normal documents more than a day after they were created. In 1900 Abbé René Graffin, a Frenchman, invented the first copying process that allowed unplanned copies of any docu-ment.[89] This process used a large camera and light-sensitive paper to produce a negative photographic image (see fig. 2.17). It was known in the United States by 1910.

At this time, photocopying technology was quite slow and expensive but was still valuable for some uses. In 1911, the Taft Commission on Economy and Efficiency evaluated the use and economics of the Photo-stat, a brand of photocopier, as part of its investigation of efficiency in office methods.[90] At a cost of $500.00, the apparatus was expensive. Since it had a top speed of about one copy every minute, a Photostat machine could only make about five hundred copies in a day. Moreover, the commission calculated the average cost per copy as $.098, or $98.00 per thousand, two orders of magnitude higher than the $0.56 per thou-sand that the Commission calculated as the cost for carbon copies. While point-of-origin copying was clearly preferable, the commission found the Photostat cost-effective as an alternative to retyping compli-cated documents (especially tables) or redrawing diagrams. Trial runs in five different government agencies showed that it saved from 56 percent to 83 percent of the cost of retyping or redrawing.[91] As one bureau head pointed out, without this machine some things simply would not have been recopied because of the cost and time involved.

Although the report does not indicate how widespread photocopying was in business, large firms and firms working with many drawings and diagrams had probably already reached the same conclusions about the technology, since the commission consistently imitated rather than led business in office technology. A 1926 manual on office appliances,

2.17 A photocopying machine from the 1920s. *(William Henry Leffingwell, ed.,* The Office Appliance Manual, *1926.)*

though clearly predisposed toward almost all labor-saving devices, is probably exaggerating only slightly when it says that "the development of the photocopying machine has demonstrated its place in the conduct of modern business. Consequently, its acceptance is general and the ways in which it is used are of many score."[92]

Photocopying filled an important niche in the range of technologies for reproducing documents. It made possible, for the first time, the reproduction of unplanned copies of any document at any time after the document had been created. The technology was too slow and too expensive, however, to be used as extensively as photocopiers have been used since 1960 when Xerox revolutionized photocopying technology with the plain paper copier. Thus, the photocopiers of the early twentieth century did not affect the internal communication system as profoundly as the mass duplicating machines had done by supporting entirely new types of internal mass communication and as carbon paper had done by making easy and routine the creation of up to ten copies of any document at the point of origin.

Innovation in Storage and Retrieval: Vertical Filing

The significance of turn-of-the-century innovations in storage systems for documents is much less widely acknowledged now than that of the typewriter, carbon paper, and duplicators.[93] Yet files of correspondence and other documents played an increasingly important role in the management of firms by the turn of the century. And while written documents were far from easily accessible in the latter half of the nineteenth century, as illustrated in figure 2.9, that situation worsened as the other technologies supported the growth in internal and external correspondence triggered by firm growth and systematization. A new storage system was clearly needed—one that would allow incoming, outgoing, and internal documents to be combined and to be arranged and rearranged to suit the company's needs.[94] Carbon paper and the rolling copier solved the problem of combining documents of any origin by replacing bound press books with loose copies of outgoing documents. Vertical filing systems in which documents were filed on edge in folders made papers easier to arrange and to use. In the 1890s and early in the twentieth century, the new vertical files helped business organize and make accessible the growing amount of written correspondence and documentation and consequently played an important role in the use and development of internal written communication.

Vertical filing of papers, a new form of flat filing which evolved from the vertical card files used by librarians, was presented to the business world in 1893.[95] Vertical card indexes for libraries had spread in conjunction with the Dewey Decimal System, introduced in 1876. The equipment for these card files was provided through the Library Bureau, an organization that Melvil Dewey founded in the same year to sell supplies and promote the decimal system to libraries.[96] Card indexes and card files gradually came to be used in business both for indexing numerically organized flat files and for compactly recording and storing various types of business information (sales office records, for example). In 1892, the Library Bureau devised guides and folders for filing correspondence on edge and had file cases designed for them. They presented that system at the Chicago World's Fair of 1893, where it won a gold medal.[97]

The debut of vertical files signaled the demise of other filing systems. However, the changeover was not immediate and universal. One expert noted that "the twentieth century had arrived before the vertical file had made any appreciable headway."[98] If that was indeed the case, vertical filing equipment made rapid headway in the early twentieth century because, by 1911, according to the Taft Commission, "vertical flat filing [had] practically supplanted all other systems" in the large companies it investigated.[99]

Like other forms of flat files, vertical files stored papers flat, eliminating the need for folding and abstracting. This type of filing apparatus, however, also had several advantages over letter boxes and horizontal cabinet flat files.[100] Most importantly, the folders allowed related papers to be grouped together and easily removed from the files for use. The movable dividers and tabs allowed the papers and folders to be arranged or rearranged as desired, with expansion possible at any point. In addition, papers were much easier to handle when they were filed on edge, since a paper or folder could be removed without lifting out papers above it. Vertical file drawers also held eight to ten times more than box files or horizontal file drawers, thus saving time by allowing the filer to open and close far fewer drawers. Moreover, the larger drawers, in which papers could be packed more closely, eliminated much wasted space. In 1909, the Shaw-Walker Company promised a 44 percent saving in wall space with the switch to vertical files.[101] While total volume saved was undoubtedly less, any space saving was significant for large filing operations.

The new equipment alone was not enough to make storage and retrieval efficient. In a textbook on filing published by the Library Bureau, vertical filing of correspondence was defined as including the *organization* of papers in the files, as well as the filing apparatus itself: "The definition of vertical filing as applied to correspondence is,—the bringing together, in one place, all correspondence to, from or about an individual, firm, place or subject, filed on edge, usually in folders and behind guides, making for speed, accuracy and accessibility."[102] This definition suggests two elements of organizing materials in vertical files: bringing together incoming, outgoing, and internal correspondence; and organizing the merged correspondence by some system that would make it readily accessible.

While vertical filing equipment was not necessary to combine incoming, outgoing, and internal documents, the new organizational scheme

Before   *After*

An example of "Y and E" System Service, in the offices of The Fox River Butter Co., Chicago, makers of Meadow-Gold Butter. Photograph show the old file room and the new—as arranged by one of our service men; the records kept in the old and the new filing rooms being identica

2.18   The transformation wrought in the office by merging various storage systems into a single vertical filing system. *(Catalogue, Yawman and Erbe Manufacturing Co., "Filing Equipment," ca. 1920. Hagley Museum and Library.)*

was frequently introduced with the new equipment. Merging correspondence into a single storage system first became possible in the 1880s with the loose copies produced by the rolling press copier and carbon paper. Indeed, some railroads apparently merged their correspondence in box files a decade before vertical files appeared.[103] On the other hand, some companies adopted vertical files in the 1890s but used them for incoming correspondence only, leaving copies of outgoing correspondence in press books.[104] As the above quotation suggests, however, vertical files were presented and apparently generally used as a means of bringing together in a single folder the previously scattered documents on a given subject or company, whether incoming, outgoing, or internal. Figure 2.18 illustrates the dramatic transformation such a merging could bring about.

More complex was the decision of how to organize the merged system. The flexibility of vertical files allowed companies to abandon the chronology of the press book in favor of a more functional arrangement. They still had to decide, however, which arrangement was most functional. The relative merits of various filing arrangements were much debated in the early years of the twentieth century, as the Taft Commission Report and the various texts on filing indicate. A 1918 bibliography on office methods lists twenty-eight different sources just on filing correspondence, and more on filing cards and other types of records.[105] Four major schemes were available, all with their advocates: numerical, alphabetical, geographical, and subject-based decimal.[106] Other schemes combined more than one of these major schemes.

Numerical schemes, used earlier by the railroads for flat files, continued to have advocates. With vertical files, folders rather than boxes were assigned numbers.[107] To retrieve a document, the searcher went to the alphabetical card index first, looking up either the name of the correspondent or the subject to find out the file number. Only with that number could the correct file be located. The numerical system allowed the company to create new files at will, generally at the end of the series, and it allowed them to group related documents into a file. It complicated the retrieval process, however, by requiring two steps: consulting the index and going to the files. Many of the railroads using this system converted to subject-based decimal filing (described below) in the early years of the twentieth century.[108]

Alphabetical files, unlike numerical ones, required no separate index; they were frequently referred to as self-indexing. The system required a folder for each person, company, or subject. These folders were then arranged alphabetically, using drawer labels and tabbed dividers to guide the filer (see fig. 2.19). Any correspondence from, to, or about that person, firm, or subject was filed in the folder, with the most recent in front. Such systems were very easy to use for correspondence with and about customers and suppliers. Internal correspondence concerning the operations of the company could be filed alphabetically under a department head's name, the department's name, or a subject heading filed alphabetically among the names. As long as the filer did not forget the subject headings, this system worked well and, because of its simplicity, steadily gained in popularity.

Geographical filing was most useful for companies organized by geographical districts. Retail sales organizations or railroads, for example, might divide their correspondence by geographical units. Folders within these geographical units were organized alphabetically, or by some other scheme. Some companies had a geographical file for sales correspondence and an alphabetical or subject-based file for general and internal correspondence.

Subject-based decimal filing was adapted from the Dewey Decimal Classification scheme for libraries. In decimal systems, all possible subjects for internal or external correspondence were divided into ten categories, which were in turn subdivided into ten more, and so on. In 1902 William Henry Williams, working at the request of L. F. Loree, president of the Baltimore and Ohio Railroad Company, designed and published a decimal classification system for railroads entitled the *Railroad Correspondence File*.[109] In it, the ten main subject classifications were such things as general matters, executive and legal matters, finance and accounting, and roadway and structures. Many, but not all, railroads adopted it, often to replace numerical systems.[110] Some other large companies with extensive internal and external correspondence—the New England Telephone and Telegraph Company, for example—developed similar decimal systems suited to their needs.[111] Such systems claimed the advantage of keeping materials on related subjects near each other. Furthermore, they were considered self-indexing, since no separate card

Vertical Filing Systems

Left Tab Alphabetical Guides.

INDIVIDUAL NAMES ALWAYS CONSPICUOUS

Left Tab Alphabetical Folders for
Miscellaneous Letters and Papers.

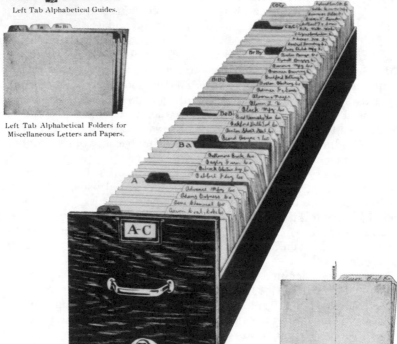

Right Tab Folder.

For supplies used in the Safeguard Method of Vertical Filing, see prices on page 131.

In Ordering Goods from this Page Please Telephone or Address Department G-2

2.19 Alphabetically organized vertical files with the requisite equipment. *(Catalogue, Hoskins Office Outfitters, Philadelphia, ca. 1912. Hagley Museum and Library.)*

index had to be maintained. On the other hand, the filer had to memorize the subject categories to use such a system. Moreover, if new and unforeseen topics arose after the closed decimal system was devised, they could create problems.

All four of these systems coexisted during the early twentieth century, but the trend seemed to be away from the straight numerical, which required a separate index, and toward the self-indexing alphabetical or decimal, sometimes combined with geographical. The Taft Commission's survey of business filing systems assessed the situation in this way:

> The numerical file, supported by one or more card indexes, is gradually being supplanted by the file arranged either alphabetically when the business is mainly with individuals or subjectively [by a decimal system] when the correspondence is upon a variety of subjects rather than with the individuals, but in both cases always upon a self-indexing basis.
>
> There were several instances of changes from the numerical to the alphabetical, or subject, basis, and in the case of each such change, the statement was made that time and money were being saved by the change. In no case did we find that an alphabetical, or subject, file had been supplanted by a numerical file.[112]

The report also described two systems that used both letters and numbers to identify a file, but also on a self-indexing basis. The Taft Commission favored the decimal-based subject system for government use. Many companies preferred the simpler alphabetical schemes. By 1938, one expert in business filing observed that alphabetical filing was clearly the most common system.[113]

The advent of vertical filing affected both external and internal communication, though the latter more profoundly than the former. By increasing the efficiency of storage and retrieval, vertical filing systems enabled companies to handle increased correspondence of all sorts without becoming bogged down. In 1909 the Shaw-Walker Company guaranteed that its vertical filing equipment and alphabetical filing system would cut labor costs by one-third over any other system.[114] While this claim is suspicious in comparison to competing systems of the early twentieth century, it is probably reasonable in comparison to the older modes of storing and retrieving documents, still in existence in many companies in 1909. The ability to handle external correspondence efficiently facilitated the growth of that correspondence and the business it represented.

More significantly, the new filing systems affected the function and nature of internal communication, as well as its quantity. As one early twentieth-century expert on filing systems asserted, "It will already have become evident that it is impossible to sever the problem of finding a good practicable filing system from the whole problem of business organization. This is particularly true of all businesses large enough to be carried on departmentally."[115] First, vertical filing systems organized by intended use rather than by origin and chronology allowed companies to create an accessible corporate memory to supplement or supersede individual memories. Accessible internal correspondence and records encouraged tighter administrative control over the growing companies. When gathering all of the information on a given subject was too time-consuming or expensive, executives might have made decisions with inadequate information. Systems that made it easy to consult documentation encouraged its use. Thus, vertical filing systems made internal communication a more effective tool of systematic management.

The merged filing systems may also have encouraged the development of certain types of internal communication by guaranteeing that such communication would survive and be accessible.[116] Previous dual systems for outgoing and incoming correspondence did not have a very good way of handling internal written communication; consequently, much internal communication, especially lateral exchanges to coordinate activities and argue points of view, may well have taken place orally or, if written, been relegated to the wastebasket after it was read. The new vertical filing systems assured that internal correspondence, like external, could serve a documentary function beyond its initial communicative function, thereby creating an added incentive for individuals to record their positions on an issue.

The more decentralized the files, the more they seemed to encourage internal correspondence. Most of the books on vertical filing advocated centralized filing departments that handled all of the firm's files.[117] They saw filing as a functional activity, like accounting or sales, that would benefit from specialization and systematization. Moreover, they felt that efficiently managed centralized filing could give departments throughout the facility faster service than decentralized filing could. Nevertheless, the new duplicating and filing technologies allowed proliferation of files. Thus, while many companies undoubtedly had such a central filing department, the general tendency seems to have been centrifugal. Geographical dispersion, of course, necessitated separate files. But soon departments in the same facility also maintained separate files.

In the early years of the twentieth century, the growth and departmentalization of the firm probably encouraged the spread of local files. While centralized files of customer correspondence might make sense, the nature of internal correspondence in a growing firm seemed almost to demand decentralized files. Each unit of the organization to receive downward directives and announcements would feel the need to file them close by for ready reference. Most departments or subunits of departments would not want to send reports up the hierarchy without also keeping copies to document their performance and positions. Moreover, the lateral correspondence coordinating activities (or airing disputes) between or within departments could best document each side of an interaction if it were stored in the sender's files, as well as in the recipient's.

If internal communication encouraged proliferation of files, evidence suggests that the spread of local files in turn encouraged internal communication. The relationship between decentralized files and increased internal communication was frequently given as a reason for restricting local files, as this passage from a 1913 article in the *Railway Age Gazette* illustrates:

> Probably the most effective means of reducing the number of letters written is the elimination of correspondence, as far as possible, between the head of a department and his subordinates, or staff officers. This is facilitated and made practicable by combining all of the files of the various staff officers with that of the department head. The principle of consolidating files to effect economies of time and the elimination of unnecessary letters, has been in use by many of our largest commercial enterprises, as well as on several railroad systems, for a number of years.[118]

The author, who had chaired a committee investigating methods of handling correspondence in the Pennsylvania Railroad, went on to assert the success of his method in that railroad. He claimed that correspondence was reduced on average 20 percent by the consolidation of files in twenty-one divisions, a significant reduction in divisions producing a total of 2,700,000 letters (internal and external) a year before the change.

While the accuracy of these estimates is open to question, they suggest the potential impact of the filing system on internal communication. A system that allowed decentralization thus supported and even encouraged internal written communication, and certainly guaranteed that more of it survived. Whether that outcome was bad (as the Pennsylvania Railroad assumed) or good is another question. Undoubtedly, the increased correspondence contributed to the corporate memory, a necessary element of systematic management.

Thus, vertical filing systems, like typewriters and duplicators, responded to, facilitated, and encouraged the development of internal written communication.

Conclusion: Communication Technology and Internal Communication

In the decades surrounding the turn of the century, an insatiable desire for efficiency created an office revolution since unequaled until the advent of the desktop computer. The major developments traced above were accompanied by many others. In 1911, the Taft Commission sponsored an exhibit of labor-saving devices for office work.[119] The records from this exhibit contain box upon box of brochures advertising such devices as Amberg Filing Cabinets, Gem paper clips, the Dictograph

Turner telephone system, the Burroughs Adding Machine, dating stamps, and many others. By 1926, the National Association of Office Appliance Manufacturers had put out an enormous tome cataloging and describing all of the available types of office equipment.[120]

These appliances facilitated many different types of paperwork and communication. Some of them were designed to aid the financial and cost-accounting processes of companies. Adding machines, for example, were developed in many varieties, such as those designed to combine typing and adding and to work on bound ledgers. Other devices, such as paper folders and envelope sealers, were used for mass mailings to external customers. Check protectors and check writers facilitated financial transactions. Pneumatic tube systems allowed rapid credit checks in retail stores. The devices became highly specialized as businesses searched for efficiency in their increasing paperwork. The most ubiquitous and influential of the devices, however, remained those such as the typewriter, file cabinet, and duplicator that could be used for many types of communication.

New technologies contributed to the specialization of office skills and consequently created an opportunity for applications of scientific and systematic management to the office as well as to the factory floor. David Lockwood has argued that "the actual division of tasks very often preceded mechanization, but machinery has speeded up the trend" of office workers becoming more like factory workers with specialized repetitive tasks requiring minimal thought.[121] In the past, clerks in manufacturing businesses had performed a very diverse array of firm-specific activities. Now the office, with its increasing force of workers and its new technology for "producing" written documents, had its specialized workers just as the factory did. As the typists, filers, and other clerical workers handled the increased paperwork needed to systematize production and other functions, they themselves became potential targets of the same systematization. And to the extent that office jobs were systematized, yet more internal communication was generated.

The net effect of all of these technologies on internal written communication was to facilitate and encourage its growth, at the same time that it spurred the development of the technologies. Without these new technologies, systematic management's dependence on extensive written communication might have imposed costs too heavy to be worth the resulting savings. Different methods might have been developed to manage the large companies. With these technologies, many avenues of control through communication were opened up.

Chapter 3
Genres of Internal Communication

The genres[1] or generic forms of modern internal communication—including the report, the in-house magazine, the committee meeting, and the ubiquitous memo—are so much a part of work life today that it is hard to imagine the business world without them. Yet, except for basic financial records and letters bridging distances, there was very little written communication within firms before the late nineteenth century, and oral communication was predominantly informal and undocumented. Many of the now-familiar genres of internal communication first developed or evolved significantly during the late nineteenth and early twentieth centuries.

The new types of internal communication did not grow out of classical rhetoric, as did many earlier forms of business writing.[2] Nor did they grow out of contemporary theories of business communication, which lagged far behind practice. In fact, most texts on business English ignored the development of internal communication well into the twentieth century, continuing to focus solely on letter writing to external audiences. Some of the new genres reflected the influence of other organizations or professions, such as the military and engineering, but this influence was based as much on common goals and needs as on direct imitation. Ultimately, new forms evolved from old primarily as a practical response to the demands of growing companies that were systematizing their management. The technologies of producing, copying, and filing documents also played a role in shaping certain aspects of their formats and occasionally their functions.

The genres of internal communication served as managerial tools for controlling the growing businesses. Circular letters, manuals, and other types of downward communication developed to aid executives in imposing system on people and processes. The late nineteenth-century developments in duplicating technology that radically lowered the cost of mass duplication encouraged companies to use them more widely. Routine and special reports—frequently consisting of tables, forms, and graphs, as well as prose—evolved to aid in drawing information up and sometimes across the hierarchy. The differentiation of the memorandum from the letter reflected a growing recognition of the importance of internal correspondence and an attempt to make it more practical and efficient to use. Finally, managerial meetings evolved to formalize and document multidirectional oral communication needed to ensure cooperation and maintain morale among foremen and lower level managers.

Downward Communication

As companies grew, the old oral channels of communication became increasingly attenuated. Moreover, word-of-mouth management did not provide the control and efficiency desired by a new breed of systematic managers. They developed a series of formal vehicles for communicating down the hierarchy, including circular letters, manuals, and individual instructions. A final genre of downward communication, the in-house magazine, developed later to build the worker loyalty that executives learned was necessary for cooperation and efficiency.

Communicating Policy: Circular Letters and General Orders

Written announcements of rules, policy, and personnel changes issued by an executive to large numbers of subordinates emerged in the nineteenth century as a major mechanism for establishing and documenting consistent procedures. As one systematizer put it early in the twentieth century, "The large business establishment of today is a living manifestation of the efficacy of the written order."[3] These downward mass communications were differentiated from letters quite early, taking on such designations as circular letters, bulletins, and general orders.

Such directives had antecedents in military communication, in certain forms of external communication, and in earlier printed lists of company rules. Orders or circular letters, issued (in printed form) by the U.S. Army beginning in 1813, allowed a military commander to announce a change in the chain of command, to give orders, or to establish procedures for all or some of the men under his command.[4] Another influence on widely disseminated announcements, suggested by the frequent use of the term *circular* or *circular letter* for such documents, is the advertising circular, a form of external communication. In the nineteenth century business world a circular, according to the Oxford English Dictionary, was "a business notice or advertisement, printed or otherwise reproduced in large numbers for distribution."

Yet a third forerunner of downward mass communication in American firms appeared in the unsystematized factories of the first half of the nineteenth century. Some early textile factories posted a list of printed regulations for employees (see fig. 3.1). These rules, set by the owners and intended to be enforced by the foremen, covered only general issues such as starting and stopping times and standards for on-the-job behavior. Consequently, they did not need frequent updating; once printed, they remained posted for years. Nor did they play a very important role in governing the workers. Daniel Nelson has pointed out that, "with the

REGULATIONS

TO BE OBSERVED BY ALL PERSONS EMPLOYED BY THE

LAWRENCE MANUFACTURING COMPANY.

The overseers are to be punctually in their rooms at the starting of the mill, and not to be absent unnecessarily during working hours.

They are to see that all those employed in their rooms are in their places in due season, and keep a correct account of their time and work.

They may grant leave of absence to those employed under them when there are spare hands in the room to supply their places; otherwise they are not to grant leave of absence except in cases of absolute necessity.

All persons in the employ of the Lawrence Manufacturing Company, are required to observe the regulations of the room where they are employed. They are not to be absent from their work without consent, except in case of sickness, and then they are to send the overseer word of the cause of their absence.

They are to board in one of the boarding houses belonging to the company, and to conform to the regulations of the house where they board.

The company will not employ any one who is habitually absent from public worship on the Sabbath.

All persons entering into the employ of the company are considered as engaged to work 12 months.

All persons intending to leave the employment of the company are to give two weeks' notice of their intention to their overseer; and their engagement with the company is not considered as fulfilled, unless they comply with this regulation.

Payments will be made monthly, including board and wages, which will be made up to the second Saturday of every month, and paid in the course of the following week.

Any one who shall take from the mills, or the yard, any yarn, cloth or other article belonging to the company, will be considered guilty of *stealing*, and prosecuted accordingly.

These regulations are considered a part of the contract with all persons entering into the employment of the LAWRENCE MANUFACTURING COMPANY.

JOHN AIKEN, Agent.

3.1.   Posted list of rules from the Lawrence Manufacturing Company, ca. 1842. *(Reproduced courtesy of the Kress Library of Business and Economics, Baker Library, Harvard Business School.)*

exception of regulations pertaining to the length of the work day and activities that would be obvious to outsiders (drinking or smoking on the job, for example), the shop rules were largely what the foreman made them."[5] In these early factories, word-of-mouth management was more important than written rules.

The railroads were among the first businesses to adopt the general order or circular letter for regular use in communicating with employees. Faced with geographical dispersion and unique safety problems, many of these companies initially printed lists or small booklets of rules for their employees.[6] These rules were probably intended to be as permanent as those of the textile factories, and may initially have been modeled on them. The rapid expansion and technical evolution of the railroads, however, required relatively frequent additions to or changes of rules and managerial structure. Both military orders and advertising circulars may have influenced their adoption of frequently issued circular letters. Certainly many early railroad engineers were trained in the military and thus would have been familiar with military models.[7]

Ultimately, however, the railroads' widespread adoption of circular letters was a practical response to their needs for consistent and hierarchical communication to ensure their safety and efficiency. One of the first crises in railroad management, the accidents on the Western Railroad described in Chapter 1, established the need for printed standard rules and a system for disseminating and acknowledging any changes in or exemptions to them.[8] Between major updates of the rule books, the circular letter or general order provided a method for communicating new or changed regulations, as well as for announcing changes in managerial personnel. Railroads soon learned that the improved control afforded by frequent directives could also be used to increase efficiency.[9]

In the late years of the nineteenth century, when other businesses grew in size and complexity, they also discovered the value of the circular letter for issuing general announcements or orders. Their motives were usually consistency and efficiency more than safety. The Bell Telephone Company, for example, discovered the necessity of uniform written orders within a year of its founding. While initially the company corresponded and negotiated individually with each agent, inconsistencies in policy quickly caused trouble. To remedy that situation, in November 1877 the company issued its "Instructions to Agents, No. 1," in which it established a uniform policy of rentals.[10] When the New England Telephone Company was licensed under the Bell Telephone Company soon after, it also isued such instructions (see fig. 3.2). Even a company whose business was facilitating oral and generally informal communication needed formal written orders for efficient operation.

At that time, creating multiple copies still required rewriting a document or having it printed. While railroads and a few widely dispersed companies such as telephone and telegraph firms were willing to accept the delays and expenses of printing to secure safety or efficiency, other

New England Telephone Company,

43 SEARS BUILDING, BOSTON.

INSTRUCTIONS TO AGENTS.

No. 1.

The New England Telephone Company respectfully announces that it has received from the Bell Telephone Company the exclusive right to rent Telephones in New England.

Parties using any other than the Bell Telephone will be liable to prosecution as infringers of the Bell patents.

The Telephone has ceased to be a novelty, and has become a well-known and recognized instrument for business and household purposes. The necessity for furnishing Telephones for trial has passed away, and hereafter the rental must be paid in advance; and every Agent will be held personally responsible for all rentals not collected within thirty days after the Telephones have been received by the Lessee.

The annual rental for Telephones shall be Ten Dollars each, payable in advance; not less than a pair of Telephones must be used at each station, except as hereinafter specified.

For social purposes, single Telephones may be used at each station. By "social purposes" is meant the use of Telephones as a matter of convenience between private houses; between a house and private stable; a house and office, etc., etc.

For house uses, a discount of fifty per cent may be made, and the use of single Telephones allowed at each station.

By "house use" is meant all places where Telephones are used in one building, or one group of buildings, as, for instance: several buildings in the same yard used by the same party, or, in fact, where Telephones substantially take the place of speaking tubes. College lines may be included in this class.

The Magneto Bell Calls may be sold for Fifteen Dollars each, or rented for Five Dollars each per annum. Agents are instructed to sell rather than lease when possible.

These rates do not apply to the use of the Telephone in mines, for which use terms will be furnished hereafter.

We enclose a copy of the form of lease which we have adopted, and which is to be used by all our agents.

The agent making the contract will sign his own name as agent under the word "accepted," and every agent will be responsible for all acts of any sub-agent within his district.

All applications for Telephones from officers in the United States service must be referred to Hon. Gardiner G. Hubbard, at Washington, D. C., unless such officer has power to contract for the payment of the rental.

No agent has authority to rent Telephones to be used for the transmission of messages for hire.

Every Agent will be required to remit the advance, which is Two Dollars on each Telephone and Three Dollars on each Bell-Call, on receipt of instruments, and no one will be considered a suitable person to act as Agent or Sub-Agent who cannot pay the advance required for a sufficient number of Telephones and Bell-Calls to supply every customer without delay. To do this, they will inform the Company from time to time what number will be required in their Agency, so that the Company can have one month's notice of their wants. The Agent will deduct the advance from the rental of the instrument on which the advance is made. It will not of course effect the commission of the Agent.

Every Agent will be required to forward monthly accounts of every Telephone received and the disposition made of it, according to a form to be prepared by the Company.

The Company delivers its instruments free of charge at the shipping place in Boston. All freight charges thereafter will be assumed and paid by the Agent.

Circulars will be furnished to Agents for $3.00 per thousand. Space is left on them for the insertion of Agent's name; and, as they have been prepared with considerable care, they will be found the most advantageous means of bringing the Telephone before the people.

No agent of this Company has any authority to reduce rentals, or allow discount.

The agents of the New England Telephone Company are especially urged to use their best efforts for the introduction of the Telephone into the District Telegraph System.

Managers of District Companies should be furnished with Telephones for trial, and their use fully explained.

For the instruction of such Agents as are not acquainted with their use in this system, we give the following instructions:

When not in use, the Telephones are hung up, out of circuit. When to be used a knob is pressed ringing a bell at the Central Office,—the Telephone taken down, which brings it into circuit and the order given.

In houses where the signal-box has not been introduced, a connection is made from the Telephone to the ground, and in this circuit a push-button is inserted. When communication with the Central Office is desired, a pressure of the button rings a battery-bell at the Office.

Where the District Telegraph System has not been introduced; or where the Company does not desire the Bell Telephone, a District Telephone Company should be organized, and metallic circuits constructed running from the Central Office to various parts of the city.

The advantages of the Telephone over the District system are apparent. The message is sent through the Telephone to the office, its receipt acknowledged, thus saving much time and expense, and making it useful for many purposes, instead of for two or three calls.

In small towns a family can be put in connection with the physician, or tradesman, and every man with his own place of business through a Central Office. Conversation can be carried on in the same way. There are few villages too small to sustain such a Company.

The District business of the New England Telephone Company in Boston will be conducted by the Telephone Despatch Company, E. T. Holmes, Manager.

GARDINER G. HUBBARD, *President.*
THOMAS SANDERS, *Treasurer.*
GEO. L. BRADLEY, *Gen'l Agent.*

Boston, Feb. 15, 1878.

3.2. The New England Telephone Company's first general order to all agents.
(Courtesy of AT&T corporate archives.)

companies with less pressing needs might have been deterred by time and cost from issuing notices or circular letters very frequently. Within a few years, however, the evolution of carbon paper, stencil copying, and other early duplicating technologies traced in Chapter 2 had cleared the way for rapid, convenient, and inexpensive duplicating of typed or hand-written documents. These innovations were particularly important for manufacturing firms, whose needs were somewhat different from those of the railroads. Because the hierarchy on railroads tended to be rela-tively shallow but very broad, those firms usually needed many copies of circulars. In manufacturing, however, because the hierarchy was often deeper and more diverse, groups of employees who needed to receive similar instructions varied greatly in size and were frequently smaller than comparable groups in railroads. Circulars might be issued from almost any level of the hierarchy to as few as five people or as many as several hundred. New duplicating processes removed a technological and financial barrier by creating a range of inexpensive options suited to these needs, from carbon paper for a few copies to mimeograph for hun-dreds. Printing remained a practical option when thousands of copies were needed.

By the early twentieth century, such mass directives were widely used. A 1909 text on handling and filing correspondence, for example, mentioned both "general orders affecting all departments" and "bulle-tins . . . issued by a department manager to all employees under him" as types of communication that had to be handled by filing depart-ments.[11] A 1910 text on business correspondence defined the genre as a tool for achieving efficiency:

> There is however, a class of interdepartment communications
> which may better be termed general or multiple orders. The general
> or multiple order is one issued by the general manager to the heads
> of the various departments under him, who are to receive the same
> general instructions. A general letter of this class will be one re-
> questing the observance of certain general conditions[, such] as one
> asking them to secure better efficiency from the various employees
> under them, one calling attention to the lack of system, the excess
> of red tape, general slackness of the house conditions due to some
> cause within their control, etc.[12]

Although secondary sources reveal much less about the form and style of such general orders than the case studies do, they suggest that the orders underwent several changes from letter form and style, dictated by their use. While some continued to use letter salutations such as "Gen-tlemen:" at the beginning, frequently these greetings were replaced by headings to facilitate filing and later reference.[13] Often these headings included numbers, as in figure 3.2, "Instructions to Agents, No. 1." In many cases, standard elements of these headings were eventually printed onto special stationery for this purpose.[14] Some longer instruc-tions or circular letters also differed from letters in providing subheads to aid the reader. In an even more radical departure from letter form, some announcements had an attached receipt to be signed and returned by each recipient.[15]

In style, the directives aimed for clarity and precision to guarantee uniform compliance. For railroads, clarity was certainly more critical than courtesy or elegance. Between 1881 and 1889, the *Railway Age* magazine contained numerous debates on the exact meaning of different sets of rules, especially important when trains used more than one company's tracks. As one such passage pointed out, "The importance of having rules for the movement of trains worded so explicitly and plainly that there can be no doubt as to their meaning is self-evident."[16] Later, manufacturing firms apparently aspired to the same goal, but for the purpose of guaranteeing efficiency more than safety.[17]

Thus, the demands of immediate use and later reference helped shape the form and style of circular letters or general orders. Their popularity led some early twentieth-century managers to explore ways of supplementing them with a more permanent compilation of rules and procedures.

Embodying a Comprehensive Organizational Memory: Manuals

The flow of directives documented rules and policies, contributing to a corporate memory. Nevertheless, each document was essentially fragmentary—it recorded only a piece of the total management system. Rule books or manuals attempted to provide a comprehensive corporate memory. Such compilations were not entirely new to the railroads, in which rule books for train movement often preceded circular letters. In most manufacturing firms, however, manuals appeared in the early twentieth century, and they were new enough to require explanation: "The record of business routine operations is known as a manual, and it is surprising to see how universal is its possible use," said one management expert.[18] Rule books became so popular that one publishing company put out a model office manual to illustrate what they should include.[19]

In their use, the modern rule books reflected preoccupations of the early twentieth-century systematic managers. Manuals were seen as critical for recording policies and establishing lines of authority.[20] As the following description of a primitive manual illustrates, they reflected an attempt to transcend the individual and create an organizational memory:

> When the production department was organized certain instructions were issued in the form of typewritten memoranda [i.e., notes]. While these were necessarily somewhat voluminous, they were planned to be permanent and to be in such complete form that a new man in the department, after reading the instructions would have a thorough understanding of the duties and responsibilities of it.[21]

Moreover, the manual also played a role in assuring "an analysis of all office work, the adoption of adequate, modern systems, and continuous organized control of clerical operations and office service to customers."[22] By compiling all information about procedures in one place, the manual made those procedures easier to analyze and to see as a whole than did separate bulletins and circular letters, thus encouraging "the adoption of adequate, modern systems." Additionally, the manuals provided a mechanism by which top management exerted "continuous organized control" over those under them, organizing everything to work together in the most efficient manner they could devise.

Manuals took either of two basic forms: bound or loose leaf. Bound (and generally printed) rule books, like those used earlier by railroads, captured a comprehensive picture of a firm's rules and procedures at a given point in time. Since companies and methods evolved over time, however, new versions had to be issued periodically. Thus, bound rule books frequently gave way to loose-leaf manuals designed to reflect a dynamic, rather than static, organizational memory. When the material on one page was out of date, a new version of the page, rather than a whole new manual, could be issued to update it. Loose-leaf manuals provided a compromise between a bound manual and a flow of circular letters. An authority on management gave the following rationale for the loose-leaf form:

> While the material in the manual is the important thing and not its form, nevertheless it is convenient, where the firm is large, to publish the manual as a looseleaf book and small enough to slip into the pocket. Thus changes may be made from time to time without re-publishing the whole book and because of its convenient form it can be carried around and thus be at hand when required.[23]

This handy, loose-leaf manual, he went on to point out, "should be fully indexed to facilitate its use." Moreover, he continued, rather than having manual pages printed, "In small companies it is convenient to typewrite, mimeograph or blue print the manual." In fact, one office systematizer even suggested that "the practice of issuing bulletins concerning various subjects can be discontinued, and in their place additional or revised manual sheets issued."[24] Such manuals had evolved in form from bound and printed rule books in ways that made them more convenient and usable as comprehensive but flexible repositories of organizational memory.

Giving Specific Orders: Notes and Forms

Although the general rules and policies of firms were communicated through circular letters and manuals, many day-to-day orders and instructions still needed to be communicated to specific employees. The railroads had determined early that any communication concerning train movement should, for the sake of safety, be in writing. By the early twentieth century, systematizers advocated written orders for manufacturing firms as well, though usually for efficiency and dependability more than for safety. One expert said, "As to the form that an order should take, the only satisfactory form is the written order."[25] Another stated, "An executive should never give an important order verbally. The most efficiently organized offices in the country recognize this principle."[26] In justifying their positions, both cited the problems inherent in depending on an individual's memory as well as the potential for friction or conflict over the specific terms of an instruction. Written orders conveyed and documented those terms so that they could be referred to later.

Either notes or forms could be used to communicate specific instructions, depending on the routineness of the instruction and the extent to which a firm was systematized. The notes or memoranda were simply another type of internal correspondence, to be discussed later in this chapter. Some systematic and scientific managers adopted printed or duplicated forms, with standard elements and space for filling in specific details, for conveying routine instructions as efficiently as possible. Forms were widely used as part of the upward reporting system; they could also, one forms expert pointed out, be used to "provide the means for carrying out decisions and policies."[27] As such, they provided an efficient and impersonal way of communicating specific instructions within a common framework.

Certain types of orders having to do with production flow were closely linked with the scientific management movement. According to Frederick Taylor, for example, each worker should be given a daily instruction card, a form on which specific instructions for the day were provided.[28] By the early twentieth century, many companies used shop orders and job tickets to convey specific production orders on the factory floor (see fig. 3.3).[29] These forms, often attached to the materials as they progressed through the production process, served in a dual capacity as part of the upward flow of accounting information as well as part of the downward flow of instructions. They usually bore no resemblance to older genres such as letters, but were designed purely to guarantee efficiency.

Forms could also be used for communication from a company to its agents, though this use was not fully accepted. One expert in business correspondence discussed the delicate issue of a main office communicating referrals to a sales agent via letter or form:

> Where diplomatic requirements call for it, the letter should be
> used, if the agent would feel slighted or would not give as close
> attention to a form notification as to one sent by letter. Usually the
> agent can be made to understand that it is not necessary to write
> him the same letter every time a prospect is turned over to him and
> will realize the labor-saving side of the uniform form notification.[30]

TO				WORKING ORDER NO.
Jno. Wilson				*5328*
DEPT. *M.S.*	SEC *2*		ASSIGNED	

YOU ARE HEREBY AUTHORIZED TO DO THE FOLLOWING WORK, VIZ.

5   N.B. Reservoirs, (C.L.)

REQUISITION GRANTED FOR THE FOLLOWING STOCK, VIZ.

12	*lbs. Sheet C.*
½	*" Rosin.*
1	*set stock bolts.*
	nuts & washers.

DATE *11/9/03*   *C.H. Andrews —*
                                                      SUPT.

3.3.   A foreman's shop order from his superintendent. (System, *May 1904. Courtesy of the Baker Library, Harvard Business School.*)

Even the form recommended for this situation resembled a letter, beginning with "Gentlemen: The following inquiry is referred to you" and ending with a (printed) signature. It was essentially a compromise between the letter and the form.

In a period during which "system" was the universal catchword, forms filled an important role in systematizing certain types of downward communication. They conveyed, as economically as possible, specific instructions that would otherwise have been conveyed orally (thus leaving no record) or in an individually composed note (thus taking more time). They were simultaneously mass communication and individual communication.

Building Morale: The In-house Magazine

If the circular letter, manual, and form were used to transcend the personal, the in-house magazine or shop paper had the contrasting goal of re-creating the personal feeling offered by the traditional firm. Firm growth and depersonalization had eroded the workers' loyalty, as the labor unrest of the postwar period made clear, and efficiency was impossible without worker cooperation. Early in the twentieth century, influenced by the corporate welfare movement and by their own labor problems, many companies started shop or company publications designed to reinject a personal element into the workplace (see fig. 3.4). The purpose, content, and form of such in-house magazines became the subject of articles such as "Fostering Plant Spirit through a Plant Paper" and "The Shop Paper as an Aid to Management" in magazines aimed at managers.[31]

In-house magazines attempted to humanize the workplace through their content and approach. The editor of one such paper noted: "Of course everything in all the subjects treated must be kept human and handled in as personal a way as possible, and the pictures must be alive and full of people doing things. But we need still more to . . . humanize our magazine with concentrated personality."[32] In particular, he suggested two mechanisms designed to "bring out clearly the personal acquaintance angle": a Who's Who column with articles about prominent executives and managers, and a series of articles on long-time employ-

3.4.   Some of the many in-house magazines being published by 1919. (Industrial Management, *March 1919.*)

ees. Both were to deal with the subjects' home lives and personal interests as well as their roles in the company. The editors of other shop papers had similar suggestions. In smaller plants, all employees could be invited to submit their news for publication in the paper, and even large-circulation magazines could include informal profiles of interesting individuals, whatever their position in the company. News of employee clubs and activities was also a staple of shop papers.[33]

In addition to humanizing the workplace, the shop paper could help educate the workforce. As one writer on the subject pointed out, "In these times when skilled labor is so scarce and so independent, many a harassed manager is driven to educate what labor he can obtain."[34] The education sometimes focused on technical issues: "Articles and illustrations of each operation in turning out the product manufactured render understanding more easy." In addition, the papers often attempted to educate employees in values such as safety.[35] Sometimes the values promoted were more clearly self-serving on the part of management. One contributor to a series of articles on shop papers suggested presenting a "printed talk" from the boss, noting that "it is not the boss, however, who speaks, but the editor. His position permits of a little straight talk, addressed to both the boss and to the men."[36] In the example that followed, however, the editor simply conveyed the management's point of view—that if the workers were not efficient, the firm would not be able to compete and the workers would lose their jobs. Another author admitted, "Despite the fact that the nominal editors are often employees, yet the papers themselves are very evidently the expression of the management end of industry."[37]

The mix of materials in shop papers was important to their success. The educational function could not be too prominent if they were to succeed in humanizing the workplace, raising morale, and getting management's point of view across:

> The general conception of the ideal plant newspaper is pretty well defined[:] it should carry lucid articles on efficiency, personal betterment, shop news and personals, with the aim of securing cooperation.
> However it rarely secures this object for the simple reason that the conventional factory paper is so dry and top-heavy that it is not thoroughly read, much less assimilated.
> But, if one puts out a paper light and breezy enough to hold attention all the way through, and eliminates flat-footed sermons, then it is possible at intervals to convey sugar-coated admonitions and advice that get under the hide of the reader and stay there.[38]

A successful in-house magazine had to "sugar coat" its attempts to instruct. The personal items, of course, helped leaven the educational element, as did cartoons and jokes. "The employees' magazine should be half full of laughs," said one editor.[39] Photographs also helped capture the interest of readers.[40] All of these entertaining elements helped boost morale at the same time that they made palatable the more direct attempts to educate workers in managerial values.

Because shop papers were generally issued only once a month and often had large distributions, many were printed rather than duplicated. Moreover photographs, which could be used only in printed papers, added greatly to their appeal. As one author noted, "Nearly all employers who have undertaken to print them have felt that in thus developing a plant spirit it is worth while to issue a paper which will command both the respect and the admiration of the average worker."[41] Their form, as well as their content, indicated that these papers were viewed as serving an important purpose. They improved morale and cooperation, thus indirectly reinforcing control. Although their ostensible function differed greatly from that of other genres of downward communication, their underlying aim was similar.

Upward Reporting

If downward communication was critical to implementing executive plans and decisions, upward communication was critical to formulating them. Reports were the major formal mechanism by which managers and executives at all levels acquired information about what went on at lower levels, information on the basis of which they made decisions for the future. Reports were of two basic types: routine or periodic reports, which were issued at regular intervals to provide information on normal operations; and special reports, which analyzed (usually in response to a special request) a specific problem, opportunity, idea, or physical entity.[42] Although both types of reports supported the aims of systematic management, the routine report became a particularly important managerial tool, assuring a regular flow of data to be used for monitoring efficiency.

Reports were not a new phenomenon, although before this time they were rather nondescript in form. As the author of the earliest known American textbook on report writing said, reports "have been made since time immemorial."[43] In the mid-nineteenth century in America, reports occurred in several situations. When sales agents located in different cities supplemented their quarterly financial accounts with frequent letters about events and competition, they were reporting. The military required subordinates to report information and analyses to superiors. And in the few companies in which ownership and management were separated, managers reported to owners or their representatives. In general, these reports were not particularly distinctive in form: they were simply extended (but not necessarily well-organized) letters, sometimes accompanied by financial accounts in traditional double-column form. In the late nineteenth and early twentieth centuries, the form of business reports evolved significantly to fulfill the greater demands placed on them as firms grew and as the philosophy of systematic management gained in popularity.

Railroads and the Evolution of Reporting

The evolution of the report genre, like that of circular letters and manuals, began early in the railroads. Routine annual reports, both of superintendents to directors and of directors to stockholders, were present from the beginning of the industry's development.[44] These reports kept key constituencies informed about progress and were important in maintaining financial support, but they did not play a critical role in the daily operation of the railroads. In the 1830s and 1840s, railroad annual reports were generally designed as letters with opening salutations and complimentary closes, though occasionally they were simply titled and signed. Within the body of such letters, basic financial information was often presented in balance sheet form. This tabular format was sometimes extended to exhibits of rudimentary operating data such as the number of miles of track. Tables would become increasingly important, and the letter form would disappear as routine reports proliferated.

As the function and frequency of reports changed, the format evolved in response. In the Western Railroad, initial innovations in report form came about when, following the series of accidents mentioned in Chapter 1, the directors demanded that monthly reports be made at several levels as a way to accumulate and learn from past experience. The monthly reports written by the president to the directors over the ensuing six months (the rest of the series did not survive) showed an evolution in form toward consistency and convenience.[45] The first of these handwritten letter reports included long prose discussions of several issues, as well as brief financial statements. By the sixth one, the financial statements within the report had become more extensive (adding various categories of monthly expenses) and had taken on standard tabular formats. The red-ruled tables, clearly modeled after the ruled ledgers used for financial accounts, were more readily accessible to the directors than figures embedded in prose. In addition, the tables provided consistent data from month to month, enabling directors to compare one month's statements with the next. Finally, the consistent tabular format no doubt eased the president's task in writing these reports.

Further changes in report form resulted from Daniel McCallum's far more extensive use of reports to improve efficiency on the New York and Erie Railroad. The hourly reports on the position of various trains, for example, were telegraphed to the General Superintendent, "the information being entered as fast as received, on a convenient tabular form."[46] The frequency of the reports and the desire to save time both in telegraphic transmission and in reading and analysis at the receiving end made the use of printed tabular forms desirable. These most basic reports permitted constant monitoring of train movement and a level of control not previously possible. The telegraph and the tabular forms worked together to speed the flow of data without imposing excessive time burdens on senders or recipients.

Moreover, McCallum continued, routine reports provided the basis for further analysis resulting in "statistical accounts kept in this office." The many operating facts were combined and analyzed to provide data comparable over time and across divisions, and thus useful in evaluating the performance of individuals and of equipment. Again, tabular statements were the mechanism for making such data accessible. For example, McCallum mentioned that an "experiment . . . has furnished data for determining the effects of the ruling grades and curvature on each Division, and from this data a tabular statement giving the amount of load each engine is capable of moving over the several Divisions has been prepared, and we shall be enabled hereafter to observe whether they work up to their capacity." While the New York and Erie was ahead of its time in its reporting system and in its use of the telegraph to transmit some reports, evidence suggests that by 1880 railroads commonly used the telegraph to transmit daily car-reports.[47]

Once again the railroads led the way in developing a genre of communication as an efficient mechanism of control. By the early twentieth century, the broader business community had also accepted the importance of upward reporting in the systematic management of large enterprises. As one expert in business communication put it, "In many large corporations the departments are joined together by a veritable network of reports."[48] Moreover, each department itself was held together by such a network, "because the [department head] must often base his action upon the reports of his subordinates." This system of reports, he continued, appeared to some as red tape, but it "is really a means to efficiency, provided it is properly handled. It enables the company to base its actions on an exact knowledge of facts; and to fix responsibility on the different members of the organization. The indirect benefits in improving as well as measuring the efficiency of men are often considerable."[49] Reports created a corporate memory of facts and provided the basis for evaluating individuals.

Manufacturing firms followed the lead provided by the railroads in adapting the report form to the specific needs of their preparers and users. As an authority on business organization put it, "The purpose and source of any report should decide the nature, the form and arrangement of its contents."[50] And, indeed, that appears to have been the case: "Reports are, of course, of many kinds. They range from the brief analysis of one man's activities for a single day to the complete summary of a great corporation's operations for a whole year—or even long periods of years. Some are composed of tables of statistics; others are filled with passages of interesting narrative and description."[51] In each case the users chose from or adapted the available models, often with the goal of assuring efficiency for writer, reader, or both. While some reports continued to consist mainly of prose, many were designed as tables, printed forms, or graphs. Still others combined prose with one or more of those three special techniques.

Efficiency through Tables and Forms

Other firms, like railroads before them, soon discovered that tables were efficient formats for routine statistical records and reports and for data within longer prose reports. Compilers of reports saved time because they did not have to embed the statistics in prose, and readers saved time because figures were more readily accessible in tabular form. Tables could be created by hand or on the typewriter. Initially typing tables was a laborious process, but the task was made easier by the introduction of tabs on typewriters around the turn of the century, evidently in response to the need created by increased use of tables.[52]

Tables were also valued because they facilitated the comparisons of data so critical to systematic managers. In describing executive profit and loss reports, an expert on business organization stated that "the data in the report should be so arranged as to permit the ready comparison of those items which show the relative standing of the different branch houses of the firm."[53] This goal was attained by setting up a table with columns for each branch house, so that statistics could be compared across the columns. The shift from descriptive to comparative data required for the systematic management of large enterprises clearly encouraged the use of tables as reports or components of reports.

The step from tables to printed forms for periodic records and reports was a small one, and one that the railroads had already made. Forms of all sorts—on paper or on cards, printed or duplicated, in tabular or prose form—became popular among progressive manufacturing managers in the early decades of the twentieth century. Magazines such as *System, Factory,* and *Industrial Management* were full of examples of forms (see figs. 3.5 and 3.6), often portraying them as the very essence of system.[54] By 1925, a book devoted entirely to the design and use of forms had been published.[55]

Forms certainly simplified and systematized the jobs of both compiler and reader. From the point of view of the compiler, forms made "clerical work easier than would be possible if the blank sheet of paper were used."[56] Instead of writing or typing out a new prose or tabular report for each period, the compiler had only to fill in the blanks. From the point of view of the user of the report, the form guaranteed consistency and, if well designed, accessibility of content. To find a particular piece of information, the user simply looked in the same place each time. Moreover, sets of forms could be designed to work together.[57] The information from several forms was often combined, reduced, and recompiled on a new form at the next organizational level. The form report shown in figure 3.6, for example, was the result of progressive consolidation of many individual sales reports and shipping reports.

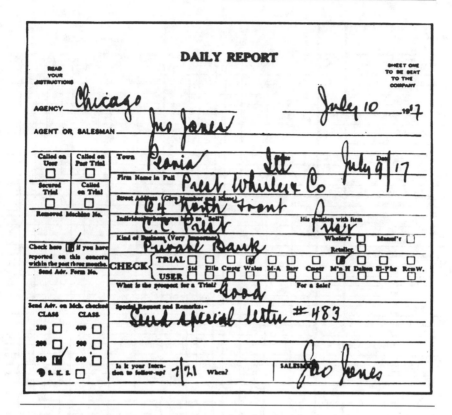

3.5.   A nontabular report form. (Industrial Management, *February 1917.*)

Many of those who wrote about forms during the period suggested ways of improving their design. On the most general level, one system-atizer noted, "Forms should be simply designed, and their use easy to learn."[58] In keeping with this principle, and with the general trend to-ward standardization characteristic of the period, he further suggested a standard, three-part format (identification matter, data, and special in-structions) "to secure a general logical rule for all forms, so that one clerk can readily use the forms of another department than his own, should the occasion arise." Another expert suggested more specific de-sign principles:

> The matter contained on the form should be arranged with refer-ence to its sequence; the sequence of entry or computation and the sequence of matter to be transcribed from another form.
> On any sheet the spaces should be such that the entries are made first in the extreme left side spaces progressing across to the right

ARTICLE	PRODUCT-ION	SALES						SHIPMENTS					
		QUANTITY			VALUE			QUANTITY			VALUE		
		TO-DAY	THIS MONTH	THIS YEAR	TO DAY	THIS MONTH	THIS YEAR	TO-DAY	THIS MONTH	THIS YEAR	TO-DAY	THIS MONTH	THIS YEAR
Standard No. 0													
No. 1													
No. 2													
No. 3													
Hot Water No. 1													
No. 2													
Units													
TOTAL INCUBATORS													
Colony Breeder													
Stove Breeder, Sr.													
" " Jr. 1													
" " Jr. 2													
TOTAL BROODERS													
K D Hoovers													
Rex "													
Jr. Portable Hoovers													
Home "													
Units "													
TOTAL HOOVERS													
M 112 Cabinet													
212 "													
312 "													
412 "													
M 120 "													
220 "													
320 "													
420 "													
T 13 "													
25 "													
33 "													
45 "													
T 16 "													
25 "													
35 "													
45 "													
Special													
12-qt. Container													
20-qt. Container													
TOTAL CABINETS													
4-gal. Can													
8-gal. "													
12-gal. "													
20-gal. "													
TOTAL CANS													
12 Tub													
20 "													
TOTAL TUBS													
Parts													
Truck Bodies													
Arro Kars													
Cedar Chests													
GRAND TOTAL													

Form for Report Giving Details of Sales and Shipments,

3.6. A tabular report form consolidating data from other reports. (Industrial Management, *January 1921.*)

in order that the hand will not cover information already written
down.

Information to be summarized or transcribed should be placed on
the right side so that the sheets may be handled with the left
hand.[59]

Figure 3.7 shows yet another writer's advice on how to revise a form to
minimize the compiler's work.

Other design suggestions in the management literature related to ease
of use by the recipient of the report. Each form should have a specific
and descriptive name—not "Monthly Report," but "Monthly Report of
Sales by Salesmen," for instance—to make it easier to request.[60] Moreover, forms should be designed to help the executive compare various
data. For example, "The same standard form embracing the same data
can be used in preparing the report upon the estimated sales and expenses as that employed for showing the actual sales and expenses. . . .
The advantage of the two reports to the executive for purposes of comparison, one showing the results desired and the other the actual accom-
plishment affected by the selling division, needs no commentary."[61]
Similarly, the revised form in figure 3.7 benefited the user of the report
(as well as its compiler) by standardizing the reasons for absence into a
few key categories, which could then be compiled and readily compared.
Like tables, forms should help the executive monitor and compare data
flowing up the hierarchy.

Finally, some recommendations for form design were intended to address practical matters of storage and retrieval as well as to minimize
printing costs. Color schemes were frequently suggested as a way to
facilitate handling.[62] One authority recommended that information by
which the form might be filed or retrieved be positioned where it could
best be seen, depending on whether the completed forms were stored in
vertical files or in ring binders, for example.[63] Another suggested providing blanks for approval signatures on the form when necessary.[64] Cost-
conscious systematizers advocated using paper of the smallest standard
size consistent with the information on the form and with its intended
filing system, as well as the cheapest weight suitable for the form's
useful life.[65] In general, experts recommended that forms be standardized so as to save time and money.[66]

Form reports, then, had evolved far from the letter format used for
most reports in the mid-nineteenth century. They took their shape from
the use to which they were put, as well as from the value placed on
efficiency and standardization by the systematic management movement. In practice, of course, many forms were haphazardly designed.[67]
Moreover, as the author of the book on forms pointed out, "Complaints
of 'too much system' and 'red tape' in an office are almost invariably
traceable to the existence of unnecessary forms. . . . Any serious attempt to improve the operation of an office or the so-called paper work
of a shop must have as one its main objects the elimination of unneces-

MR._____   DATE_____

On the above date employees failed to register or were absent as indicated by check marks. If failure to register, give time IN or OUT in column checked. If absent, state reason for absence, and whether COMPANY BUSINESS, EXCUSED, or NOT EXCUSED. Sign this sheet and return it to payroll department immediately. Irregularities will be left unadjusted on payroll until this sheet is returned properly signed.

Clock Number	Name	A. M.		P. M.		Night		Absent		Remarks
		In	Out	In	Out	In	Out	A. M.	P. M.	

WHY WRITE THE REASON UNDER "REMARKS"—

In this daily absence-or-failure-to-register report it was necessary for the clerk to write out in longhand under the remarks column the reason for the absence. Obtaining the separate totals for the various excuses given was rather difficult

---

NOTICE OF ABSENCE OR FAILURE TO REGISTER

MR._____   DEPT. NO._____

RETURN TO DEPT. NO._____   DATE _____

On the above date employees failed to register or were absent as indicated by check marks. If failure to register, give time IN or OUT in column checked. If absent, state reason for absence, and indicate by a check whether at OWN EXPENSE, COMPANY BUSINESS, EXCUSED, or NOT EXCUSED. Sign this sheet and return to payroll department immediately. Irregularities will be left unadjusted on payroll until this sheet is returned properly signed.

Clock Number	Name	A. M.		P. M.		Night	Absent			Remarks			
		In	Out	In	Out		A. M.	P. M.	Sick	At Own Expense	Company Business	Excused	Not Excused

—WHEN A CHECK MARK WILL SUFFICE?

But when the report was simplified and redesigned a check mark placed under the proper column indicated the reason. This not only decreased the clerical work, but made the totaling of the separate excuses comparatively simple

---

3.7.   How to simplify a form report. (Factory, *August 1917.*)

sary forms."[68] In spite of these dangers, however, tables and forms made it feasible and reasonably efficient to collect, analyze, and transmit up the hierarchy the vast amounts of data needed for systematic control.

Graphs for Effective Display of Data

Yet, the very efficiency of reporting via forms and tables often created problems for the ultimate recipients at the top. As companies grew and as the data from many routine reports moved up the hierarchy, the amount of information rapidly became overwhelming. How could an executive handle all that information, even consolidated on tabular forms, and get anything out of it? In the early twentieth century, graphs came to be widely accepted as a useful technique for reporting large amounts of data in a readily accessible form.[69] One systematizer, Carl Parsons, articulated both the problem and this solution in 1909: "[The executive] must have reports of his costs, his sales, his profits or his losses, but he must have them in such forms that he can interpret them instantly and draw conclusions for future guidance. . . . In a modern organization the executive obtains this information through a system of graphic records, a simplified summary of countless departmental statistics and itemized reports."[70] With graphs, unlike with most tables, he continued, "the impression, the fact, is obtained at a glance." Parsons added, "This point puts the graphic method of reports at once into harmony with up-to-date office systems; and makes it the most effective and quickest way of getting a grasp on details." Graphs came to be seen as a modern, efficient mode of communication.

Graphic techniques for representing empirical data were first developed in the late eighteenth century, but initially in America they were used primarily to represent demographic and trade statistics. Graphs were not commonly used to present managerial data in firms until the decades surrounding the turn of the twentieth century.[71] By then, engineers were using graphs for displaying experimental data and physical relationships, thus introducing graphic techniques into firms.[72] The quantity of data being amassed in tables and forms encouraged managers to adapt these increasingly familiar graphic techniques to operational and financial data at the firm level. Not surprisingly, advocates of systematic management were among the first to employ graphs as a managerial tool. In the mid-1880s, Henry R. Towne, an engineer and early supporter of systematic management, used the first graph of managerial data to appear in the *Transactions of the American Society of Mechanical Engineers* (see fig. 3.8), a publication that pioneered in treating managerial issues before the magazines of management had been founded.[73] From that time forward, graphs saw increasing (if still relatively infrequent) use for managerial purposes. In 1914 they came of age as a managerial tool: Willard C. Brinton published the first American book on graphic techniques for business and general audiences; and a committee of representatives from American scientific and engineering societies,

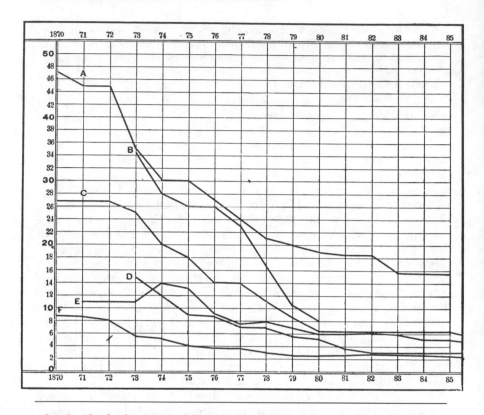

3.8.   Graph of reductions in labor cost of selected products under a special system of contract and piecework. (ASME Transactions, *1885–86.*)

chaired by Brinton, published its preliminary report on graphic standards.[74] Brinton's book was followed by others as graphs caught on in the business world.[75]

For the busy executive, graphic reports had several advantages over tabular and prose reports. First, as Parsons pointed out in a passage quoted above, graphs were often "the most effective and quickest way of getting a grasp on details." William Leffingwell, a well-known advocate of system in office management, noted several years later that "there is a growing preference among executives for statistics prepared in graphical form. . . . There is . . . no doubt that a graphical chart, correctly made, shows tendencies much quicker and impresses the mind more accurately and emphatically than do figures."[76] Certainly, as one authority on graphs pointed out, graphic display was effective for getting and keeping the attention of busy executives:

> What the executives need is accurate and easily understandable information about what is going on in the several divisions of their

own company. Typewritten reports and columns of statistics about the sections of a company in which the executive in question is not personally interested are almost always too dry and uninteresting to cause the head of the section in question to take the time and the trouble to dig out the information that he ought to have. But when this same information is presented in the form of a simple chart or graph which does not need a lot of time for its comprehension, the busy executive will take the time necessary to study something that is at one and the same time so interesting to him and so helpful.[77]

For example, the numbers reported to a middle manager on the tabular report shown in figure 3.6 were combined with other data to be presented to the general manager in the graph shown in figure 3.9. From this and similar graphs, he could more rapidly grasp the trends for each product. Moreover, Brinton pointed out, "In many cases, the graphic method requires less space than is required for words."[78] "There is, besides," he continued, "the great advantage that with graphic methods facts are presented so that the reader may make deductions of his own, while

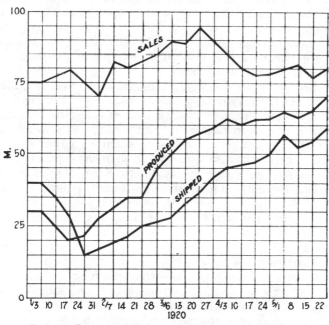

Graphic Record of Production, Sales and Shipment.

3.9. A graphic report incorporating the data in figure 3.6. (Industrial Management, *January 1921.*)

when words are used the reader must usually accept the ready-made conclusions handed to him."

Furthermore, graphs, even more than tables, facilitated the analytic comparisons so important to the systematic management of large, multi-unit enterprises. Tables might juxtapose figures from comparable business units, but the reader still had to study those figures carefully to interpret them. Parsons observed, "Business men today know that not the facts of themselves are significant, but only their relation to other facts."[79] Graphs highlighted trends over time and relationships across units so that they could be grasped rapidly:

> It is in just such problems as these, where a number of different sets of data must be compared, that curves have tremendous advantage over presentation by columns of figures. A man must be almost a genius to grasp quickly the facts contained in several parallel columns of figures, yet anyone of average intelligence can interpret correctly a chart which has been properly made for the presentation of curves.[80]

Another expert in graphics added that "charts appeal to the mind through the eye, which is the organ best adapted by nature for the comparison of quantities."[81] This comparative function might be enhanced by the use of logarithmic scales or cumulative curves. For example, Brinton showed how the graph depicted in figure 3.10, with its cumulative curves of scheduled production and actual production, could be used to highlight deviations from schedule.

Perhaps the best-known application of graphic analysis to scheduling is the Gantt chart. Henry Laurence Gantt was an early follower of Frederick Taylor (though ultimately they broke with each other) and a major proponent of scientific management.[82] Although he used various graphic techniques throughout his career, his most famous contribution is the progress chart that bears his name (see fig. 3.11). Developed in 1917 to aid in the war effort, the widely praised Gantt chart focused on progress toward scheduled goals, rather than on actual quantities.[83] One of Gantt's followers, author of an entire book on the Gantt chart, went so far as to say of it, " 'The Gantt Chart, because of its presentation of facts in their relation to time, is the most notable contribution to the art of management made in this generation.' "[84] Although that statement is certainly a partisan one, it underscores the excitement with which graphs and charts were embraced as managerial tools.

In addition to their role in monitoring trends and schedules, graphs could play a valuable role in forecasting the future.[85] One expert pointed out that "the graphic chart, by reason of the fact that over a given period of time it shows the trend or tendency of events, may be used as a means of predicting what is apt to occur in the future."[86] Simple projections of trend lines on a chart could bypass mathematical calculations. The forecast might then be compared to actual values, very much as Brinton

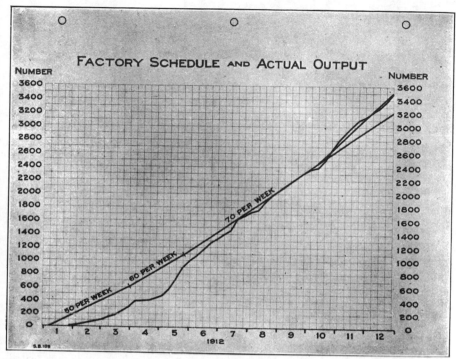

Fig. 134. Production Schedule and Actual Output of an Automobile Factory for One Year

3.10. Cumulative curves of scheduled and actual production for highlighting deviations from schedule. *(Willard C. Brinton,* Graphic Methods for Presenting Facts, *1914.)*

used a chart to compare scheduled and actual production. This forecasting capability was useful, although its limitations had to be acknowledged: "This, of course, assumes that the conditions governing future events are the same as those which governed past events. A graphic chart, per se, is not able to look into the future and tell of unforseen occurrences."[87] Such forecasting graphs were not as common as graphs of past events and were more likely to appear as part of longer, special reports than to stand alone as routine reports.

Although brief graphic reports were valued in part for their lack of bias in presenting data, graphs could also perform a persuasive function, usually as part of a longer report. Brinton explained the need for persuasive tools as follows:

> After a person has collected data and studied a proposition with great care so that his own mind is made up as to the best solution

GANTT CHART SHOWING RELATION OF MATERIAL ORDERED, COMPLETED AND ISSUED TO SCHEDULED REQUIREMENTS

ITEM	UP TO NOV. 30	DEC.	JAN.	FEB.	MAR.	APR.	MAY	JUNE	JULY	AUG.	SEPT.	OCT.
REQUIREMENTS SCHEDULE	1906 M / 1906 M	1990 M / 84 M	2074 M / 84 M	2161 M / 87 M	2251 M / 90 M	2344 M / 93 M	2443 M / 99 M	3154 M / 711 M	3273 M / 119 M	3397 M / 124 M	3525 M / 128 M	3656 M / 131 M
Ordered												
Completed												
Issued from Stores												
REQUIREMENTS SCHEDULE	2182 M / 2182 M	2342 M / 160 M	2492 M / 150 M	2647 M / 155 M	2807 M / 160 M	2972 M / 165 M	3150 M / 178 M	3948 M / 798 M	4167 M / 219 M	4392 M / 225 M	4624 M / 232 M	4862 M / 238 M
Ordered												
Completed												
Issued from Stores												
REQUIREMENTS SCHEDULE	1997 M / 1997 M	2130 M / 133 M	2266 M / 136 M	2405 M / 139 M	2547 M / 142 M	2691 M / 144 M	2844 M / 153 M	3546 M / 702 M	3737 M / 191 M	3932 M / 195 M	4130 M / 198 M	4331 M / 201 M
Ordered												
Completed												
Issued from Stores												
REQUIREMENTS SCHEDULE	268 M / 268 M	281 M / 13 M	295 M / 14 M	309 M / 14 M	323 M / 14 M	337 M / 14 M	352 M / 15 M	416 M / 94 M	464 M / 18 M	483 M / 19 M	502 M / 19 M	521 M / 19 M
Ordered												
Completed												
Issued from Stores												
REQUIREMENTS SCHEDULE	986 M / 986 M	1047 M / 61 M	1097 M / 50 M	1149 M / 52 M	1202 M / 53 M	1256 M / 54 M	1313 M / 37 M	1668 M / 355 M	1739 M / 71 M	1811 M / 72 M	1885 M / 74 M	1960 M / 75 M
Ordered												
Completed												
Issued from Stores												

GANTT CHART SHOWING RELATION OF MATERIAL ORDERED, COMPLETED AND ISSUED TO SCHEDULED REQUIREMENTS

3.11. The first published Gantt chart. (Industrial Management, *February 1918*.)

for the problem, he is apt to feel that his work is about completed. Usually, however, when his own mind is made up, his task is only half done. The larger and more difficult part of the work is to convince the minds of others that the proposed solution is the best one—that all the recommendations are really necessary. Time after time it happens that some ignorant or presumptuous member of a committee or a board of directors will upset the carefully-thought-out plan of a man who knows the facts, simply because the man with the facts cannot present his facts readily enough to overcome the opposition. It is often with impotent exasperation that a person having the knowledge sees some fallacious conclusion accepted, or some wrong policy adopted, just because known facts cannot be marshalled and presented in such manner as to be effective.[88]

Because they visualized and highlighted complex relationships for the reader, Brinton went on to state, graphs were often more effective for persuasion than tables or prose. Graphs could, of course, be manipulated to exaggerate certain trends: horizontal and vertical scales could be stretched or shrunk to achieve the desired effect. Even without such manipulation, however, graphs might be more persuasive just because they were eye-catching and easy to understand.

In part because of the susceptibility of graphs to distortion, and in part because of a desire for standardization, scientists, engineers, and managers saw a need for graphic standards.[89] The panel chaired by Brinton presented preliminary rules designed to increase readability, to minimize distortion, and to standardize methods. For example, the rule that "it is advisable not to show any more co-ordinate lines than necessary to guide the eye in reading the diagram" attempted to make reading a graph easier.[90] Another rule attempted to minimize distortion: "Where possible represent quantities by linear magnitude[,] as areas or volumes are more likely to be misinterpreted." Another rule, that "the general arrangement of a diagram should proceed from left to right," apparently aimed at achieving standardization as a value in its own right. The rules to aid readability and to standardize are analogous to the rules suggested by some experts in forms; the rules to minimize distortion were peculiar to the powers and dangers of graphic representation of data.

Graphs, then, were added to the set of techniques available to convey data up the hierarchy. Whether used alone as routine reports or as components of longer reports, graphs proved their value to the report genre.

More Efficient Prose Reports

At the same time that tables, forms, and graphs were coming to be used as reports or components thereof, prose reports were increasing in number and evolving in form and style, as well. The need for special reports increased as companies grew, adopted new methods, and entered new markets. There were, one expert noted, financial reports, engineering reports, efficiency reports, and marketing reports to be written.[91]

Moreover, longer periodic reports (such as departmental annual reports) were often still predominantly prose; tables, forms, and graphs did not allow for discussion and interpretation of data, nor for recommendations based on the figures.

As the number of prose reports increased, their form and style received more attention. In the second decade of the twentieth century, for example, report writing became part of the curriculum in New York University's School of Commerce and in the Massachusetts Institute of Technology's program in engineering administration (the predecessor of management there).[92] In 1924 Ralph U. Fitting published what he claimed was the first textbook on report writing.[93] Discussions of reports by academics, proponents of systematic management, and practicing managers revealed that the letterlike report of the nineteenth century had given way to newer forms and styles reflecting the need to transfer information as efficiently as possible.

The various authorities on reports agreed that form should follow function rather than tradition, but they did not always agree on either the function or the optimal form. George B. Hotchkiss, head of the Department of Business English at New York University, emphasized the need to pull facts up the hierarchy to executives:

> All reports, whatever their kind, have substantially the same general purpose. They aim to present facts for the information of an interested reader or group of readers; usually they give also conclusions or recommendations based on the facts. These facts and conclusions serve as a basis for someone's action. . . . The report serves as a basis for action rather than as a stimulus to action. It is not primarily an argument, though the final effect may be the same as that of an argument. The facts are almost invariably more important than the conclusions and recommendations, because if the facts are right the report has value despite the lack of logical conclusions.[94]

Since he viewed the facts as more important than the conclusion, he felt the structure of the report should emphasize those facts and not risk diverting the reader into disputing the conclusions. Consequently, he advised that "the report should usually be constructed in the inductive order; that is, with the facts first, and then the conclusions drawn from them."

While Hotchkiss was training his students to write reports with facts first and conclusions last, some of those in the trenches were giving very different advice. They emphasized not just the facts, but the efficiency with which an executive could acquire them. These managers viewed the function of special reports as supporting rapid decision making, rather than simply transmitting masses of information. In an article aptly titled "Putting It Up to the President," Lester Bernstein, supervisor of traffic statistics at the Baltimore and Ohio Railroad Company, recommended a report structure exactly opposite to that presented by Hotchkiss.[95] In a very modern-sounding admonition, he said: "Because an executive wants to know the meat of a report as soon as he can, there is no reason why he should be compelled to waste time hunting for what he wants or to laboriously thumb pages to find out the basis for something in the back of the report. Time to him is precious, therefore project the essentials just as quickly as possible in the beginning of the report."

Based on this assumption, he proposed that reports be organized as follows: "Title page, foreword, contents, recommendations, conclusions, summary, body." His reference to title page, foreword, and table of contents shows how far the report had evolved from the earlier extended letters. More importantly, however, he proposed putting the recommendations first and the facts last. The goal of this structure was to enable the reader to read as little as necessary: "Should he not agree with the recommendations, wholly or in part, he would then read the conclusions to determine on what grounds the recommendations were based. The reading of the conclusions may show him that his previous conceptions were not correct; and if he accepts the conclusions as given, there is then no need for his reading further." Bernstein sought to make the report efficient for the reader. Moreover, he emphasized the analysis over the data upon which it was based.

Although the two report structures apparently coexisted (as they still do), by 1924 Fitting could claim that some consensus was beginning to emerge:

> The increase in the number of reports has brought with it, more or less unconsciously, a greatly improved technique of presentation. A fairly uniform procedure has thus been developed for certain types. The development has been so recent, however, that as yet it has hardly been realized, either by those who read reports or by most of those who write them. Nor has it as yet affected the form of the great majority of reports on which the activity of the world of affairs so largely depends.[96]

This "improved technique," like Bernstein's report structure, was an attempt to reduce the reading burden on executives. To handle the quantity of reading demanded of them, Fitting noted, executives often found themselves turning directly to the conclusion of a long report, or handing it to a subordinate to summarize. "To avoid this way of reading reports a change has been made in recent years in the manner of presentation. Instead of one report . . . , two separate reports are submitted. The first is variously called 'brief report,' 'epitome,' or 'digested report.' "[97] His solution was to have the report author write what would now be called an executive summary for the report. The executive could read the short version, turning to the longer version only if necessary.

Thus, while Hotchkiss was most concerned with completeness, Bernstein and Fitting were more concerned with efficiency of presentation.

Leffingwell, the expert in scientific office management, focused on still another value of the evolving managerial ethos—standardization for its own sake:

> First there should be standard format—that is, all reports made by one office should follow certain rules devised for that office, as to size, type-face, style, and manner of presentation, to the end that all who have occasion to work upon or with the reports, or who use them, may contract uniform and desirable habits of work and thought in relation to this activity. In like manner, if reports are to render the greatest service to an organization, the general format to be followed must be devised with considerable care and uniformly carried out. It is not particularly important just what order should be followed—different offices may easily follow different styles— but what is important is that there shall not be several different styles of reports in the same organization, imposed thereon by different generations of executives.[98]

All of these report structures had come a long way from the leisurely letter reports common earlier.

While none of the experts spent much time discussing style, brief references indicate that report style omitted many of the frills of business letter writing. Hotchkiss pointed out that "one of the great points of distinction between reports and most other forms of business English is that the interest of the reader is taken for granted."[99] Consequently, he argued, clarity was more important than attracting the reader's attention. While reports could be formal or informal in style, depending on how well the writer knew the reader, they must always be clear. Another source commented on the need for "explicitness, literalness, and cool impersonality of statement" in report writing.[100] In general, authorities agreed that reports should be efficient and easy to follow in style as well as structure.

The report genre evolved in ways intended to draw information and opinions up the hierarchy as effectively and efficiently as possible. Tables, forms, graphs, and prose reports were all designed to lessen the load on the executive receiving the report. Tables and forms also lightened the burden on the person making the report. In other cases, the time of the source was sacrificed to that of the recipient, as with the extra synopsis for a long report or with the graph that was time-consuming to produce. Given the hierarchical structure, with fewer people and more information at each higher level, that trade-off seemed appropriate. The system of reports had to compress information progressively as it traveled up the pyramid in order to avoid overwhelming those at the top. This system was not always successful. Leffingwell pointed out that "over systematizing a business in the keeping of records is easily possible and by no means infrequent."[101] Nevertheless, the report genre evolved to support the upward flow of information that became so necessary to managerial practice.

Internal Correspondence: The Birth of the Memo

The memorandum acquired its identity later than some of the other genres of internal communication. (In fact, during the period up to 1920 a piece of internal correspondence was usually referred to as a note or letter. A memorandum or memo generally meant a note to oneself, though it was occasionally used in its modern sense.) Internal correspondence across distance had long been an accepted part of business, but in the late nineteenth and early twentieth centuries considerable correspondence—upward, downward, and lateral—emerged within facilities as a response to plant growth and systematization. The form and style of this correspondence began to diverge from that of external letters, reflecting the preoccupation with efficiency and system that shaped downward and upward communication. While custom and courtesy restricted the form and style of external letters, internal correspondence evolved in ways intended to make it more functional to read and to handle.

External letters took their form from British predecessors. In England, model letter books had long provided instruction in correspondence by offering examples of letters written to deal with common social and business situations. As early as 1698 Americans also began to produce model letter books with examples of sales letters, letters of collection, and other types of commercial correspondence.[102] The format of eighteenth- and nineteenth-century letters was much what it is today, allowing for the differences between handwriting and typing: the date and inside address were followed by the salutation, the body, and the complimentary close. In style, the letters were extremely formal and stilted by modern standards, making heavy use of certain wordy phrases especially in letter openings and closings, a style one modern commentator has described as "the 'your-esteemed-favor-of-the-sixteenth-ult.-duly-to-hand' style."[103] By the opening of the twentieth century, some authorities were beginning to rebel against this style even for external letters. One expert urged his readers, "I cannot repeat too often that the style in which a business letter ought to be written is that of a simple natural conversation."[104] Nevertheless, the older style continued to be popular in external correspondence.[105]

Authorities in business writing continued to focus on external letters even after internal written communication had begun to assume a major role in business. Well into the twentieth century, most textbooks on business English provided detailed examples and principles for writing various types of letters but never mentioned the growing volume of internal business correspondence.[106] Beginning in the early twentieth

century, however, books on office methods and a very few business writing texts noted the existence and differentiation of internal correspondence. The need for efficient handling and filing led to the initial distinction of internal from external correspondence. As an authority on
filing noted, subject-based filing demanded that "so far as practicable
each letter be confined to a single subject."[107] Since it was hard to limit
the subject in external correspondence, the limitation applied principally to internal correspondence, creating an initial divergence between
the two.

Further changes in form were designed to make internal correspondence cheaper and more efficient to type, handle, and file. Writing in
1910 about what he called "interhouse correspondence," or correspondence between different locations of a single company, one author recommended several changes in form that would make these documents
look less like letters and more like present-day memos.[108] His discussion
is worth quoting at length, for it sheds light on the underlying reasons
for the changes.

> In the first place, all unnecessary courtesy, such as "Fred Brown &
> Co.," "Gentlemen," "yours very truly," and other phrases are omit-
> ted entirely. In a business where hundreds and sometimes
> thousands of interhouse letters are written daily the saving of time
> is considerable. Next, an expensive letterhead is done away with,
> and this also is a factor in reducing expense. The blank is made
> with simply the words, "From Chicago," "From Atlanta," or what-
> ever may be the name of the town where the letter is written,
> printed in the upper left-hand corner, and underneath the word,
> "Subject." In the upper right-hand corner is the serial number of
> the letter and the words, "In reply refer to No." and "Replying to
> No." It will thus be seen that the only typewriting necessary in
> addressing a letter consists of the location of the house to which
> the letter is to be sent, a short summary of the matter contained in
> the letter for indexing purposes, the number of the letter, and date,
> with the initials of the writer, and the number and date of the letter
> which is under reply (in case there has been previous correspon-
> dence), with the initials of the former correspondent.

The traditional salutation and closing with their formal and polite style
were eliminated to save time—presumably the writer's and the typist's
time. Letterhead stationery was replaced by a cheaper and more efficient
form heading that included spaces for all the information the sender and
receiver might need to file and retrieve it.

Not all authorities had moved so far from the traditional letter and
toward the modern memo format at that time. In 1912, the Taft Commission on Economy and Efficiency noted that both foreign governments
and large American corporations were "experimenting" with simplified
and more functional forms for internal correspondence, but not that
these forms had been widely adopted.[109] A sample memorandum in a

1913 book on office management showed aspects of both new and old formats. It used a special form for internal correspondence, omitted the salutation, and even was labeled a "Memo," but it still ended with a standard letter closing.[110] Yet within a few years the existence and general outlines of the new genre were well established. A 1918 text on business English noted, "As [departmental letters] go to different departments of the house, the usual formal heading, introductory address, and salutation are needless, and a uniform system of mechanical details is adopted in each house."[111] A few years later, a committee of the New York City Board of Education, reporting on what should be taught in business English, defined the memo as a "miniature report" rather than as a variant of the letter.[112] Tradition had given way to functionality, and the memo had clearly diverged in form from the letter.

The differences extended beyond the heading and format into the documents themselves. While letters had traditionally lacked a tight structure, memos were now characterized as similar to reports in their "logical arrangement."[113] Moreover, memos, like reports, could include tables to increase readability.[114]

Style, as well as structure, responded to the demands of efficiency by becoming less traditional and formal. The 1918 text noted that "in departmental letters the absolute essential is clearness. Questions take time and delay action. . . . There is also no need of establishing personal relationship, so that courtesy should receive less consideration than directness and completeness."[115] Another treatment of the memo commented that it shared a report's "explicitness, literalness, and cool impersonality of statement" and praised a sample for its "strictly businesslike tone," which "makes for brevity and directness."[116]

In some cases, the removal of traditional polite phrases led to clarity and directness to the point of bluntness: "During the past few days some of you have been a little careless about getting to the office on time. If you arrive 5 or 10 minutes late, it will have a very unfortunate effect on the discipline of your department, and I trust that each of you will strive to overcome the habit of arriving late—a habit which, I am sure, is entirely due to carelessness on your part."[117] In other cases, the writer cultivated "cool impersonality of statement" to the point of bureaucratic style. One sample memo, for example, used the passive construction "It has been decided" instead of the more direct "We have decided."[118] This impersonality may in part have reflected a desire to sound scientific, in keeping with the management philosophies of the day. It was probably also a practical response to evolving bureaucratic politics in larger organizations.[119] In either case, the traditional polite phraseology of the model business letter disappeared. Of course, at this time letter-writing style was changing as well, but not as rapidly.

By the second decade of the twentieth century the memorandum had emerged as a genre in its own right, differentiated from external correspondence in form and style. Moreover, the differentiation, like the explosion of internal written communication itself, had its roots in the search for efficiency and system.

Meetings for Managers

One final type of formal internal communication deserves some attention here: managerial meetings or shop conferences. As Chapter 1 demonstrated, these meetings of foremen and/or middle managers had multiple purposes, including injecting a personal element into work life, promoting esprit de corps, monitoring and comparing the performance of comparable units, discussing policy, and generating ideas for promoting efficiency. Early in 1916, *Factory* magazine began a long-running monthly column on "Getting More Out of Shop Conferences," in which executives of different companies discussed the forms and techniques of shop conferences in their companies.[120] As these columns and other treatments of managerial committees reveal, this genre, like the other types of communication considered in this chapter, displayed some standard forms and styles.

Based on a questionnaire sent to twenty-five executives, *Factory*'s editors could identify two general types of shop conferences: " 'good will' and supervision conferences."[121] The former, which tended to occur less frequently, were "for general policy discussions and to promote good will and fellowship between the several department leaders." One executive writing in the column described them as follows: "A general shop conference is of considerable value in a large plant, bringing a large body of men together to receive advice, or have instilled into them the get-together spirit which every plant needs."[122] These general or good will conferences were closely related to the social clubs of the welfare movement. One executive commented explicitly that general shop conferences in his company "were the outgrowth of the very valuable suggestions brought out and the team spirit developed at our company dinners which have since 1902 annually brought together under the most pleasant auspices thirty to fifty of our responsible men."[123]

"Supervision" conferences, on the other hand, denoted meetings "where foremen or department heads, either as a whole or in groups, watch and discuss the progress of the work and ways to expedite its progress."[124] Such working sessions often took place "once a week or in some cases more frequent[ly]." These meetings could also take different forms, ranging from routine meetings to special problem-solving ones, and involving different groups of managers depending on the purpose. While the general or "good will" conference might include up to forty-five participants, most of the surveyed executives found eight to fifteen participants the appropriate size for working meetings.

Although the precise subjects, purposes, and procedures for the meetings varied, a few key characteristics were commonly found in the genre. The *Factory* questionnaire determined that all but three of the twenty-five executives surveyed said that they kept written records of the meetings. These ranged from general summaries to "a complete report of every transaction or statement."[125] In their use of minutes, these working meetings more closely resembled a board of directors' meeting than a social club. One advocate explained that the written record played a dual role, both focusing discussion and ensuring accountability:

> At each committee meeting, a stenographer should be present to act as secretary. When no record is kept, discussions are quite likely to be rambling rather than confined to a specific subject. Then, too, matters which were discussed at the last meeting have grown hazy, and if a man who is responsible for putting into operation a given suggestion has failed in his duty, there is always the chance that no one will think of it at the next meeting unless he brings it up. If, however, an exact record is kept of these discussions and full reports are delivered to every member of the committee before the next meeting, a foreman, or other member, will be very careful about making statements or promises unless he knows that they can be fulfilled.[126]

Another authority suggested a different mechanism: a folding blackboard used to record conditions and tasks agreed upon, then brought back out at subsequent meetings for comparison.[127] In either case, the written record documented the outcome of otherwise ephemeral oral communication, thus formalizing these meetings as a mode of communication.

Written and oral reports often supplemented the discussion, as well. According to an expert on industrial organization, "The basis of discussions in the majority of such meetings will lie in the consideration of the departmental records. It is in the defense of these reports by the different foremen that many new plans are suggested which are later developed to the great benefit of the company."[128] One executive interviewed for the *Factory* series revealed that in foremen's meetings at his plant, "The department expense analyses are also given to the foremen and each month the figures are compared to prove the advantage of constant discipline on the part of department heads."[129] In addition to analyzing routine reports, such committees often discussed special reports or papers prepared especially for the meetings. A rotating schedule of papers given by participants allowed peers to educate each other on their work and their methods for managing workers, improving efficiency, promoting safety, and so on.[130] For such reports, either lanterns used with slides or reflecting lanterns (which did not require specially prepared slides but could project an image from cards, papers, or books) could be used to display graphs or tables.[131] Whatever the medium, such records and reports focused the participants on specific departmental issues up for discussion.

Management committee meetings fostered multidirectional communication very different from that embodied in the written genres discussed above. Such committees represented, in some ways, a counter-force to the depersonalization of most of the strictly hierarchical downward and upward communication. Like the other genres of communication, however, such committee meetings were adapted from older precedents in the face of new demands posed by growth and systematization. As the easy, oral communication of owner to foreman gave way to standardized written reports and orders, a more formal mechanism was needed to maintain the communication necessary for promoting coordination and cooperation. These meetings also reinforced the new policies and goals of systematic management. As one executive consulted by *Factory* summed it up, "We feel that the shop conferences are growing in value and consider them most important both from the standpoint of the human element and cost-cutting aspect."[132]

Conclusion: Genre and Function

The genres of internal communication that emerged during the late nineteenth and early twentieth centuries evolved in response to new demands put on them by growth and by changing management philosophy, within the constraints of communication technologies. In the case of vertical communication, the railroads led the way in the mid-nineteenth century. Later in the nineteenth century, manufacturing firms followed that lead. The genres developed new forms, shaped by the desire for efficiency and standardization. In the early twentieth century, forms of communication that were not strictly vertical also emerged. Although the memo genre, like the report, was shaped by the desire for impersonal efficiency, in-house magazines and managerial meetings reintroduced a personal element, harnessing it to the aims of efficiency.

These genres were important mechanisms by which large firms were controlled and managed. In subsequent chapters, the case studies demonstrate more concretely the interrelationships among managerial functions, communication technologies, and communication genres.

Chapter 4
The Illinois Central
before 1887
Communication for Safety,
Consistency, and Honesty

The Illinois Central Railroad, the first of the land-grant railroads, played an important role in the agricultural and economic development of the midsection of the United States. The land grant passed Congress in 1850, and the Illinois Central Railroad was chartered early in 1851. Although it came into existence almost half a century after Scovill and Du Pont, and more than two decades after the first eastern railroads were established, it grew into a large firm with thousands of employees almost immediately. It ranked as one of America's major railroads throughout the nineteenth century.

Because of its size and the characteristics of the railroad business—a dispersed workforce, potential danger to people and expensive equipment, and the collection of large sums of money—the Illinois Central was forced to deal with issues of control and communication early in its corporate life. In its first decade it established a rudimentary formal communication system to assure safety, consistent service, and honesty. Various factors, including income from land sales and limited competition on its north-south route, shielded the Illinois Central from extreme pressure to increase efficiency. In spite of incremental refinements over the ensuing quarter century, its communication system at the beginning of 1887 was still very similar to what it had been in 1862. Not until the late 1880s and early 1890s, under the leadership of Stuyvesant Fish and in response to both increased competition and regulation by the Interstate Commerce Commission (ICC), did the company thoroughly modernize its management methods and communication system to promote efficiency.

The First Decade: Laying the Groundwork

The federal land granted by Congress was conveyed to the incorporators of the Illinois Central Railroad in 1851. Thus, the Land Department, through which the land was to be sold, was the first unit of the company to be organized.[1] The sale of land loomed large in the affairs of the corporation during the early years; one scholar has said that in its first decade, the Illinois Central "was primarily a land company and secondarily a railroad company."[2] Construction of the main line was the other major task in the first five years. Nevertheless, these challenges were, ultimately, peripheral to the railroad's main purpose.[3] Passenger and freight traffic began to run on the first segment of the track as soon as it was completed in 1853.[4] When construction of the 705-mile main line was completed in 1856 (see fig. 4.1), the company employed 3,581 men and operated what was, at that time, the world's longest railroad.[5]

From the start, the dispersion of the workforce as well as the varied tasks involved in constructing and operating a railroad necessitated a relatively complex hierarchical structure. The various functions—construction and maintenance, transportation, freight, and passenger service—had their own departments, which were further subdivided into geographical units. The organizational complexity was increased by the fact that the geographical divisions in different departments did not necessarily coincide. In 1855, for example, the Engineer's Department divided the road into twelve divisions, while the Transportation Department divided it into only three.[6] To complicate matters further, the large capital requirements of a railroad demanded that the company's financial office and its president be in the East, near the capital markets. Thus, financial management was centered in New York, while operational management was centered in Chicago. Consequently, even the most basic managerial coordination could pose logistical problems.

Of course, the Illinois Central benefited from being able to adopt basic operating procedures established by older eastern railroads and well known to its early management team. Its first president, Robert Schuyler, was simultaneously president of four other roads (he eventually paid for this overreaching ambition when a conflict of interest scandal caught up with him).[7] The first vice-president, Captain David A. Neal, was president of two railroads, the Eastern and the Reading, when he resigned to join the Illinois Central.[8] Colonel Roswell Mason, chief engineer during construction, and John B. Calhoun, who handled his accounts and finances, had both previously worked on the Housatonic Railroad. Such men helped transfer to the Illinois Central the basic techniques of communication and management developed by eastern roads, but not the newer methods then taking shape in the most progressive roads, such as the Erie.

During the first decade, downward communication established the rules and procedures needed to secure safety, honesty, and consistent service. Upward reporting drew basic financial and minimal operating information up the hierarchy to help executives assess performance, though only in the most general way. Internal correspondence bridged distances between the geographically separated officers.

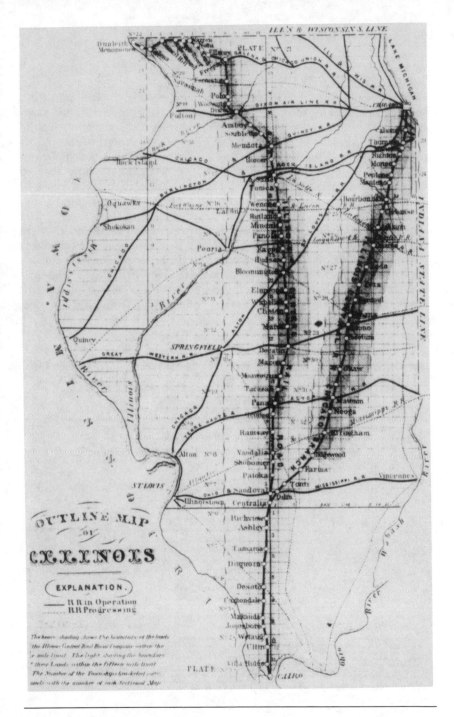

4.1. Map of completed Illinois Central main line, 1856. *(Courtesy of the American Association of Railroads.)*

Establishing Rules for Train Movement

The safety of employees, passengers, and expensive equipment demanded that rules for moving trains be identically understood and scrupulously obeyed by all employees involved in train movement. By the 1850s, the need for written rules to govern the movement of trains was widely accepted. The Illinois Central did not need to discover the lesson learned at such cost by its predecessors; its 1860 annual report boasted that the railroad had had no fatal injuries in the first seven years of operation, even though the line ran on a single track for most of its length.

In these early years at the Illinois Central, however, the rules were still limited and the method of conveying them still basic. Procedures for the movement of trains were printed on the reverse side of the current employee timetable carried by everyone involved in train movement.[9] As the 1857 timetable in figure 4.2 illustrates, while the service was still relatively limited the scheduled stops and meeting times and any nec-

---

FOR THE GOVERNMENT OF EMPLOYEES ONLY.

Urbana Section.   Illinois Central Railroad.   Chicago Division.

BETWEEN CENTRALIA & URBANA.

TIME TABLE, NO. 6.

TO TAKE EFFECT SUNDAY, AUGUST 23d, 1857, AT 12 O'CLOCK, NOON.

FROM URBANA, GOING SOUTH.						DISTANCES FROM CHICAGO	STATIONS.	DISTANCES FROM CAIRO	FROM CENTRALIA, GOING NORTH.					
Freight		St. Louis and Cairo Express		St. Louis and Cairo Express					St. Louis, Cincinnati and Chicago Express		Cairo and Chicago Express		Freight	
TIME	REMARKS	TIME	REMARKS	TIME	REMARKS				TIME	REMARKS	TIME	REMARKS	TIME	REMARKS
5.45 A.M. Leave		3.25 A.M. Leave		3.20 P.M. Leave		128	Urbana	238	5.30 P.M. Arrive	3.25 A.M. Arr. Meet Pass			4.45 P.M. Arrive	
6.40		3.48		3.55 Meet Freight		137½	Tolono	227¾	5.05		3.00		3.55 L've / 3.49 Arr. Meet Pass	
7.07		3.59		4.07		142½	Pesotum	222½	4.55		2.48		3.17	
7.52		4.19		4.39 Meet Pass		150	Ia. Cent. Crossing	215	4.39 Meet Pass		2.30		2.32	
8.37		4.37		5.00		158	Okaw	207	4.13		2.11		1.48	
9.11		4.52		5.20		165	Milton	200	3.58		1.56		1.17	
10.00 Arrive / 10.15 Leave		5.15 Arr. Con. with T.H. Train going West / 5.30 Lea.		5.50 Arr. Con. with T.H. Train going West / 6.10 Lea.		171½	Mattoon	193½	3.35 Lea. Con. with T.H. Train going East / 3.10 Arr.		1.35 Lea. / 1.24 Arr.		12.30 P.M.	
11.25 Meet Frt.		6.06		6.51		183½	Neoga	181½	2.36		12.48		11.25 B. Freight	
12.46 P.M.		6.46		7.36		198	Effingham	167	1.55		12.05 A.M.		10.06	
1.25 Arr. / 1.35 Lea. Meet Pass		7.05		7.58		205	Watson	160	1.25 Meet Freight		11.48		9.28	
2.08		7.22		8.17		209	Mason	154	1.19		11.32		8.57	
2.23		7.31		8.27		214	Edgewood	151	1.10		11.23		8.39	
3.09	Mt. Freight	7.55		8.53		222½	Farina	142½	12.47		10.59		7.55 Lea. / 7.45 Arr. Meet Pass	
3.41		8.10		9.12		230	Kinmundy	135	12.31		10.43		7.10	
4.33		8.38		9.43		238½	Tonti	126½	12.03 P.M.		10.15		6.15	
5.02		8.53 Meet Pass		10.00 Meet Pass		244	Odin	121	11.48		10.00 Meet Pass		5.45	
5.36		9.11		10.21		250½	Central City	114½	11.29		9.39		5.08	
5.45 P.M. Arrive		9.15 A.M. Arrive		10.25 P.M. Arrive		252	Centralia	113	11.25 A.M. Leave		9.35 P.M. Leave		5.00 A.M. Leave	

The Clock in the office of the Superintendent, at Centralia, will be taken as the standard time. Conductors and Engineers must compare their watches daily.   See Rule 10.
All Trains on Branch must come to a full stop at the Junction and see that no Trains are approaching on Main Line.
For Rules and Regulations, see back of this Table.   Trains will not stop to receive or leave Passengers at Stations, except on Signal.   Reduce speed to 8 miles per hour over all wooden bridges and trussel works.   Study well the Instructions on back of this Table.
Freight Trains must not run on time of Passenger Trains, unless by written orders from Superintendent or his Assistants.   Trains going South have the right of Road over Trains of similar class going North.   REGULAR Trains, however, have the right of Road over all Extra Trains, whether going North or South.
Trains must be run at such speed as to enable the Mails to be exchanged at all Stations where there are Post Offices.
☞ Observe change in Rule 2, of Regulations, and Rules 1 & 2, of Directions concerning Signals.

4.2.   Above, an 1857 timetable; facing page, rules printed on the reverse side of the timetable. *(Courtesy of the Newberry Library.)*

Regulations for the Running of Trains.

Both Engineers and Conductors will be held responsible for any violation of these Rules.

1st. No Train will be allowed to leave a station before its time, as specified in the Time Table, nor run faster between stations than is required to reach the next station in season to start from it on its regular time.

2nd. All regular Passenger and Freight Trains will leave Chicago on their card time, and shall have the right of road to Calumet, for thirty minutes beyond their card time, against all delayed trains, both Passenger and Freight.

South of Calumet, trains going South have the right of road for thirty minutes beyond their card time, against all trains of same or inferior class. At place of meeting, however, when the approaching Train has not arrived, they will wait Five Minutes beyond card time, to allow for variation of watches.

3rd. Trains going North will not leave a turn-out, unless they can, *without doubt*, reach the meeting place on or before the time marked in the Time Table for the departure of the approaching train; but will wait where they expect to meet the approaching train *thirty minutes* beyond the time set for it to be there, and then proceed with caution, keeping *thirty minutes* behind its own time, until the delayed approaching train is passed.

4th. Passenger Trains will not wait for Freight Trains at the places appointed for meeting. Freight Trains wait indefinitely for Passenger Trains.

5th. Freight Trains, as between Freight Trains, will be governed by rules 2 and 3. See rule 4.

6th. Trains, on approaching a meeting point, will enter the side track at the nearest end.

7th. No engine, with or without a train, will be allowed to pass along the line without previous notice, except by the written permission of the Superintendent; and no engine or train following another train will be allowed to run nearer than *one mile* of the preceding train except at stations, when they may run up within sight, but with the greatest care.

8th. Whenever any Freight or Baggage, and Passenger Cars, are formed into one train, the latter must, in all cases, be placed on the rear of the train.

9th. The engine bell must be rung at the distance of eighty rods from each road crossing, and be kept ringing until the engine has crossed.

10th. Passenger trains will leave the termini of the road (Cairo and Chicago,) on their regular time, and proceed to the first station, all delayed trains keeping out of their way; after reaching this station they will be governed by Rules 2 and 3.

11th. If it shall be found impracticable, from any cause, for a train, in passing from one station to another, to reach the station to which it is proceeding in season, and another train is expected, then the Conductor will send a man in the direction of the approaching train, with a flag by day, or a lantern by night, to give notice of his position; and should it be necessary to back a train, a man must be sent in advance around the curves, and a sharp lookout observed.

12th. Enginemen will sound the whistle, and approach all stations slowly, pass all switches cautiously, and be sure that the switch is seen by its lever to be right. No excuse can be received for running off a switch when the lever shows it to be wrong.

13th. Engineers will not allow any person to ride upon their engines, except the Road Master, or by permission of Master of Transportation, Superintendents, their assistants, or the Master Mechanics.

14th. All persons engaged in the service of this Company are required to give notice of any obstructions on the road caused by their work or otherwise, by exhibiting conspicuously a red flag in the day time, and a lantern by night, at least fifty rods in both directions of the road from the obstruction. Conductors and Engineers are required to proceed with extreme caution until such obstruction is passed.

15th. All foremen of repairs, and men under their direction, must at all times hold themselves in readiness to aid all in their power, the passage of trains, and assist in case of accident, by conveying intelligence when required so to do by the Conductor, or by giving a prompt and willing obedience to his orders.

16th. Station agents will be held responsible for the proper position and security of their switches, and at night will either themselves attend, or see that some person stands at the switch, and exhibits a light while the trains are passing.

17th. The general direction and government of the trains, from the time of receiving their passengers and freight, until their arrival at their destination, is vested in the Conductor. He will be held responsible for its safe and proper conduct, and all the men employed on the train are required to yield a willing obedience to his orders.

18th. A brakeman must, in all cases, while the train is in motion, be on the rear car. He must see that the bell-cord is adjusted and kept in its place, and that it extends to and is attached to the rear car of the train.

19th. Station agents are required to have the doors of all cars on the side track securely fastened out of the way of passing trains. They will be held responsible for the position of the switches, and in no case will allow them on the main track, except when a train has arrived to enter the side track.

20th. Baggagemen will invariably and immediately report to the Conductor of their trains whenever any baggage has been lost or miscarried, or whenever there has been any call by passengers for missing baggage.

21st. Whenever any articles of freight are lost or miscarried, the Conductor of the Freight Train must immediately report the same to the agent of the station from which the train was sent.

22d. All trains will run with great care after rains, and slacken their speed when the track is in bad order, while passing switches, when crossing long bridges and tressel work, and when practicable shut off steam.

23d. All accidents, such as breakages, getting off the track, uncoupling of trains, killing stock, failure, in any way, of the engines, defective places in track, road crossings or bridges, etc., must be immediately reported by the Conductor to the Superintendent, and to the man in charge of the division.

24th. No brakeman will be allowed to leave his post, or to ride in a car, when the train is in motion.

25th. Strict observance of all regulations, and the greatest care and attention to their several duties, are enjoined upon all.

26th. Whenever you are in doubt, take the safe course.

27th. All trains will come to a full stop before crossing another Railroad at grade, and wait to be signaled by the Conductor to start.

28th. Avoid killing Stock. Train men will be held to a strict accountability for all Stock injured by their respective trains.

Directions Concerning Signals.

1st. One Red Flag by day, or one Red Light by night, carried upon an engine, indicates that another engine is following, which engine will be allowed all the rights and privileges of the engine bearing such signal.

2nd. One Green Flag by day, or Green Light by night, carried upon an engine, indicates that another engine is following; and the engine following such signal will be run with great care, and must be kept out of the way of all regular trains.

3rd. Two Red Flags by day, or two Red Lanterns by night, carried upon an engine, indicate that another engine or train is following, which must be treated as a part of the forward train. No train, after meeting two Flags or Lanterns upon an engine, will leave the station where it meets the two Flags or Lanterns so carried, until the flagged train arrives, or is positively heard from.

It must be distinctly understood that when such Flags or Lights are carried, the train for which they are carried must be in sight at every station. If it is not in sight, then the forward train must substitute one Green Flag or Lantern, and leave a written communication to that effect for the following train with the station agent, and then the following train must run under the second rule of signals.

4th. A Red Flag by day, or Lantern by night, waved upon the track, signifies that a train must come to a full stop.

5th. A stationary Red Flag signifies that the track is not in perfect order, and must be run over with great caution.

6th. A White Flag denotes that the track is clear and in order.

7th. Two Red Signal Lights must be exhibited on the rear of each Passenger train in the night time, and one on Freight trains, until the train arrives at its destination.

8th. One puff of the steam whistle is a signal to Brake; two puffs is the signal to loose the Brakes; three puffs is the signal to Back; five or more rapid puffs is the signal for Wooding up, or calling in Signal men, when stationed out on the road.

9th. A Lantern swung over the head is the signal to go forward; raised and lowered perpendicularly, to stop; swung sideways, to back.

SILAS BENT, Supt.

essary explanations could be printed on one side of a single sheet of paper. At this stage, the simple rules fit readily on the other side. The sheet folded into a four-panel folder, about three by eight inches, easily carried by employees. Those responsible for train movement were expected to know these rules by heart, but the timetables were available for reference if any question ever arose. Such rules were occasionally reinforced or modified in a particular case in printed circular letters issued to all affected employees.[10] In general, however, train movement was governed strictly by the rules printed on the timetables.

The Illinois Central had access to a telegraph line along its length very early in the road's operations, as the 1855 Annual Report explains: "The Company has erected a line of Telegraph from Cairo to Dunleith, which is operated by the Illinois Telegraph Company who contract to keep the line in order, the free use of it being secured to this Company, with the privilege of using 1200 miles of telegraph controlled by the Illinois Telegraph Company."[11] Although at least one historian of the railroad states that the Illinois Central used the telegraph for dispatching as well as for other business from the very beginning,[12] other evidence suggests that its role in train movement was at best minor and peripheral until 1863. The rules on the 1857 timetable (fig. 4.2), for example, include no mention of the telegraph, although they cover many other contingencies. In contrast, timetables from the late 1860s include extensive sections on telegraphic control of trains. Moreover, although the correspondence from this period referred to the telegraph's use for other business purposes, the only references to its use to facilitate train movement involved special circumstances (such as sending for cars to carry stock) rather than to routine train movement.[13]

Further evidence that the Illinois Central did not routinely use the telegraph to control train movement until the mid-1860s can be found in reports and correspondence concerning a second telegraph line erected in 1863. The railroad had only very limited and undependable service from the original telegraph line at this time because, due to wartime demands, it was "fully engaged by the Government and the public."[14] In fact, in early 1863 William H. Osborn, president of the railroad, complained to the president of the telegraph company operating the original line that "to get a reply from any point to any message I may send takes an average of six hours."[15] Although Osborn was not referring to train dispatching, this evidence suggests that the wire could not be counted on for routine dispatching as long as railroad messages competed with public and governmental use.[16] Finally and perhaps most conclusively, in a letter written at the end of 1863 after a storm had knocked down many parts of the new telegraph line, Osborn complained, "Now that our trains are moved by Telegraph, this disarrangement causes the most serious consequences."[17] The language of this statement clearly implies that the situation was a recent one.

It seems highly likely, then, that the telegraph was not used for routine dispatching until 1863. In the first decade, train movement was still governed primarily by the rules printed on the timetables. While lacking the speed and flexibility of telegraphic dispatching, these printed time-tables conveyed identical rules for train movement to all relevant employees in a form that could easily be consulted. The timetables were essentially a portable organizational memory concerning train movement. The communication vehicle, however, restricted the number and consequently the flexibility of the rules. Because only a limited number of rules would fit onto the back of a timetable (though as service grew, so did the timetables), rules for moving trains more efficiently under a variety of special circumstances such as breakdowns or weather delays could not be included. The basic rules would always provide safety, but they might do so at the cost of efficiency.

Communicating Other Regulations and Procedures

Executives of the Illinois Central also recognized the need to promulgate standard rules, procedures, and chains of command for operations beyond those directly connected with the movement of trains. They needed to guarantee adequate service to customers as well as uniform and honest handling of finances throughout the organization. The size and dispersion of the workforce led management to follow the precedent of eastern railroads in formalizing downward communication about such matters quite early.

Printed circular letters and notices—issued either from departmental headquarters in Chicago (by the general superintendent or department heads) or from division headquarters (by division superintendents)—established and updated procedures of all types. In these early days many procedures needed to be established, so circulars frequently provided several related rules and instructions, rather than a single one. In 1854, for example, the general superintendent issued a one-page circular letter entitled "Regulations adopted by the Illinois Central Railroad Company in regard to FREE PASSES."[18] Once basic rules in a given area had been communicated, subsequent circular letters could provide additions and changes. At the beginning of 1859, the land commissioner issued an instructional circular on land sales to all station agents. A few months later he issued another circular letter, updating the specific terms for land sales but otherwise reaffirming the previous instructions.[19] In both of these cases, consistent and honest handling of financial matters was the goal of the regulations.

In addition to establishing rules, circulars were also used to reinforce existing rules and reprimand lapses from them. For example, an 1857 circular from the master of transportation included this statement: "Complaints have been repeatedly made to this office of the neglect in the past of Baggage Masters of Pass[enger] Trains to deliver articles placed in their charge for points along the Road. A repetition of this will be followed by the dismissal of the person found to have neglected his duty."[20] Here the goal was to assure consistent service to customers. Circulars were also used to give special instructions for specific situations, such as an order to allow a surveyor to get on and off trains anywhere he desired during a specified period.[21] Finally, they were used to announce new appointments and lines of authority. Some simply announced new appointees, with no comment as to their functions, while others attempted to define, at least generally, the authority of new appointees.[22] When the superintendent of the North Division announced the appointment of a general freight agent, for example, he added, "All Agents are required to obey implicitly any orders or instructions emanating from the General Frieght [sic] Department."[23]

Company executives also showed some signs of wanting to create a primitive organizational memory of procedures and instructions. The circulars that included several rules on a given subject were obviously intended for reference. At headquarters, copies of some (but not all) circulars were pasted into bound books as a record.[24] In spite of these limited movements toward establishing an organizational memory, in this first decade no comprehensive compendium of rules existed.

As far as surviving records show, circular letters were issued relatively infrequently, perhaps partly as a result of the duplication method. Many copies of each circular or notice were needed. Since hand writing hundreds of copies was obviously too time-consuming and stencil duplicating methods were not yet available, printing was the only feasible option for mass downward communication. This process was both expensive and slow. In 1856 alone, the company incurred a "Stationery and Printing" expense of just over $15,000.[25] Once printed, the circulars were apparently distributed hierarchically, with batches sent to middle managers, who in turn sent them down the line to their subordinates.[26] (Physical transmission, of course, was available at no charge on the trains themselves.[27]) Presumably recipients of such circulars saved the relevant ones for awhile, but most such personal collections have not survived.

The basic forms of mass downward communication of rules and procedures—the timetable for train movement and the circular letter for other matters—were established at the Illinois Central during its first decade. The firm did not have to develop these genres, for other railroads had already established them. Such communication was used as a tool of managerial control to assure safety, consistency in service, and honesty in financial matters.

Reports to Stockholders and Directors

Because the Illinois Central, like other railroads, was large and capital-intensive, ownership was separated from management from the start. From the start, too, annual reports from the management to the stockholders were necessary to keep stockholders informed. Moreover, the board of directors that represented the stockholders needed even more information about the railroad's operations, frequently gained through additional special reports, in order to give good advice and make good decisions.

In 1852, following established precedent, the Illinois Central's president and board of directors began to issue annual reports. These reports initially focused on the company's financial status and land sales. The land sales were new to such reports, since the Illinois Central was the first land-grant railroad, but they were presented in a form similar to that used for passenger and freight receipts. Once the road opened for use, the reports contained more information about operations. By 1855 the annual report included letter-reports from the treasurer, the chief engineer, the general superintendent, and the land commissioner—all addressed to the president and directors—as well as one from the president to the stockholders. The separate letter-reports making up the annual report each combined prose discussions of departmental progress with traditional balance sheets and more modern tables of figures (e.g., revenues of various types, or miles of track open for use).

As the company's operations became well established in the latter half of the 1850s, the Illinois Central again followed precedents in providing certain basic data that could be used to compare the road with others or with itself in previous years. Starting in 1856, the first full year of operation on the entire main line, the annual report provided the operating ratio (operating expenditures as a percentage of gross receipts) for the line each year. As a result of weak performance in the ensuing three years (blamed on crop failures and a financial panic), however, the general superintendent's section of the 1859 annual report cautioned readers about using that figure as the primary measure of the line's performance:

> Allow me here to remark, that the question of the expenses of operating the Illinois Central Railroad, compared with the earnings, so far as the profitableness of the property is concerned, is one that cannot be fairly measured by the ordinary rules governing other roads of this country. This road was built on the basis of a munificent land grant, and a traffic to any considerable extent, from many portions of the line adopted, was merely prospective, depending upon the sale and development of the lands through which the location was made. In constructing the road and organizing the operative department, it was done, not only with the view to business already offering, but also for the purpose of carrying out one of the original objects of the corporation—that of furnishing facilities of transportation for the large and fertile bodies of land donated for the construction of the line, thereby rendering them marketable. In doing this, it has not been necessary to run a greater number of trains, or employ a greater number of men than will keep up a daily communication over the entire line. It is, therefore, not exactly consistent with a clear statement of the case, to compare the expenses with the traffic receipts alone, as in ordinary lines. Due importance

should be given to the enhanced value of a property out of which
the construction indebtedness is expected to be met.

From early on, then, the Illinois Central argued that because of the land
grant, its success could not be completely gauged by standard measures
of railroad efficiency. Nevertheless, the company continued to report
the operating ratio as well as a few more figures that could be compared
across lines, such as receipts (but not expenses) per ton mile. Beginning
in 1856 and 1857, it also included its previous year's performance in each
category as a benchmark for comparison.

Annual reports to stockholders, while technically internal reports to
the owners of the railroad, were in reality public documents of little use
in daily managerial decision making. The directors, who were involved
in the major executive decisions on behalf of the stockholders, needed
more frequent and more detailed information than what was given in
the annual reports. Some of that information was communicated by
letter to individual board members or orally in New York board meetings
attended by the president or vice-president.[28] In addition, the 1856 Annual Report notes that the board deputed some of its powers to an executive committee of leading operating officers in Chicago, so that questions of local management "outside of the ordinary routine" could be
handled locally. Occasionally executives sent this committee long letters that served as special reports. In the spring of 1856, for example,
President Osborn wrote a ten-page letter to that committee explaining
the steps that J. C. Clarke, new superintendent of the Northern Division, had taken to control the insubordination, inefficiency, and chaos
then prevalent at division headquarters in Amboy, Illinois.[29] Even such
an extended letter-report retained the format and loose organization of a
letter, showing no development from traditional formats to differentiate
it from an external letter or to make it easier for the committee to read.

Although both annual and special reports were issued by top executives to directors, stockholders, and the executive committee, these
reports played little part in the day-to-day management of the road and
showed no real innovation in form or function.

Primitive Routine Reporting System

More interesting for my purposes than annual reports and special
reports to stockholders and directors are routine reports designed to
carry information up the company hierarchy for use in decision making.
A system for gathering and reporting the information necessary to guarantee honest handling of money was established immediately, and some
of this information was compiled in a form that allowed limited comparisons over time. Nevertheless, very little reporting useful in evaluating daily operating performance was established until the late 1850s. In
1858, when financial problems drove the top executives to focus briefly
on controlling costs and improving efficiency, they established a primi-

tive system of reports for collecting and monitoring data on operations and expenses. Even then, however, their reporting system was not a very sophisticated managerial tool.

From the start, the company handled large amounts of money. It paid money out for the expenses incurred in advertising the land it needed to sell, and in building and then operating the line. It took money in from land sales, and, once the trains started running, from passenger and freight service. From the start, also, the company established standard financial accounting and auditing procedures to assure honest handling of this money.

Any monetary transaction created a record of how the money was spent or from what source it was collected. For example, flows of cash such as those from stations along the line to the local treasurer, and in turn to the main treasurer's department, were accompanied by brief statements of cash receipts.[30] Moreover, the accounting department recorded various earnings and expense figures on a monthly or annual basis in large bound volumes that spanned many years.[31] Beginning in 1855, for example, annual earnings were compiled by source (passenger, freight, mails, storage and dockage, etc.) into huge tables that eventually covered two decades. Total monthly earnings (uncategorized by source) were also compiled over a period of years. Similarly, annual operating expenses were recorded under 28 different accounts each year, as well as by month in an uncategorized form.

While 28 fell far short of the 144 accounts used by the Pennsylvania Railroad at this time,[32] such an organizational repository of data for comparisons over time was a first step in establishing effective control through reporting. But these centrally compiled and maintained volumes were not in a form useful to operating managers in their day-to-day decision making. The ponderous volumes were probably compiled in the treasurer's office in New York, and thus were not accessible to operating managers in Chicago. Even if a duplicate series of volumes existed in the Chicago office, however, managers wishing to consult the series would have had to go to a central location; the volumes were far too bulky to be circulated. Moreover, most of the data was compiled annually. A few statistics were maintained on a monthly basis, but during this first decade none were compiled more frequently than that. Even with these few, there was undoubtedly a considerable time lag between the end of the month and the time that they were recorded.

The more critical issue, then, is the extent to which data reached operating managers in timely routine reports useful as tools of managerial control. Top managers of the Illinois Central clearly were at least aware of (and in fact felt that they were imitating) Daniel McCallum's 1854–56 innovations in managerial methods on the Erie Railroad, although they apparently did not fully understand the role that systematic reporting played in those methods.[33] In his 1856 letter-report to the executive committee, President Osborn praised James C. Clarke's actions as division superintendent by saying that Clarke "has been systematizing his accounts, and is keeping a check on his master machinist, the consumption of oil, etc."[34] In fact, Osborn asserted, Clarke's operation was "fully as systematic for all practical purposes as the plan adopted by the Erie Railroad." If Clarke did gather systematic information on a variety of operating issues, however, it apparently stopped at his own desk, rather than serving as part of a comprehensive upward reporting system. Before 1858, no systematic reports appeared in the bound volumes of either his incoming or his outgoing correspondence.[35] He reported to Osborn only via frequent but irregular and unsystematic letters that conveyed a mass of unorganized information.

The only indication that any of the Illinois Central managers really understood the principles behind the innovations the Erie had implemented was an 1855 letter from a middle-level manager to his subordinate: "Will you please have kept a report made at the end of the month to this office of the number of miles run by each engine and whether passenger or freight trains and the number of gallons of oil and pounds of waste furnished the same, with the name of the Engineer[,] for publishment and comparison. The roads East have adopted this plan and find a saving of 30%."[36] This letter expressed the fundamental importance of recording and comparing statistical information as the basis for improving efficiency, a principle clearly enunciated by McCallum. Such a monthly analytic report was a beginning, but the key comparative figure drawn from it, costs per mile for each engine, left much to be desired as the major measure of operational efficiency. It did not distinguish between the engine's and the engineer's efficiency, for example, nor did it take overhead costs into account. Moreover, the report was still to be made monthly, rather than more frequently. This attempt to set up a standard of comparison for train performance is virtually the only evidence of systematic gathering of data for management monitoring at this time. Pre-1858 management of railroad operations at the Illinois Central seems to have been largely ad hoc, in spite of pretensions to a more progressive system.

In the wake of the Panic of 1857 and the company's consequent near brush with insolvency, however, upper managers—particularly George B. McClellan, vice-president, and J. C. Clarke, now master of transportation—became concerned with reducing costs on the line.[37] In 1858, they devised a series of daily, weekly, and monthly reports to carry financial and some operating information up the hierarchy.

Beginning in July 1858, Clarke wrote daily letters to Vice-President McClellan in Chicago, who forwarded them to President Osborn in New York.[38] At first, these letters only listed the number of loaded and empty railroad cars received and forwarded at each station. The letters continued into the fall, at which point McClellan began compiling information sent up the line to the Chicago office into more comprehensive daily letters to send to Osborn in New York.[39] They included the "Daily report of local Treasurer," sometimes referred to as the "Daily cash statement"; the "Daily report of business at Telegraph Stations"; and the "Daily Commercial Letter," also referred to as the "daily statement of business at this station," which continued the listing of cars received and forwarded at each station. The first two items were essentially extensions of the financial reporting system for tracking cash received, and the third part provided operational data. Although there is no evidence that these car reports were then subjected to the analysis necessary to make them useful for monitoring and control of operations, at least the information was being collected. These letters began a daily flow of financial and operational data that would continue for decades.

Initially these letters were handwritten and inconsistent in form. Parts of them were fairly standardized (the list of stations with their business, for example), while other parts varied in form and even in the items included. Such letters were tedious to write out and, because of the inconsistencies, inefficient and sometimes even inappropriate to use as the basis of routine analysis.

In addition to these daily reports, a series of monthly reports and at least one weekly report were devised at the same time to pull information on expenditures up the hierarchy. In September 1858, McClellan issued a circular on handling monthly accounts.[40] The newly devised procedure required the person in charge of a set of accounts to close all supply vouchers by the twenty-eighth of each month and to make out "an approximate statement of the total expenditures in your department—estimated for each month—including Vouchers paid & unpaid & the amt. of Pay Rolls" on the last day of the month. The final figure for estimated monthly expenditures was telegraphed to the Chicago office immediately, and the payrolls, vouchers, and statement were sent by train soon after.

This new procedure for estimating expenses gave McClellan and the New York executives regular information on each department's expenditures, though in a form more useful for financial than for operational control. The monthly reports allowed them to track financial demands on the company so that a sudden increase in expenses would not go undetected until it caused serious problems with cash flow. The system for closing accounts on the twenty-eighth and telegraphing estimated totals on the final day of each month gave them timely information for each department, bypassing the delays that would have ensued had they waited for the exact total, which followed several days later. This new procedure clearly was intended to give top management better control over expenditures and cash flow by enabling them to see any increases as they were occurring.

The monthly estimated expenditure reports, however, still fell short of real cost accounting. In 1857 J. Edgar Thomson of the Pennsylvania Railroad had created a system of monthly expense reports divided into 144 categories. The expense figures were supplemented by detailed op-

erating data to enable extensive analysis.[41] The Illinois Central's estimates did not break down the expenditures into any categories, nor
provide other operating information. Thus, they did not provide a mean-
ingful managerial tool for accurately and fairly evaluating the performance of various managers or parts of the line. On the basis of these
accounts, top management could still do no more than exhort the department head to reduce expenses if they were high.

In a few cases, additional operating information came to supplement
the financial information in routine reports. By at least early 1860, one
division engineer promised the chief engineer a "tool and material distribution" for the month as soon as possible after sending the vouchers
and payrolls.[42] A breakdown of expenses by categories and projects
would allow the chief engineer better control over the expenditures than
he could achieve with only the vouchers and payrolls. But the surviving
monthly reports from the division engineer to the chief engineer during
this period carried this and other operational statements (such as a list
of passes issued during the month) only sporadically.[43] The irregularity
of these segments of the monthly reports would certainly have limited
their usefulness as managerial tools. Only by routinely monitoring such
items could the chief engineer have maintained tight control over the
division engineers. Perhaps continued pressure to reduce costs would
eventually have led to such routine monitoring. But the Civil War and
its aftermath were to bring increased traffic and profits, as well as to
turn the company's sights southward.

The Illinois Central's routine reporting system, as it had developed by
the end of the firm's first decade, was a better aid for studying revenues,
sales, and even cash flow than for understanding costs and their relationship to operations. This priority was also revealed in the annual reports
of this period, which gave *revenue* per passenger mile and per ton mile,
but said nothing about *cost* per passenger or ton mile.[44] Although much
of the information necessary for analyzing expenses and ultimately for
comparing managerial and employee performance was already being collected in the process of accounting for all cash flowing into and out of
the company, most of that information was not incorporated into the
daily, weekly, and monthly operating reports. Thus, these routine reports were still a primitive management tool in comparison to those of
the Erie and Pennsylvania railroads.

Internal Correspondence to Bridge Distances

In addition to the upward reports and the downward circular letters
and notices, there was much general correspondence—downward, upward, and lateral—between pairs of individuals. During the first decade,
correspondence served more often to span distance than to document
information or action for future reference. Except for time-sensitive exchanges that were carried on via telegraph, most of this internal correspondence took the same form as standard external letters.

Much internal correspondence conveyed exhortations, orders, or instructions to a single employee. At high levels, such letters often urged that the recipient cut expenses for whatever function he was overseeing.[45] Because the routine reports did not give executives precise cost breakdowns by division and type of expense, the admonitions to cut costs tended to be vague, urging general reductions of expenses but not suggesting how they ought to be made. A few downward letters conveyed more specific instructions.[46] Other downward letters stated a new policy or procedure in general terms and asked the recipient to convey specific orders to those below him, presumably by writing and distributing a circular letter or notice.[47] A few of these letters, particularly the exhortations to cut costs, may have had a documentary function, recording the sender's efforts for future reference. The majority, however, seem intended primarily to bridge distance.

Much internal correspondence was also upward or lateral, reporting informally on happenings, coordinating activities between functional areas, and exchanging information or views. For example, during the initial years of construction, Chief Engineer R. B. Mason frequently wrote or telegraphed the president, sometimes more than once a day, from locations along the line.[48] He described progress, announced agents hired for newly completed stations, informed the president about negotiations for right of way, and asked for instructions. The master of transportation, stationed at the main Chicago office, frequently wrote to the master of machinery, headquartered at the Weldon Machine Shop, to coordinate the activities of the two departments.[49] (Although line-staff relationships were not yet very clearly defined at the Illinois Central, Master of Transportation Clarke and his deputy made their requests for action in ways that presumed their ascendancy over the master of machinery.) During 1859, Vice-President McClellan, whether in his office in Chicago or traveling up and down the line, corresponded frequently with a variety of company officers including the president, the treasurer, the auditor, the master of machinery, the chief engineer, the general purchasing agent, and the superintendents of the Chicago and Northern divisions.[50] All of these upward and lateral letters seem intended primarily to bridge distance, rather than to document exchanges for future reference.

Further evidence that correspondence was not primarily documentary lies in the fact that the storage system was not designed to facilitate such a purpose. During its early decades, the company depended on press books for copying and storing outgoing correspondence, although the dispersion and mobility of internal correspondents required some variations from the typical use of this technology in manufacturing firms. Rather than having a single, centralized series of company press books, the Illinois Central had separate sets for various offices or officers.[51] It was obviously necessary to use different sets of press books at different locations. A single series of Chicago office press books, for example, held all nonconfidential correspondence from the ranking officer in that city (and his secretary) to the New York office.[52] Most of the press books that have survived were tied to individuals rather than to offices, however, with a single press book series (or even volume) often containing one man's outgoing letters written over a period during which he held more than one position.[53] Thus, when an individual moved from one position to another, he often took his press books with him, rather than leaving them with his successor, a factor that worked against establishing organizational memory. Moreover, certain officers who traveled a great deal had more than one set of press books, making it even more difficult to locate a specific letter. For example, during the late 1850s, when Vice-President George McClellan frequently traveled up and down the line in a private car, he carried with him a separate press book that overlapped his office press books for the same time period.[54] All of these factors undoubtedly discouraged frequent references to previous correspondence.

Incoming correspondence was also frequently kept in personal, rather than company or office, collections at this time. There was no standard system for storing such correspondence. In some collections, letters have been folded and annotated for storage in pigeonholes.[55] In others, the letters were pasted chronologically into bound books of blank pages or of adhesive stubs.[56] Storage in bound books anticipated flat filing in not requiring that letters be folded and annotated, but imitated press books in permanently fixing the letters in chronological order, thus making them impossible to reorganize more functionally. These books were sometimes, but not always, indexed in the back by correspondent.

The entire storage system, then, was an eclectic combination of traditional methods, many of which had been adapted to suit the company's geographical spread and the taste of individual correspondents, but none of which had been adapted to improve access to a given document or to emphasize organizational over individual memory. Moreover, the format used in the letters had not yet evolved to facilitate the storage and retrieval of internal correspondence. No subject lines or other identifiers were added to internal letters to make them easier to locate at a later time. In sum, the format and methods of storage for such letters, as well as their contents, suggest that internal correspondence was seen less as documentation than as a simple medium for bridging distances.

Thus, during its first decade, the Illinois Central developed only a rudimentary internal communication system. This embryonic communication system would evolve only gradually during the next quarter century.

The Next Quarter Century: Incremental Development

Between 1862 and 1887 the Illinois Central tripled in length, from just over seven hundred miles to over twenty-one hundred miles.[57] Some of the growth came in the form of an extension across Iowa and additional feeder lines in Wisconsin and Illinois,[58] but the biggest increment, and the most important to the railroad's role in the national transportation system, was its acquisition in the 1870s of southern lines extending to the Gulf of Mexico. The firm accomplished this southern expansion through majority stock ownership (beginning in 1876), followed by long-term lease (beginning in 1882), of the Chicago, St. Louis and New Orleans Railroad, itself created by the consolidation of two other southern roads.[59] By creating a through route from Chicago to New Orleans, this extension made the Illinois Central the key north-south route in the center of the country. Although the company fully controlled the southern line, until 1890 the main and southern lines were operated separately, with different (but interlocking) officers and different rules and procedures.

Although the Illinois Central's growth was enough to increase the complexities of managing the company, it was not great enough to precipitate thorough systematizing of the firm. It grew only moderately compared to the great east-west trunk lines. Moreover, its growth was partially financed through its land sales. One historian of the company has pointed out that land sales paid for most of the original construction and enabled the company to pay good dividends through the 1860s.[60]

Neither did the firm face the intensity of competition encountered by the trunk lines. The war, during which it served a key role in transporting troops and equipment, brought plenty of business.[61] After the war, its land sales and freight business from the fertile Illinois farm country up to the Great Lakes made it prosperous. The east-west trunk lines began siphoning off some business in the 1870s and, combined with various other problems, created financial difficulties for the firm. It responded primarily by expanding to the south to pull in new business, with only a minor effort at systematizing to reduce costs. As one analyst has stated the company's strategy, "In the vicious world of cutthroat enterprise, cutting costs was at best a stopgap measure retarding, but not turning away, inevitable failure. Only by expanding trade to increase earning power could a railroad survive its competition."[62]

Thus, a combination of factors—slower growth, war profits, funds from land sales, lower levels of competition, and the opportunity to expand south—limited the pressure on the Illinois Central's management to systematize administrative procedures for greater operating efficiency.[63] Consequently, this quarter century was a period of incremental development, not of major reforms such as those in the major eastern lines.

During the quarter century, several aspects of the communication system evolved incrementally. The telegraph assumed a more important role. Downward communication underwent minor changes in form, while upward reports became slightly more analytic and took on different formats. Correspondence showed only a hint of evolution in format to suit an expanded function. Further major developments in function and form were not to occur until after 1887. Control was still maintained as much through personal, ad hoc management as through systematic, impersonal communication.

The Increasing Role of the Telegraph

Beginning at around the time that the company erected a second telegraph line along its right of way exclusively for its own use, other signs also indicated that the Illinois Central was making a major managerial commitment to the telegraph. In 1862 (the year in which the company made its first appropriation for constructing the new line) it hired a man who, over the next four years, would be its first train dispatcher and superintendent of the telegraph.[64] Moreover, operating expenses associated with the telegraph went up from their previous level of around $10,000 to around $15,000 in 1863 and $25,000 in 1864.[65] That expense category, which covered salaries of the superintendent and operators as well as maintenance of the line, would continue to rise.

As the company grew by construction and acquisition, it assured full control of a set of wires for the entire rail line. By at least the mid-1870s, the railroad had a special agreement with Western Union for telegraphic services along large parts of the line, giving that telegraph company the right to operate wires along the Illinois Central right of way in exchange for the railroad's exclusive use of certain wires, with free telegraphing up to a specific value and half rate in excess of that value.[66] By 1886, the railroad owned and operated (for itself and for public use) 550 miles of wire, and Western Union owned 3,986 miles set apart for the sole use of the railroad.

After 1863, the railroad regularly used telegraphic dispatching to handle schedule disruptions. The ability to move trains by telegraph provided enormous gains in flexibility and efficiency. The company was clearly concerned, however, that these gains not come at the cost of safety. Thus, from at least 1869 on, the rules printed on the backs of timetables included an extensive section on "Movement of Trains by Telegraph."[67] These rules strongly emphasized safety precautions. At every stage, the rules required that telegraph messages be fully written out and elaborately confirmed in writing. These regulations comprised almost a third of the rules printed on the timetables.

During this period, the telegraph was also used to speed routine reporting along the road and, to a lesser extent, to New York. Several inputs to the daily reports were telegraphed up the line to Chicago.[68] These telegraphic reports guaranteed timely and accurate information for the daily letters. At the same time, the reports were vulnerable to disruptions along the telegraph line, a common occurrence at that time. When weather or other factors caused breaks in the telegraph line, some segments of the daily letter had to be omitted. While the daily reports from Chicago to New York (which frequently ran three to four pages) traveled through the mail, by at least the 1880s certain key financial figures in the weekly and monthly reports (e.g., earnings figures for each division) were telegraphed from Chicago to New York in advance of the written report.[69]

The telegraph also played a prominent role in nonroutine communication within the company. On the operational level, for example, agents and conductors were expected to telegraph the division superintendent about any loss or damage to freight in transit, in addition to writing a report.[70] On a higher level, the negotiations concerning the southern lines required extensive use of telegrams between New York, Chicago, and various points along the southern lines.[71] The telegraph lines clearly were used heavily, because by 1884, even though the company had a monopoly over most of the telegraph lines it used, the general superintendent had to remind officers and employees to restrict their use of the telegraph to time-sensitive matters to avoid delaying important items.[72]

Because of the railroad's special relationship to the telegraph, the Illinois Central received enormous amounts of free telegraphing. The company spent large amounts on operator salaries and wire maintenance, but very little on per-word charges. Thus, minimizing words to save money never seems to have been an issue for the Illinois Central, as it was for manufacturing companies. The terse style so often associated with telegrams was less common in those of the railroad.[73] Moreover, the firm's lack of concern for economizing on words is also suggested by the various cipher codes they adopted beginning sometime before March 1883.[74] These codes were designed primarily for secrecy, not economy. Although a few code words substituted for more than one normal word, as when "Ibex" in the wired weekly and monthly reports meant "Illinois Earnings first week," most code words replaced a single normal word, as when "mooning" in one code meant "month."[75] An 1893 letter distributing a new code to top officers made clear that secrecy was the primary purpose of the code: "I send you herewith copy of new standard cipher code, which is to be used in the dispatching of important messages between officers of this company, in cases where it is desired that the content of the message should not be known to others than the person to whom it is addressed."[76]

The Illinois Central Railroad came to depend heavily on the telegraph for routine and special communications on business affairs, as well as for train dispatching. The railroad's special relationship with the telegraph enabled the offices and officers all down the line as well as in New York to communicate quickly and inexpensively.

Circular Letters as Vehicles of Systematization

During the quarter century from 1862 to 1887, management continued to depend primarily on the circular letter and its variants to convey rules, regulations, and information to employees at lower levels. The uses and forms of the genre evolved in response to and as a vehicle of slowly increasing systematization and documentation of procedures. Nevertheless, neither the essential nature of downward communication nor the relationship with employees that it implied changed radically during this period.

Circular letters were, by their nature, fragmentary. They generally announced a single new procedure, rule, or modification; where they covered more than one item, they still were generally restricted to a single page with only a few closely related rules. No comprehensive rule book had yet been compiled. In fact, the only surviving evidence of movement away from the fragmentary and toward the more comprehensive—a four-page circular of rules on the handling of freight cars traveling across more than one railroad line—came as a result of coordination with other railroads.[77] For the most part, downward communication continued to be fragmentary.

The fragmentation was partially countered by an increasing emphasis on more complete and permanent storage of circulars for documentary purposes. Although only a few of the circular letters issued in the 1850s and 1860s have survived, beginning in the 1870s, the scrapbooks of circulars became much more inclusive, although still far from comprehensive. Such scrapbooks appear to have been started in the 1870s by at least two different offices: the General Freight Office and the General Superintendent's Office.[78] These volumes were clearly an initial attempt to create an organizational memory of the rules that had been established. Still, most new employees would have had to depend on superiors and peers, rather than on a unified, comprehensive, and accessible rule book to learn essential rules and procedures.

A new trend toward numbering circulars reinforced the increasing emphasis on storage. Beginning in 1873, Freight Department circulars were numbered consecutively, and within a few years circulars from the General Superintendent's Office also began to be numbered.[79] This numbering enabled recipients to be sure that they had received all the circulars issued. More importantly, it set the circulars in context. Further, the numbering, which made referring to a circular easier, implied that the circulars should be saved for future reference. The idea that systems should be independent of the individual, however, had not yet been established at the Illinois Central. When a new general superintendent took over, he began numbering again from the beginning.[80]

In spite of the limitations imposed by their fragmentary nature, beginning in the 1870s the circulars were sometimes used to standardize procedures. Although many circulars simply announced new appointments or gave instructions for specific occurrences (e.g., state fairs, fruit trains, etc.), an increasing number reinforced and systematized procedures. An 1872 circular from the treasurer to station agents, for example, established a new procedure for remitting to the local treasurer the cash received on freight or passenger accounts.[81] Another circular from the general freight agent on handling way-bills attempted to enforce as standard policy a practice already in existence, but not invariably followed.[82]

Many of the procedures established by the circulars involved increased documentation of transactions through forms and other primary records. One circular from the General Freight Office distributed copies of a new form to be used to standardize and document a certain type of freight transaction with customers: "In future verbal directions for the shipment of car-load freight will not be accepted. Shipping tickets (sample herewith enclosed) will be furnished in order by the Division Superintendent. These you will distribute to your shippers, and for every shipment of car-load freight you will require one of them filled in by consignor, and preserve the same carefully on file."[83] Another such circular announced a new procedure for having freight of certain types transported at "Owner's Risk Released," requiring the owner to sign a release form, which was then preserved for future reference.[84] Yet another circular was sent with "blanks to be filled up with ink, and attached to all Loss and Damage claims sent to this Department."[85]

As the examples suggest, most of the transactions documented at this point involved external parties, either directly or indirectly. A few subjects for increased documentation related only to internal matters, however, and thus were desired primarily for managerial purposes. For example, one circular from General Superintendent E. T. Jeffery announced an addition to the daily report to document transportation of company materials:

> On account of unnecessary delay of cars containing company material, it is found desirable to have a daily report of same.
>   You will therefore on all reports sent to the Car Recorder, note against such cars "Company's material"; and as the correctness of the report will depend upon the making in every instance of this notation, no omission must be allowed to occur.[86]

Here the circular was used to announce standardization of an internal procedure through a system of routine reports, the ultimate goal of which was to improve efficiency by reducing unnecessary delays. In another case, a circular distributing a new tariff or rate list asked that agents acknowledge the receipt of the new tariff on the transmittal circular and return it.[87] Thus, the notice itself became a form of sorts, used to document receipt of the standard rate list.[88]

The expanding functions of circulars were revealed in their format as well as their content. Although the numbered circulars usually still included letter salutations and closings, they were differentiated from external letters in format by the prominent heading, "Circular Letter #XX." Numbering such documents also caused General Superintendent J. F. Tucker to differentiate between "Circular Letters," which were statements of rules or policies, and "Notices" (unnumbered), which announced appointments of individuals to positions.[89] This distinction also implied that individuals were transient, while rules formed a more permanent record.

Changes in the technology of document duplication also affected the form and function of downward communication at the Illinois Central. The railroad clearly needed an inexpensive and rapid alternative to printing, for in 1876, the very year in which it was patented, the Edison Electric Pen found its way into the railroad offices.[90] In spite of the primitive nature of this early stencil device, which produced messy-looking handwritten documents that contrasted strikingly with printed ones, the Superintendent's Office used it frequently through the end of the decade. Although it was not used for those circulars or notices to be posted in public view, the company did use it during the first year or two after its appearance for those to be seen only by employees.[91] After that period, it acquired a more specialized use: filling in parts of printed forms.[92] Notices of the appointment of new agents, undoubtedly the most commonly issued form of notice, were printed up in large quantities with blanks left for the name of the station, the effective date, and the names of the old and new agents. Then the Edison Electric Pen was used to make a stencil master, which was in turn used to fill in the blanks on an adequate number of copies for each appointment. These routine notices were thus even further differentiated in appearance from the more important circulars announcing new policies. And certainly this use reduced the onerous printing bills associated with announcing changes in the hundreds of station agents working for the company.

By 1885 the Freight Office's need for a neat alternative to printing had led it to adopt another new duplicating technology, the hectograph.[93] Freight rates were changed frequently, and each time new rate lists were issued to all agents.[94] Using a hectograph in the Freight Office, rather than sending the rate circulars to be printed, was faster as well as cheaper. And although the hectograph duplicating process itself was messy, the final products were neater and more readable than those produced with the Edison Electric Pen.

The Illinois Central's extensive use of mass downward communication made it a likely candidate for duplicating technologies as they became available in the 1870s and 1880s. The railroad's rapid adoption of such technologies for certain uses was dramatic evidence of its need. While the company continued to use printing for announcements of new or changed rules and for anything that might be seen by the public, it adopted less expensive and more rapid duplicating technologies for relatively transient internal announcements. This use of new technologies had the side effect of further differentiating the format and appearance of internal communications, and consequently of continuing the trend by which their form was gradually coming to reflect their function.

During the quarter century beginning in 1862, downward communication at the Illinois Central began to introduce elements of standardization and documentation within the company. Even this incremental progress was, however, limited in scope to the main line. It did not extend to other railroad lines acquired or controlled by the Illinois Central. Much later, in a ten-year study of the railroad's progress between 1886 and 1896, Stuyvesant Fish was to note that, in part because of "the growth of the System by the absorption of a number of small railroads in different States," in 1886 the railroad system as a whole was characterized by an "utter lack of uniformity of standards and methods" and "an absence of standard rules for the movement of trains, government of employees, train orders, train signals, &c."[95]

On the main line, the railroad had progressed beyond ad hoc oral management but had not yet achieved any more comprehensive embodiment of organizational memory such as rule books or employee manuals. This intermediate stage in the development of downward communication reflected the firm's relationship with its employees. Though certain procedures were routinized and systematized, and bureaucratic processes were playing a slowly increasing role in the company, the Illinois Central still retained a large element of personal influence and paternalism. Relations had not yet come to be defined primarily by a system, rather than by individuals.

In writing to two superintendents concerned about discontent among the engineers in 1883, First Vice-President Clarke began groping toward some principles that he felt ought to govern the relationship between the firm and its employees.[96] He argued that "it is due to every Employee in the Service—high or low position—to have fair and Even handed Justice done to him," and that "No officer controlling men should have any favorites, or permit his personal feelings to enter into his official duties." These statements certainly reflected the systematization and depersonalization of management that was begun in part by the mass downward communication of the circular letters. By stating such previously unarticulated principles as "Seniority in Service should have preference," Clarke was beginning the task of standardizing relations that had previously been handled inconsistently.

These developments, like the changes in downward communication, forwarded a gradual and still far from complete systematization of procedures. But in 1886, according to Fish's report written ten years later, "The authority, duty, responsibility and rights of subordinates were nowhere defined," and consequently "Conflicts of authority, called 'friction,' were of constant recurrence."[97]

The 1870s: Function and Form in Routine Reports

At the same time that downward communication was gradually developing in function and form, so was upward reporting, although it was to see even more radical development after 1887. Almost from the beginning the company had collected and made public certain very basic analytic figures and comparisons; nevertheless, it did not begin to incorporate analysis and comparisons into frequent reports that were useful to operating officers for routine decision making until the 1870s. The 1860s, during which first the Civil War, then good freight business and lucrative land sales kept profits high, created no incentive for radical changes in managerial methods. The Panic of 1873 and the depression that followed it, however, along with increased competition for the Chicago grain trade, drove down traffic rates and, consequently, dividends.[98] These problems created pressures toward efficiency and cost cutting, and thus probably precipitated the limited changes in reporting during this period.

From very early in the company's existence some basic data had been recorded on an annual and, in some cases, a monthly basis in the large bound accounting volumes. These volumes were the source of some of the items included in the annual reports, including the operating ratio and such performance measures as earnings at each station and costs per mile for each locomotive. Moreover, a brief monthly statement of land sales, earnings from traffic, and a few other items, which was published about a week after the end of every month, included a comparison of the earnings from the current month and those from the same month of the preceding year.[99] Nevertheless, the weekly operating reports established in the first decade did not include such comparisons. Consequently, day-to-day managerial decisions in Chicago were made without the benefit of routine comparisons.

Around 1870, weekly reports began to include estimated earnings for the separate operating divisions of the road (the main line, the Iowa Division, and later the Southern lines) for the current week and for the corresponding week of the previous year.[100] (Estimates were used instead of actual figures to avoid the time lag necessary to get accurate actual figures.) This comparison gave officers a rough indication of how the line was doing each week. With it, they could monitor the financial performance of each major section of the line and spot drops in earnings as they were occurring. The comparison did not break down the earnings in such a way as to indicate why a drop might have occurred, but at least it alerted officers to problems.

Further changes were introduced by 1875, probably as a result of the Panic of 1873 and its aftermath. The weekly estimated earnings for each division were subdivided into freight, passenger, and miscellaneous.[101] Moreover, the weekly reports began to include comparative figures for freight movement, operational data useful in interpreting any rise or fall in freight earnings.[102] Still, the absence of expense data that could be directly related to the earnings made it difficult to draw comparisons between the divisions, or to assess whether the earnings from freight or passenger traffic were profitable in relation to their expenses.

The company's Accounting Department also introduced some innovations into monthly records and statements in 1875, when William K. Ackerman became general auditor. He changed the system for gathering, recording, and reporting data on expenses in ways designed to help the executives monitor them. For the first time, officers responsible for each of the separately operated major divisions could see its monthly earnings, expenses, and operating ratio at a glance.[103] In the centrally maintained accounting volumes, these figures were also now recorded, along with a cumulative total updated each month.[104]

Although these innovations increased the amount of information available to managers each month (but not, apparently, each week), the Illinois Central was still far behind the most progressive railroads in developing its reporting system as a managerial tool. Beginning in the late 1860s, according to Chandler, other railroads were learning to allocate costs to freight or passenger traffic in order to figure cost per ton mile and per passenger mile.[105] These cost accounting figures could be compared across operating divisions to evaluate managerial performance. The Illinois Central had not even begun to experiment with such measures. Moreover, even the operating ratio was only available for segments of the road operated essentially as separate railroads. Such figures were not available, for example, for each of the Traffic Department's divisions.

The limited progress the company had made in developing the routine reporting system precipitated some changes in the storage system. The daily letters from Chicago to New York had initially been press copied into the ranking Chicago officer's letter book, mixed in with other correspondence and indexed only by recipient.[106] By 1866, the Chicago office had started a separate series of press books for daily letters to New York.[107] This innovation made the daily letters easier to find. It also made them part of the organizational memory of the office, rather than of whatever individual happened to be ranking officer. These separate press books were not indexed, however, since all letters went to the same recipient. The monthly and weekly statements were sometimes, but not always, copied into the press books of daily letters.

The comparisons added to the weekly statements in the 1870s placed new requirements on the storage system, since they required ready access to the values for the preceding year. By 1875, it had become clear that the existing system was not completely adequate to this task. In one case, the daily letter noted, "The comparative statement of freight movements for the third week in November 1874–1875 cannot be prepared, owing to our not being able to find the statement for last year."[108] The letter went on to say, "We are trying to find the book in which the daily entries were made last year, but have not yet succeeded." Obviously the system did not provide an accessible organizational memory. Such incidents apparently prompted measures to improve accessibility, for the press volume that began in November of 1875 included all weekly and monthly statements and, for the first time, indexed them by *subject*.[109] The first attempts at providing regular comparative analysis, then, illustrated how critical the storage system was to such efforts.

Format, too, could affect the accessibility of data needed for routine comparisons. In 1877, Ackerman and his staff introduced a series of format changes that improved the accessibility, consistency, and efficiency of the daily and weekly reports. Until this time, the daily letters had been handwritten, though relatively consistent in language and structure. During 1877, he and his office converted successive parts of the daily letter into printed forms that had only to be filled in. First, the list of thirty-eight railroads from which freight loads could be received and to which loads could be delivered, one page of the normal three- to four-page daily report, was converted to a printed form with blank spaces for the number of loads.[110] Soon the weekly comparative report of estimated earnings was also converted to a form.[111] A few months later, the opening page of the daily letter, including the number of cars loaded and unloaded at various points and with various cargoes, was also converted into a form, leaving only the local treasurer's report of receipts handwritten.[112]

These forms served as a systematizing device that improved accessibility, consistency, and efficiency. The forms guaranteed that the same categories of data would be collected each day or week, and that they could be found in the same place in each report. Moreover, it was much more efficient for the report compiler to fill in the numbers than to write out the whole report each day. Since the forms were printed in copying ink, the form itself, as well as the numbers, was copied in the press books. Ironically, the forms of the daily (but not weekly) report retained the old-fashioned letter heading, salutation, and closing—relics of earlier years. Nevertheless, the medium for producing this "letter" had changed in response to the evolution of management methods in the company.

The Mid-1880s: Further Advances in Reporting

After the period of innovation in comparative reporting in the mid-1870s, routine reports progressed no further in analytic sophistication until the mid-1880s. The southern lines brought new business and prosperity to the company in the early 1880s, in spite of falling freight rates and rising expenses. After a drop in dividends from 10 percent between 1865 and 1873 to a low of 4 percent in 1877, the dividend rate rose again to 10 percent in 1884.[113] Still, rising costs worried at least one of the top officers, who initiated further changes in reporting in the mid-1880s. Stuyvesant Fish, destined to become president in 1887, joined the company as second vice-president in 1883.[114]

As the vice-president located in the New York office and in charge of overseeing financial affairs, Fish wanted more and better financial information. Two months after taking his position, Fish had already changed the form of the weekly and monthly financial statements, making, according to Ackerman, "a decided improvement certainly on the old form."[115] The new form, which could be used for the weekly or monthly statement, still provided estimated freight, passenger, and miscellaneous earnings for each division of the road, as well as for the total line. But instead of using the two or three separately operated segments of the company as the divisions, it divided the road into seven operating divisions, providing earnings data (but still no expense data) for each of them. This segmentation enabled him to track and compare the earnings for each division of the Traffic Department. (Since the report did not provide comparable expense data, however, it could not be used to compute operating ratios, a more legitimate comparative performance measure.) Moreover, in addition to providing corresponding totals for the previous year, it had a section that highlighted the increase or decrease from that previous period. These improved forms, which were later bound together in volumes, replaced the old prebound accounting books.[116]

Some of the changes Fish wanted were not so easily made. Later in the same month Fish questioned President Ackerman about why the weekly estimated earnings for the current week were compared to *estimated* earnings rather than *actual* earnings from the previous year, pointing out that monthly estimated earnings were compared to actual earnings from the corresponding month of the previous year.[117] Ackerman replied that actual earnings were computed only on a monthly basis. That practice seems to have continued for the next few years, but in 1886 actual earnings began to be computed on a weekly as well as monthly basis.[118]

Within two months of the request for weekly actual earnings, Fish was requesting still more financial information. Since the end of 1882, when the Illinois Central had taken over the operations of the Chicago, St. Louis and New Orleans Railroad on a long-term lease, that line had become the Southern Division of the Illinois Central. When Fish wanted information on the expenses of the Southern Division only, Ackerman replied that "our accounts are not kept in such a way as to show" that.[119] At the same time that earnings were being figured for more divisions of the railroad, expenses were being figured for fewer—only for separately operated lines. "To keep separate the sums paid on these accounts for the Southern Division and the rest of the line," Ackerman continued, "would involve great trouble and considerable expense, without any compensating advantage." Ackerman, who had risen through the ranks in the Illinois Central[120] and who was well schooled in older methods of management, did not see better statistical control as a "compensating advantage." The improving dividend rate undoubtedly bolstered his confidence in the older methods.[121]

Nevertheless, further correspondence involving both General Auditor J. C. Welling and First Vice-President (soon to be President) James C. Clarke reveals that Fish traveled to Chicago to discuss the problem with Welling. Welling wrote to Clarke with his view of the matter, which was in line with Ackerman's initial statement:

To keep a record showing just how much money each local Treasurer should remit to New York, would require us to open accounts for Southern Division, similar to those kept at New Orleans, prior to 1st January 1883 [the date on which operations for the southern lines were consolidated with those of the Illinois Central itself].— This would be impracticable and I know you do not want it done. . . . Any radical change in the manner of dealing with these matters would involve much labor and additional expense.[122]

Clarke himself seems to have been more sympathetic to Fish's desire for better information on the performance of the Southern Division. He forwarded Welling's letter to Fish, adding a note at the bottom asking Fish what conclusion he had reached when he was in Chicago and commenting that without additional information, "It will always be a matter of uncertainly as to what Southern Div. is doing in way of net results."

The final resolution was a compromise that satisfied at least Clarke. After meeting with Welling and General Superintendent E. T. Jeffery, Clarke decided to route large expenses back to the Southern Division headquarters for payment, thus enabling them to be identified as expenses of the Southern Division.[123] Smaller, miscellaneous payments would still be merged with those of the rest of the line. Clarke justified this compromise as follows: "You can see if the auditors office in Chicago undertook to divide up the Earnings & Expenses in detail as was formerly done before consolidation of the lines, it would require a considerably increased force in the Clerical branch of the audit office. I think we shall get at all matters in a general way that will be satisfactory."

In terms of statistical control, this compromise was not really adequate. Without an accurate account of Southern Division expenses, the company could not know either the most critical financial performance measure for the division, the operating ratio, or the most critical figure for internal assessment of managerial performance, the cost per ton or passenger mile. At this point, however, the Illinois Central was still not calculating these figures. This compromise was not optimal, but it gained Fish some additional information.

Fish's belief in statistical analysis as the basis for managerial decision making was clear even in this period before he became president, when he did not have the power to order all the changes that he wanted made in the reporting system. In some cases, he compensated by subjecting existing data to more complicated analysis. In May 1885, when he was treasurer and Clarke was president, he wrote a long letter-report to Clarke analyzing the drop in earnings the company had experienced in the past year.[124] He showed that earnings had fallen much more seriously on the feeder lines than on the "Trunk Line" from Chicago to New Orleans. As a result, he argued that "It is on this Trunk Line, with the large volume of business already established, that we must look to make net earnings." Still, the reporting system at the Illinois Central remained primitive compared to those of more progressive railroads. Initially cushioned by revenues from land sales and later managed by individuals more concerned with extending the road south to bring in more business than with cutting costs, the Illinois Central was still behind in its managerial methods.

Changes in Nonroutine Reports

The quarter century from 1862 to 1887 also saw a few changes in special reports and in reports to stockholders and directors, although these were neither as extensive nor as significant for purposes of internal managerial control as those in routine reports.

Special reports, which were used occasionally for investigating some unusual problem or opportunity, altered slightly in format to become easier for readers to digest. An 1876 report by E. T. Jeffery surveying compensation practices at other railroads and recommending a system for the Illinois Central was a traditional, chronological letter-report.[125] Its recipient had to read through all eighteen pages to extract the recommendations. President Ackerman's 1882 report to the board of directors recommending the construction of a branch road into an area with coal deposits, however, showed signs of more concern for the readers' time.[126] Although he used a traditional letter format without subheads, Ackerman made his recommendation near the beginning of the compact, two-page report, then presented various arguments to back it up. Attachments provided additional detail. It is not clear, however, whether other managers also adopted this more modern report structure.

Finally, some interesting developments took place in the annual reports to stockholders. These reports were produced for a basically external audience, and thus reflected different pressures than those evident in the purely internal reports. For the most part, they continued to follow the general pattern established in the railroad's first decade. Internal correspondence reveals, however, a growing self-consciousness about how this information would affect the perceptions of stockholders and the public. In one case, for example, Ackerman asked Clarke to look over one section of the annual report "and point out anything that is erroneous or sounds extravagant."[127] Precision and impression were equally important.

This self-consciousness was even more pronounced when it involved regulation or potential regulation. In the early 1870s, the state of Illinois, against the bitter objections of the companies affected, began regulating railroad rates within the state.[128] The heightened awareness created by regulation encouraged at least one minor change in the form of the annual report:

> As to Abstract G, this was adopted in its present shape in 1876 after
> a good deal of consideration, and owing to our relations with the

State in reporting to them the gross earnings on which the tax was due. The discrepancies that had been shown in our annual reports and the reports to the state had been commented upon. Of course these were easily explained, but it was then thought best that the annual report should in some way show the same figures as we returned to the State, and the first part of Abstract G has now shown this for several years past.[129]

The addition of this abstract appears to have been the first, relatively minor effect of governmental regulation on the Illinois Central's reporting system.

In the 1880s, as the clamor for federal regulation of railroads in the public interest began to rise,[130] the company became even more concerned about the effect of the information it revealed. For example, in response to an inquiry from Fish about the annual report, Clarke, then president, stated his rationale for limiting the information given out: "Again we do not want to show at present the exact amount of earnings on the various branches of the Southern Division, nor on the main line. We would prefer to keep them altogether at present—certainly until we get through with this clamor for regulation of rates &c."[131] In a letter about additional information to be given out at the annual meeting of stockholders, Clarke announced, "I do not propose to give the public so much information about our affairs as has, here to fore been done. Our traffic and resources from which our revenues is [sic] derived is our own business."[132]

Such defensiveness, far from helping the railroads avoid federal regulation, probably accelerated its arrival. In terms of changes in the reporting system, however, it is enough to note that the consciousness of popular and governmental response to publicly reported information shaped the Illinois Central's presentation of that information during this period

Internal Correspondence: Little Change

During the quarter century preceding 1887, internal correspondence altered very little in purpose and format. The evidence suggests that in most cases correspondence was still seen more as an incidental consequence of the physical distances between individuals than as a tool used to document an interaction. Only rarely does a letter seem to have been written more to document a point of view than to bridge distance.[133] Furthermore, this period saw only a slow and tentative evolution in format and storage methods. Unlike circular letters and reports, internal correspondence did not, in general, become easier to read. Only a few of the top officers, who were involved in extensive correspondence between Chicago and New York in the mid-1880s, adopted subheads to make the letters more functional.[134]

Progress was equally slow in methods of storing and indexing correspondence to make later reference to it more efficient. While the special press books of daily, weekly, and monthly reports were created and came to be indexed by subject in 1875, internal correspondence remained mixed with external correspondence in the main set of press books. Similarly, the volumes into which incoming letters were bound also mixed internal and external correspondence. These volumes were still generally indexed only by name, not a very useful system for letters from frequent internal correspondents. In 1885 a few subject entries began to appear in the index of the bound volumes of President Clarke's in-letters.[135] Only in 1887, when Fish became president, did subject indexing become thorough enough to be very useful.

Moreover, the company was slow to adopt technologies that would aid in producing and reproducing correspondence. The first use of the typewriter in the Illinois Central occurred in the New Orleans office in 1882 or 1883.[136] It was not until 1886 that the first typed letters emerged from the Chicago and New York offices, and not until the 1890s that typed letters became more the norm than the exception in the company.[137]

Similarly, although mass duplicating technologies such as the Edison Electric Pen had been adopted immediately for certain types of downward communication, analogous innovations for regular correspondence were less readily adopted. There is evidence that the company had experimented with rolling press copiers and carbon copying by the early 1880s.[138] Nevertheless, these new copying technologies were used only very rarely.[139] The company apparently did not explore the possibilities of using unbound copies as the basis of a more functional or flexible system for storing internal correspondence.

Clearly, efficiency in the format, the storage system, and the technology of production and reproduction for correspondence were not considered high priorities by the Illinois Central in this quarter century. These factors support the presumption that internal correspondence was not yet widely used for documentation and reference. Instead, it played essentially the same distance-bridging role in 1887 that it had played in 1862.

Conclusion

In the period from its origin up to 1887, the Illinois Central responded primarily to the imperatives of safety, consistency, and honesty by establishing a relatively primitive formal communication system. The timetables and circular letters communicated the rules and information necessary for safety, for consistent service, and for honest money handling throughout the dispersed railroad. This downward communication was functional, if fragmentary. The upward reporting system drew a limited amount of information up the hierarchy. Initially it was descriptive only, but by the end of the period it was subjected to a limited amount of comparative analysis. Still, the reporting system was quite primitive and had not yet become much of a force toward efficiency. Internal correspondence still functioned primarily to span distances. Changes in the technology of communication were adopted primarily as they facilitated the communication functions already established.

Although the Illinois Central's internal communication system was far ahead of those used in contemporary manufacturing firms, it lagged behind those of railroads such as the Erie that were seeking efficiency. Initially cushioned by income from its land sales and the lesser competition on its north-south route, the Illinois Central did not face early pressures to improve efficiency. Later, when competition became more significant, the road was run by managers more interested in raising revenues than in cutting costs. Although the depression of the 1870s prompted minor improvements designed to provide better statistical control, the southern expansion soon brought increased business, thereby probably decreasing pressure toward efficiency. A series of events that took place in 1887, however, initiated major changes in the communication system.

Chapter 5

The Illinois Central after 1887
Communication for Compliance and Efficiency

In 1887, two events with profound implications for the Illinois Central's internal communication system occurred: the U.S. Congress passed the Interstate Commerce Act, and Stuyvesant Fish (see fig. 5.1) became president of the railroad. Federal regulation imposed reporting requirements that, as a side effect, necessitated extensive changes in the company's internal reporting system. Ironically, in light of Fish's opposition to regulation, these changes reinforced the modifications he was initiating to improve managerial control.

Fish's ascent to the presidency allowed him, and events pressured him, to pursue his campaign to make internal communication a better instrument of managerial control. His own interest in doing so had already manifested itself during his vice-presidency. Financial pressures, in addition to the regulation already noted, also pushed him to develop better managerial tools. The prosperous days of the late 1860s and early 1870s were long gone, and the improved dividend rate of the mid-1880s was giving way in the face of competitive rate cutting.[1] It became increasingly necessary to cut costs and improve efficiency.

All aspects of internal communication changed during this period. Fish continued to improve the reporting system as an instrument for monitoring and control. In addition, he standardized and documented rules, procedures, and equipment, introducing new forms of downward communication in the process. Internal correspondence also underwent changes in form, function, and technology. For all of these reforms, he had precedents in his own industry, as well as reinforcement in the systematic management philosophy that was just beginning to emerge in manufacturing during this period.

Impact of the Interstate Commerce Act

The Interstate Commerce Commission (ICC) created by the act was intended to assure fair rates and to prohibit rebates, pooling, and other such practices. In its early years the ICC had very limited powers and

5.1.   Stuyvesant Fish, who systematized the Illinois Central Railroad. *(From* History of the Illinois Central Railroad Company and Representative Employes, *Railroad History Company, 1900.)*

enforcement mechanisms to achieve these goals.[2] Nevertheless, it had one seemingly innocuous but in fact significant power over the railroads. The act authorized the ICC "to require annual reports from all common carriers subject to the provisions of this act, to fix the time and prescribe the manner in which such reports shall be made, and to require from such carriers specific answers to all questions upon which the Commission may need information."[3] The legislation spelled out many types of information that the commission might request, including such items as dividends paid, earnings and receipts from each branch of business, and operating expenses.

This federal reporting system enabled the commission to monitor—and thus to exercise some influence over—the railroads that were required to report to it. The reporting requirements of the ICC forced companies such as the Illinois Central to collect and make public more and different statistics than in the past. Moreover, in order to avoid duplication of effort, the Illinois Central began using these statistics, in place of or in addition to statistics the company had previously gathered, as part of its own internal reporting systems.

Illinois Central executives were initially hostile to federal regulation. At the beginning of 1887, shortly before the Interstate Commerce Act passed but when its passage seemed inevitable, outgoing president Clarke complained to Fish: "Stockholders may have to go without dividends until this insane crusade against Rail Road property finds a more healthy sentiment in the public mind and a higher standard of intelligence is exhibited in State and National Legislations, on this particular species of private property."[4] Even at this point, however, he admitted, "The fact is RR managers are in a great degree more responsible than any others for the hostile legislation that has been and is being enacted against RRds."

In the months that followed the passage of the act, Fish, now president, seemed to alternate between resigned compliance and revolt. In August, at the same time that he was having the first required ICC report prepared, he wrote to the company's general solicitor to seek his legal advice as to whether Congress had the authority to require such a report. The solicitor replied that he thought the ICC's request would hold up in court, noting that thus far, "there has been no refusal to furnish this information."[5] Nevertheless, when Fish submitted his first report to the ICC in October, he registered his disapproval in the cover letter. After announcing that at this time he was not challenging the ICC on the legality of its demands, he stated, "In submitting this report I do not wish to be understood as intending to waive the right of the Company to raise [this issue of legality], if it should see fit to do so hereafter."[6] In a December letter to General Manager E. T. Jeffery he rose to new heights of antiregulation rhetoric, asserting that "the usurpation by Congress of the right to inquire respecting matters not delegated by the states is but the beginning of an Inquisition which unless checked will result in Confiscation or Revolution."[7]

Although he obviously resented the ICC reporting requirements, Fish fulfilled them, and even began changing company procedures to make doing so more efficient. On the same day that he denounced the ICC in the letter to Jeffery, he ordered General Auditor Welling "to have our accounts conformed as near as may be to the requirements of the Interstate Commerce Commission and kept in such a manner that on June

30, 1889 we can make a report for the twelve months then ended, and thereafter close our fiscal year on June 30."[8] In a letter he wrote to the secretary of the ICC three years later, disputing a specific request, Fish pointed out how much he had already done to meet the ICC's requirements:

> It should be borne in mind that not only has this Company furnished every other detail asked for by you . . . but it has also, in order to meet your views, changed its Fiscal Year and its method of classifying accounts, notwithstanding that its use for over thirty years of a different system had proved perfectly satisfactory to its stockholders and to the State which gave it being and has so large an interest in its receipts.[9]

Indeed, such changes in an accounting system long in place were not trivial. They required adjustments in recording and reporting procedures at every level of the firm. Even the annual reports were modified to match ICC reporting categories.[10]

Although the changes imposed a bureaucratic burden on the company and served as a vehicle of unwanted external influence over it, they also helped Fish rationalize the system of accounts and reports to improve his own internal control. For example, the Illinois Central had lagged behind more progressive railroads in using cost accounting measures such as cost per ton or passenger mile as management tools. Such measures of cost, while clearly useful for comparing managerial performance, were more difficult to figure than earnings, since they required some system for allocating indirect costs between freight and passenger traffic. Although as early as 1859 the company's annual report included receipts or earnings per ton mile, as late as 1888 the company still could not compute cost per ton mile.[11] It did not compute this key measure for internal assessment until 1890, when it was forced to do so by the ICC for that agency's use in determining fair rates.

Even then, the Illinois Central fought the ICC's use of cost per ton mile in determining rates. In 1890 Fish began two years of correspondence with the commission about its method of dividing expenses between passenger and freight service.[12] The Illinois Central was not alone in its opposition to this method of determining "fair" rates. Whether or not they used cost per ton or per passenger mile internally, most other roads appeared to agree that the division of costs between passenger and freight traffic was too arbitrary to make this measure the sole determinant of rates. Railroad accountants and ICC commissioners debated the issue at a convention of railroad commissioners held in the spring of 1892, with the former attacking the practice and the latter defending it.[13] Within a year, the ICC abandoned its attempt to separate freight expense from passenger expense.[14] In tracing the evolution of the ICC's use of statistics a few years later, the Committee on Railway Statistics of the

National Convention of Railroad Commissioners described this change as follows:

> A third step in the perfecting of statistical methods consisted in the abandonment of the attempt to secure the cost per ton per mile and the cost per passenger per mile; that is to say, in the abandonment of the separation of operating expenses into expenses incident to the freight and to the passenger service. This was done because it was believed by the committee and by the convention that the results attained rested in too large a degree upon estimates to be worthy of confidence. It was not, however, intended to abandon the principle that statistics would be the basis of reasonable rates.[15]

Although the Illinois Central (and other railroads) won this particular round with the ICC, it did so only after revising its reporting system to enable it to compute those key measures, thereby acquiring useful information for internal monitoring. In many other matters, the railroad was forced to permanently adapt its methods of record keeping and reporting to those of the commission.

In spite of many complaints about the ICC reports it had to submit and the information it was thus forced to make public, the Illinois Central quickly discovered that the ICC reports provided, for the first time, the basis for detailed financial and operational comparisons between the Illinois Central and other railroad lines. In 1890, at the same time that Fish was complaining so vociferously about required federal reporting, he asked his auditor to get copies of the ICC reports submitted by its competitors and to carry out a thorough comparison of the Illinois Central with those lines.[16] Up until this time, the operating ratio was the basic point of comparison between lines, but it was a limited and, because of differences in methods of capital accounting, unreliable measure. The ICC reports, Fish realized, provided a ready source of carefully defined and consequently comparable statistics on many aspects of railroad operations. He was willing and anxious to use them to highlight areas of strength and weakness in the Illinois Central's performance. Thus, even though Fish saw ICC regulation as a threat, he seized opportunities to benefit from it, as well.

Improving Statistical Reporting

The required ICC reports helped shape the Illinois Central's internal information gathering and reporting system; in addition, they helped Fish forward his own agenda as president of the company. When he took over the presidency, he continued his campaign for more precise reporting and better analysis of statistics as the basis for monitoring and control, frequently over the resistance of subordinates.

His insistence on better reporting was partially, though certainly not entirely, rooted in geography. Although the first several presidents had operated out of New York to stay in touch with the financial markets so crucial to the company in its early years, presidents beginning with John Douglas (whose term began in 1865) had been located in the main offices in Chicago.[17] When Fish rose from the financial vice-presidency to the presidency, however, he chose to continue operating out of the New York office.[18] This shift in the location of the president required some adjustments in the reporting system. The daily letters and weekly and monthly summary reports had previously been compiled and studied in the president's office in Chicago, press copied for Chicago use and records, then mailed to the New York office where they were used by the financial vice-president and the board members as needed. During Fish's presidency, however, the letters were prepared in Chicago in the office of the general manager or of the vice-president and sent immediately to the president in New York.[19]

To keep informed of the company's performance, the president wanted more rapid transmission of the daily reports to New York. Within a year, a new policy was instituted for speeding all Illinois Central mail between Chicago and New York. Each day, all items were sent to New York in a single special delivery packet, "the object being that all communications from this office shall reach the New York office at one time, and in as speedy a manner as possible."[20] Special delivery lowered the average transit time from two days to one, an important gain in keeping Fish informed about the line's daily performance.[21]

Even one full day was still a significant lag for Fish. During the summer of 1889, Fish introduced a request for a private telegraph line between the Chicago and New York offices into routine contract negotiations with Western Union.[22] Such a wire would have provided easy access and complete privacy for executive communication between the two locations, allowing daily reports to reach New York shortly after being compiled in Chicago. Unfortunately, Western Union balked at the terms suggested, and the Illinois Central decided to drop the issue in order not to jeopardize the negotiations as a whole.[23]

Perhaps even more important than speed of transit, however, was the usefulness of the reports for evaluating performance. Because Fish was located in New York, he was more dependent than his immediate predecessors had been on reports and correspondence as sources of operating information. Although distance was certainly not the only factor in Fish's campaign to improve the quality of routine reports, in this period of growing competition and falling rates, it increased his incentive to make them as useful as possible.

Initial Efforts

The early years of Fish's presidency saw significant changes in the nature and increases in the amount of statistical information flowing up the hierarchy to improve efficiency and cut costs. In the fall of 1888 he corresponded with a Mr. Bruen in the Vice-President's Office in Chicago about revising the form and content of the daily and monthly letters.[24] For example, he asked that a line summing up the number of loaded and unloaded cars for all divisions be added to the daily letters. Initially this line was to be added by hand at the bottom of the forms, then when new forms were made, the line was to be added to the forms. He also asked for a new monthly letter, patterned on letters that were sent from the New Orleans office when it was still operated as a separate company, summarizing several aspects of the business not yet reported on regularly. In particular, he wanted "a comparative Traffic statement showing daily averages of (1) 'Line' Business, that is to say, the whole system, (2) with connections (to the railroads) and (3) also with certain important stations."[25] In this case, he requested that the office compile such statements retrospectively for each month of that year, so that trends could be observed for the entire current year and monthly comparisons could be made in the following year.

The evolution of routine reports did not end here. Fish seems to have instilled Bruen with his passion for improving statistical reporting, because two years later Bruen, then working in the general manager's office, suggested further additions to the monthly reports.[26] The new information was useful in analyzing the line's competitiveness in shipping several different crops from various points and in pinpointing problems in car movement along the line. Both were useful to the president and general manager in assessing the performance of subordinates.

In some cases Fish's immediate request was for a special report with additional statistical analysis, although he frequently expected the data to be available on a more routine basis if needed. In the summer of 1888, freight rates were dropping, and an extra report for the first six months of the year, necessitated by the ICC-mandated shift to fiscal year reporting, revealed slumping profits. Fish requested that General Auditor Welling send him figures for revenue per ton mile for the preceding six months, along with a comparative statement of traffic for the first six months of 1887 and 1888.[27] In addition, Fish requested "anything which will show whether rates have been reduced more on one part of the system than another. . . . What I want to know is the relation of rates on different 'Lines' and their tendency, or rather their comparative downward tendency." He wanted a statistical analysis similar to the one he had performed in 1885 to discover that earnings had dropped less on the trunk line than elsewhere in the system. He went on to ask Welling about routine use of such information: "What figures of this sort do you furnish the Traffic Department? Or the General Manager?" At the same time, he made a related request to General Manager Jeffery. In asking Jeffery to account for "the reasons for the unsatisfactory results of the first six months' operations of the current year, and the causes for the greatly increased operation expenses," he requested the tons of freight moved one mile and the cost per ton per mile for moving it.[28] Clearly, Fish was responding to worsening overall financial results by trying to gather more data on revenues and costs to help him locate the reasons for the decline.

The responses of Welling and Jeffery to this and ensuing requests for data and analyses reveal both the internal resistance to Fish's new management methods and the ultimate outcome of the conflict. Welling, who as general auditor was used to dealing with the railroad through statistics, eventually bought into the system and worked with Fish to institutionalize data collection. Jeffery remained an advocate of ad hoc management. When Fish's methods of management exerted increasing control over Jeffery, restricting his previous autonomy, he eventually resigned in disgust.

Welling and the Institutionalization of Statistical Control

Welling's initial response to Fish's request for data was surprise: "We have never attempted to figure the rate per ton per mile except at the end of each year."[29] He went on to express a polite and regretful negative:

> It would not be possible under the circumstances, to furnish the figures for the first six months of 1888 and 1887 in less than sixty days, and would require eight or ten good clerks more than we now have. I am sorry that we cannot give you the information promptly as desired, as I can readily see it would be desirable to see the figures.

The two months and ten clerks were enough to get Fish to rescind his request, at least temporarily: "Of course the game isn't worth the candle."[30] But he went on to suggest that "it would be well, however, to ascertain how the other Railroad Companies get at these figures, which they certainly do more frequently than once a year." A few days later, Welling sent Fish a table of earnings per ton of freight on three divisions of the Illinois Central for the two six-month periods.[31] Although this information was not exactly what Fish had requested, it was related and certainly useful to Fish in determining the source of problems on the railroad. This incident demonstrates that Welling's opposition was less philosophical than practical: compiling new statistical reports required more resources than seemed to him worth it.

Fish continued to press Welling for more and better information. For example, in the following year Fish had Welling compile a "statement showing the Gross Earnings and Net Earnings of whole Line excluding Iowa . . . by months during the years 1883 to 1888 inclusive; also statements covering the same time, showing percentages both for Gross and Net Earnings."[32] Although much of this information was available, it had to be pulled together in a useful form. Fish was going beyond noting simple comparisons of one month to the corresponding month in the previous year, the normal baseline for comparisons, to examining trends over several years. Fish also instructed Welling to reorganize the accounting system to divide earnings into three divisions: the main line and the southern and Iowa branches.[33] At every turn, Fish sought to do more with data the company had already collected through the accounting and reporting systems and to modify those systems to collect more and better data. At every stage, Welling expressed reluctance to expend the time and resources, but eventually gave in.

This struggle between Fish and Welling to improve statistical control promoted, as a side effect, the spread of documentary correspondence. Both Fish and Welling began to write letters intended to document their positions, not just to communicate across distance. When Fish ordered Welling to revise accounts in accordance with ICC requirements, Fish reinforced his oral order by documenting it: "I wish to confirm the verbal instructions heretofore given you."[34] Welling began to document his own actions, as well. When Welling received Fish's oral orders to separate the accounts by divisions, Welling replied in writing, "I have your verbal instructions as to the division of Earnings between Main Line and Branches. I understand them to be, as follows."[35] After repeating those instructions, he ended by saying, "To avoid any possible misunderstanding about this in the future, will you be good enough to confirm it at your convenience, and if I am not entirely correct in every particular, please advise me."

Although Fish managed to acquire from Welling many of the special reports, statistical statements, and changes in methods of keeping accounts that he sought, ultimately Fish had to address the underlying issue: Welling's reluctance to devote the necessary money and clerical force to obtaining statistics. To ensure the ultimate success of his campaign to use statistical reporting to monitor and evaluate the company's performance, Fish had both to gain the willing cooperation of individuals such as Welling and to institutionalize the necessary bureaucratic staff and procedures.

Fish addressed both those issues in a set of major organizational changes put into effect at the beginning of 1890. First, he co-opted Welling's mild but significant opposition by making him the first company-wide comptroller.[36] Although Welling, as general auditor, was already the head of the Accounting Department, the new title reinforced a widened mandate, particularly in the matter of statistical record keeping and reporting. According to the Code of Rules issued in 1890, his duties included the following:

> The Comptroller shall have supervision over all the accounts and account books of the Company, and shall see that the system of keeping the same is enforced and maintained.
>
> He shall direct as to forms and blanks relating to accounts in all departments, and no change shall be made without his written consent.
>
> He shall furnish the President, for the information of the Board of Directors, in time for its stated meeting in each month, a statement of the earnings, expenses, etc., to the end of the previous

month, and shall keep books and records for the purpose of furnishing other desired statistics.[37]

Fish appears to have persuaded Welling to buy into the new management
methods in part by putting him in charge of collecting, analyzing, and
reporting much of the needed statistical data.

At the same time, Fish expanded the function of the Accounting Department and institutionalized increased data collection and analysis by
creating an auditor of disbursements to join the already existing auditors
of freight receipts and of passenger receipts.[38] All along, the company
had monitored receipts more thoroughly than expenditures. This new
unit within the Accounting Department created the forms and reports
to collect detailed information on all disbursements, classified by type
of expense and separated by division.[39]

According to the ten-year report on the period 1886–96, the new office
radically improved the accounts for disbursements or expenses:

> The system of disbursement accounts, prior to 1887, was very sim-
> ple, no attempt being made to obtain for permanent record any de-
> tails of expenditures of the various departments; no check was
> made by the Accounting Department of record of time of employees
> kept by them, etc., that only being done which was absolutely nec-
> essary to keep the accounts correctly and furnish information as to
> the general results. This method was followed until, on account of
> the acquisition and construction of new lines, the number of em-
> ployees and the amount of the expenditures being gradually in-
> creased, it was deemed expedient to adopt a more elaborate system,
> with a view of obtaining a better check upon the expenditures, and
> for the purpose of obtaining for record and file, in a more perma-
> nent manner and for ready access, the details of them. Towards this
> end, on January 1, 1890, the office of Auditor of Disbursements was
> created. Between that date and July, 1890, a new system of disburse-
> ment accounts was adopted, which went into effect on the latter
> date.
>   The new system provided for several improvements over the
> old. . . . The details of every item of expense is [sic] obtained and
> filed in the office, from which almost any information desired, con-
> cerning operation expenses, may be readily obtained without refer-
> ence to the various departments, as under the old system.[40]

Before, the data only showed general (predominantly financial) results;
now they were broken down in such detail as to provide insight into
operational results. Moreover, the new office created a permanent and
accessible file of information, a corporate memory of expenditures from
which "any information desired, concerning operation expenses" could
be compiled. Indeed, once this information was readily available, it was
used to analyze operational problems. In a letter written by Second Vice-
President James T. Harahan to General Superintendent Albert W. Sullivan just a year later, Harahan was able to support his complaints about

increased expenses by both stating the total increase and identifying the divisions creating the problem.[41] Thus, Fish created a permanent office with a mandate to gather statistics on expenses and let that office institutionalize new forms, reports, and procedures.

From this time on, the friction between Fish and Welling seems to have disappeared. Within the year, Welling was made first vice-president as well as comptroller.[42] Two years later, when he reported to Fish, who was in England on a long vacation, Welling used some of the statistics he had been so reluctant to collect and monitor earlier as a way of defining an area of concern in operations.[43] Fish had succeeded in converting Welling as he institutionalized his new system of managerial control.

Jeffery and the Defeat of Ad Hoc Management

The friction between Fish and General Manager Jeffery was not so easily resolved. Jeffery's objection to Fish's new managerial methods was much deeper and was finally resolved only by Jeffery's departure from the company. Ultimately, this conflict was a clash between ad hoc management and systematic management.

Jeffery, like Welling, had received an inquiry from Fish for statistics to help him analyze the poor performance of the railroad in the first six months of 1888. Jeffery's response was a long, descriptive letter that did not answer Fish's questions about specific figures such as cost per ton mile.[44] He argued, as Welling had earlier, that the cost of providing them was prohibitive: "I regret that I cannot give the tons of freight moved one mile and the cost per ton per mile for moving it, but to make the necessary computations would require a corps of clerks for a period of six or eight weeks. The expense and time required, made me abandon the idea of presenting the figures." Even though a few weeks later Jeffery stated that "we cannot hope to do business against our competitors unless we modernize all our methods,"[45] he clearly did not understand the modern methods. Cost per ton mile was a basic cost accounting tool already used by many railroads.

In two related instances during the following year, Jeffery again showed his inability to go beyond the simplest descriptive reporting methods to apply statistical tools to managerial tasks. First, Fish requested from Jeffery a list of capital improvements that "in this and future years permanently increase the Net Earnings or permanently diminish the Expense of Operation to an extent exceeding 6 percent per annum upon the cost of such outlays."[46] In attempting to look at return on investment as a financial measure, Fish was ahead of his time, for that measure was not fully developed as a statistical control tool until three decades later.[47] In a follow-up letter prodding Jeffery to respond, Fish restated the question in simpler terms for Jeffery: "Will the proposed construction pay a fair return on the money invested without any regard to the source whence the money is to come?" Although the reminder finally elicited a reply, Jeffery made it clear that he had no idea what Fish wanted. He simply recommended various improvements he felt needed to be undertaken, giving the estimated cost of each but not even attempting to estimate monetary benefits. The idea of justifying improvements on the basis of financial figures was clearly alien to him.

Later that month, Fish asked Jeffery for an analysis of the major capital acquisitions anticipated or recommended, along with the person who recommended the project and its anticipated return on investment.[48] This time, since Fish asked for financial return in the more familiar form of expected earnings on the properties, Jeffery was able to give at least rough estimates.[49] But his reply also revealed the lack of systematic analysis with which he had approached the initial recommendations for the projects. For many of his proposed acquisitions, he could locate no written analysis or recommendation, and for others the recommendation was based on outdated information or on a simple desire to protect the Illinois Central's rates, regardless of cost.

Jeffery was certainly concerned about the railroad's bad performance in 1888 and 1889, but rather than using statistical data to analyze the problems and potential solutions, he simply berated his subordinates. Although he, like Welling, had adopted the practice of documenting his position in writing, his position was a simple one of threatening the subordinate. In one such communication he said, "I feel somewhat pained at having to write this letter, but the interests of the Company require that I put the subject to you in plain language, and that I insist that there be either a change of men or a betterment of roadway."[50] Similarly, in a letter castigating the general superintendent for the line's performance, he announced: "This state of affairs cannot be permitted to continue, for it must have but one result, and that will be a change in the management of the Company."[51] And indeed, such a change was soon to come, though not in the form Jeffery would have predicted or desired.

Jeffery was reputed to be a good operating officer, but his reputation was based on an old-fashioned concept of management. Ultimately, as Fish tightened up control over the company, Jeffery was not able to adjust as Welling had. The break came, however, not in a conflict with Fish himself, but in a conflict with Edward H. Harriman, the famous railroad financier who served as a director and vice-president of the Illinois Central, and who was acting president while Fish was in Europe during the summer of 1889.[52] Before he left, Fish had been attempting to rationalize rates for the first time, using the data now collected and analyzed by Welling as the basis for determining remunerative rates. Consequently, he established a rule that all officers obtain his approval for any rate reductions. When Harriman attempted to curb Jeffery's tendency to reduce rates without approval, Jeffery resigned in protest. In a letter to Harriman in response to the news that Jeffery had resigned, Fish wrote, "Pride and incapacity to submit to control were the real sources of his trouble."[53] Ad hoc management had been pushed out by impersonal, systematic control.

Further Changes in Reporting

Regardless of whether individual upper-level managers such as Welling and Jeffery adapted to the new ways, the upward reporting system under Fish continued to evolve in the 1890s, collecting and analyzing increasing quantities of statistical information to improve monitoring and control. By 1894, for example, the reporting system was being used to support forecasting of revenues.[54] Weekly returns of freight loads moved at each station were transmitted to headquarters via telegraph, where they were compiled and summarized on a tabular form. This information was used to forecast future revenues and to interpret weekly expense information more accurately.

In addition to the routine reports, special statements and reports were also compiled frequently during the 1890s. Statements comparing various quantities over periods of time or among divisions or even railroads appeared more and more often.[55] These reports were requested to address a problem area or simply to give company executives a better idea of the line's performance on some dimension. Within a very few years of when Fish took over the presidency, routine and special reports, frequently statistical, were commonly used as managerial tools in the company.

Although the content and function of reports evolved rapidly during the first years of Fish's presidency, by the 1890s form had begun to evolve in response. As special reports increased in number, the reading burden on the recipients increased, as well. The format of an 1892 report from Superintendent of Machinery Henry Schlacks and Engineer John F. Wallace to Fish shows efforts to make the organization and format more efficient for Fish to read.[56] The report began with recommendations on the immediate project, leaving details of their research, a discussion of future implications, and proposals from several contractors for the project for subsequent sections. Each section of this typed report had a subhead to help Fish and other readers find a desired section.

The graphical depiction of statistical data in special or routine reports also appeared at the Illinois Central by the 1890s. Tables had been used to display statistical data from the railroad's earliest years. As the quantity and complexity of statistics increased during Fish's presidency, however, Fish and other executives were faced with absorbing and monitoring it. Graphs could potentially make such information more immediately accessible.

The first regular use of graphs in the company was probably around 1890. At the end of 1894, an outside consultant named Raymond Du Puy proposed to introduce a system of graphic reports at the Illinois Central.[57] In the first of a series of letters concerning this proposal, Fish mentioned that "we have had something of this sort in one or other of

the Departments of our service for several years past." In a letter to
Welling, Fish noted, "I presume Mr. DuPuy's [*sic*] idea is somewhat
akin to the plan Mr. Wallace has of showing expenses in his Department
by means of blue prints."[58] Unfortunately these blueprints have not
survived. At this time, however, graphic reports were apparently still
unusual in the firm.

The correspondence surrounding Du Puy's attempt to introduce his
system of "graphical statistics" suggests that neither Fish nor his fellow
officers fully understood either the power or the use of graphs at this
point. In fact, the correspondence reveals a tendency to confuse the
graphic presentation method with the accounting and statistical methods used to collect and analyze the data. An endorsement letter from the
president of *Railway Age* magazine, which referred to Du Puy's "system
for the improvement of railway accounting methods," could have engen-
dered this confusion.[59] Nevertheless, Fish initially seemed to understand that Du Puy was simply selling a system for "showing railroad
accounts and statistics graphically."[60] In response to an inquiry from
Fish, Welling revealed his similar understanding in saying that Du Puy's
system "does bring out in a striking way the facts in relation to the
movements of loaded and empty cars. I have no doubt it would work
equally well in the matter of expenses, if properly used."[61]

In later correspondence, however, both Welling and Fish confused statistical and presentational issues. Based on Welling's statement and on
Fish's initial meeting with Du Puy, Fish asked him to work up examples
of his "Graphical Statistics" using Illinois Central data.[62] Once these
examples were worked up, however, Welling criticized the way the statistics were used and interpreted as much as the form in which they
were presented:

> I did not say so to Mr. Dupuy, but I do not think his diagram has
> any practical value. He takes the mileage of loaded cars and emp-
> ties, north bound, treating two empty cars as the equivalent of one
> loaded car and applies the "equivalent loaded car miles in the direc-
> tion of the greatest traffic" to the total wages of men handling
> trains in both directions. The basis of comparison is wrong and the
> results, as shown by him might be very misleading. I do not see
> what good use could be made of the figures, whether tabulated or
> shown graphically, as proposed by Mr. Dupuy.[63]

Welling also commented that no matter what figures were used, the
graphs would cost more to display:

> Altogether, I think we have on [our present] system a very complete
> and satisfactory set of reports. These are furnished at minimum
> cost. In case we should introduce Mr. Dupuy's method, or some
> other similar method, this work would necessarily have to be done
> and the cost of diagrams, whatever they might be, would be so
> much additional. To certain classes of statistics, doubtless good use

might be made of Mr. Dupuy's methods, but I do not think it could be profitably availed of for current statistics.

Although he had by this time come to view the cost of the accounting and reporting system as minimal, he saw the additional cost of graphs as an unacceptable addition. On the basis of this critique, Du Puy was told that his services were not needed.[64]

Du Puy made a second unsuccessful attempt to sell his services to the Illinois Central nine months later. He wrote to Fish reminding him of his earlier claim that his system would reduce expenses and proposing that he be hired to test this assertion on a new line that the Illinois Central was acquiring.[65] Once again Fish turned him down, because "we would not want to have the method of accounting on that Road different from the one which we have on the Illinois Central." This letter suggests that Fish was not distinguishing between methods of collecting and methods of presenting data. Du Puy wrote one final time to set Fish straight, saying, "I wish merely to say that I fear you mistake the use and reason of Graphic Statistics. No question of accounts enters into the method. The information given from them is for the every day use of the operating officials, supplied to them in a form constantly at hand and impressive in its blunt indifference to figures."[66]

While Du Puy's explanation did not get him the job, it may have helped clarify the purpose of graphs in Fish's mind. Within two years the company was experimenting on its own with new uses of graphs. The ten-year report (1886–96) issued in 1897 and intended primarily for the board of directors, stockholders, and other external audiences used graphs to demonstrate the growth of the company.[67] As figure 5.2 illustrates, the graphic techniques used were not yet very effective. Nevertheless, the graph revealed at a glance the upward trend in the business. If a reader gave it a bit more attention, it showed that railroad freight business was increasing at a faster rate than was transportation of freight by water. This graph functioned as a dramatic persuasive device aimed at external audiences, rather than as an analytic tool for management.

In the following year the company continued to experiment with graphs, this time primarily for internal use by top executives. In this case, the graphs compared the Illinois Central's performance with the performance of other railroads over time. Fish had General Manager and Assistant Second Vice-President Wallace plot a series of statistics on freight traffic for the Illinois Central itself, for a recently acquired but separately operated southern line, and for a composite of other railroads over a five-year period.[68] Figure 5.3 shows only a few of the many graphs he prepared on a large blueprint. Blueprint copies were distributed to Welling and to Second Vice-President Harahan, and arrangements were made to extend the graph each year.[69] These graphs were intended to help the top executives monitor key traffic statistics and visualize how their railroad compared to others. To alleviate the difficulty of following

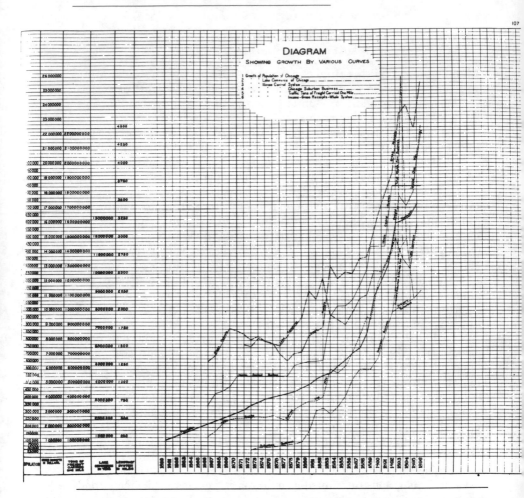

5.2.   Graphs from the Illinois Central's ten-year report, 1886–1896. *(Courtesy of the Newberry Library.)*

the many lines on each graph, Wallace added color to the blueprints by hand.

By this time Fish evidently felt that graphs were worth the effort necessary to produce them, for a few months later he had similar graphs created for his own use out of several tables from the vice-president's annual report.[70] As far as surviving evidence indicates, however, graphs remained rather special efforts at the Illinois Central up to the turn of the century.

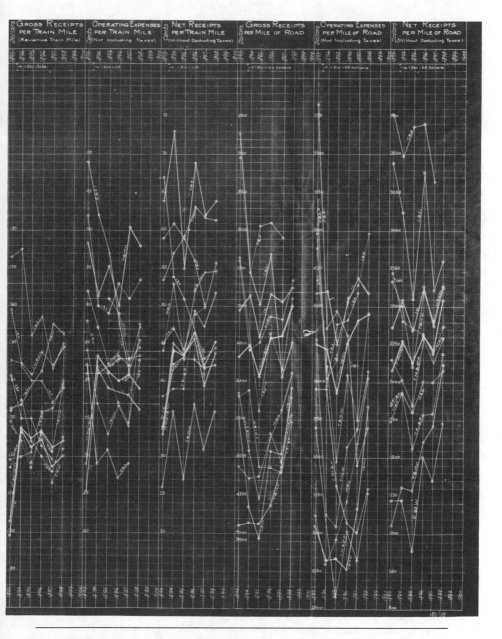

5.3. Five-year graphic comparison of the Illinois Central to other railroads,
May 1898. (Courtesy of the Newberry Library.)

By 1900, the Illinois Central had doubled in length since Fish took over as president in 1887 (see fig. 5.4).[71] Statistical reporting in the Illinois Central continued to evolve in content and form through the turn of the century. In addition to his own work on internal statistical reporting and presentation, Fish also sought to exchange managerial methods and statistical results with other large railroads in order to gain further improvement. In an exchange of correspondence and graphical comparisons with William Mahl, comptroller of the Union Pacific Railroad, another Harriman-controlled line,[72] Fish advocated such cooperation: "I believe firmly that we can all learn a great deal and be put in position to save a great deal for our respective companies, by going thoroughly into comparisons of this sort."[73] In reply, Mahl pointed out that all the large lines shared this need for better internal control and consequently for cooperation in sharing methods:

> I concur fully in your views that we all can learn a great deal from each other, and believe that the railroads will have to do so if the large systems which have been formed during the last five or six years are to survive under the stress of unfavorable conditions. Organization must now supply the means of controlling wisely these great enterprises, and common sense business methods of accounting will be an important factor in this control.[74]

Mahl's "organization" was similar to the "system" being advocated in manufacturing circles in this period. Both depended heavily on a reporting system that gathered, analyzed, and presented the statistics needed for control

Compiling an Organizational Memory

Meanwhile, Fish was also instituting a dramatic reform in downward communication. Recognizing the fragmentary nature of the existing system of circular letters, he committed himself to establishing a comprehensive organizational memory.

The first and most basic element of Fish's campaign to create comprehensive downward communication was aimed not at labor but at management. The Code of Rules, adopted by the board of directors in December 1889 and taking effect on January 1, 1890, established responsibilities of and relations between departments and officers.[75] In his ten-year report (1886–96), Fish recounted the origin of the code:

> One of the first subjects which I had the honor to bring to [the board's] attention was the lack of a Code for the government of the Officers and Employees. Ten years ago, though the Officers were men of experience and ability, there was an absence of system in the organization as a whole. The control by the Officers over the Employees was personal and paternal. The authority, duty, responsibility and rights of subordinates were nowhere defined. Conflicts

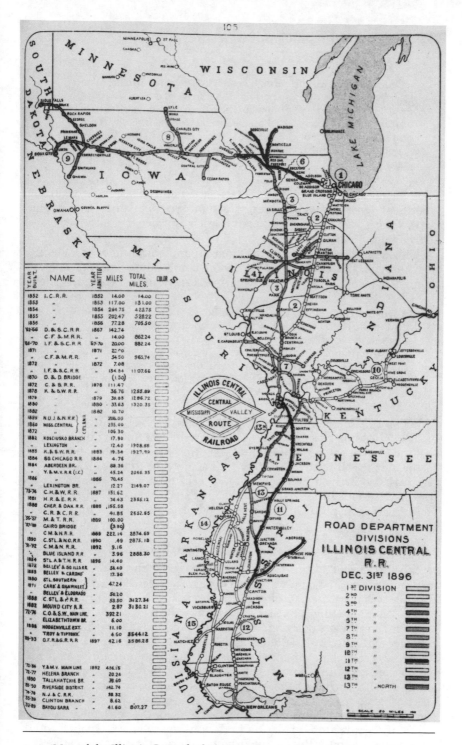

5.4. Map of the Illinois Central, 1896. *(Courtesy of the American Association of Railroads.)*

of authority, called "friction," were of constant recurrence. The matter had the serious attention of the Board and of its Committees for more than a year. In 1889, the subject was delegated to a Committee of Officers, and on December 16, 1889, there was promulgated, by order of the Board of Directors, a "Code of Rules for Conducting the Business of the Illinois Central Railroad Company," which has since been amended from time to time.[76]

The code defined the duties and interrelationships of each of five major functional departments: Treasury, Accounting, Operating, Traffic, and Legal. Points of possible friction between the departments were addressed in particular detail.

The code also defined how information and communication should flow through the company. Downward communication was to pass through all levels of the hierarchy in turn, except in cases of emergency:

> All instructions emanating from the Board or the President in regard to the business of any of the departments, shall be given through the head thereof.
> Instructions emanating from heads of departments shall be given through the next officer in their several departments; and this rule shall be observed, except in cases of emergency, by their subordinate officers.[77]

Although this hierarchical flow was the main path of control, "Copies of all circulars or general orders issued by the head of any department shall be sent to the head of each of the other departments, and also to the President and to the Secretary."[78] The upward communication system was also defined hierarchically: "Every officer and agent shall be responsible to his immediate superior, and shall report to him."[79] Other reporting responsibilities were spelled out as necessary.[80] Communication flows were thus a critical part of the comprehensive system governing the company.

Both the code itself and Fish's description of it in his ten-year report reveal a belief that systems, not individuals, were the key to successful management of a large enterprise. He characterized the company after the code was instituted as "an organization capable of indefinite expansion, without the slightest fear of its being disturbed by the withdrawal of any man from any position."[81] Clearly, he accepted the principle that management systems should transcend the merely personal.

The new code supplemented the fragmentary communication of the circular letters with a comprehensive description of duties and responsibilities at the highest level—a repository of organizational memory. It was not, of course, immutable. Although the code was to be "amended from time to time," however, it provided a relatively permanent compilation of organizational rules and relationships. Since it applied to all lines owned or controlled by the Illinois Central, it also transcended the geographical fragmentation previously caused by allowing railroads that were taken over to continue operation under their own rules, organizational structure, and even officers.

Within the code's framework, the departments adopted or created their own rule books to define detailed rules and procedures. In 1891, comprehensive rules on train movement replaced the limited rules printed on timetables when the Illinois Central's Transportation Department adopted the code book of standard rules recommended by the American Railway Association.[82] This code established uniform, standardized procedures throughout the Illinois Central system and even beyond. The extensive rules covered every contingency possible, removing as much individual discretion as possible in order to guarantee safety and to improve efficiency.

The ARA Standard Code Book was distributed to all employees involved with train movement. Moreover, "Conductors, Engineers, Train Flagmen, Station Agents and Operators," the general superintendent announced, were "required to pass a satisfactory examination before they [were] qualified to work under the new rules." An individual, no matter how satisfactorily he had performed his role in the past, could not work until he had demonstrated mastery of the new rules for train movement. Like the general code, this Transportation Department code was not permanently fixed. When rules were amended via circular letters, however, gummed slips printed with the new version were distributed to be pasted into the books.[83] Thus, the comprehensive rule book was kept up to date.

Other departments created comprehensive manuals of standards and procedures, as well. The Office of the Auditor of Disbursements created a pamphlet of rules for all company employees on handling disbursements.[84] In 1891 the Road Department created its own book of rules and standards for maintaining the tracks and roadbeds.[85] This rule book standardized every aspect of the road bed from culverts to guard rails. Annual inspections of the railroad line made sure that the conditions were maintained at the designated level. The rule book helped reverse the "utter lack of uniformity of standards and methods, especially in matters pertaining to the permanent way" that had existed as late as 1886.[86] Other departments also created compilations of rules and standards.

Thus, in downward communication, as in upward reporting, Fish oversaw extensive reform. He used comprehensive rule books as a mechanism for progressing from the "personal and paternal" to the systematic in relations among officers and employees. By 1897, he had instituted a standardized "Discipline by Record" system that converted even discipline to an impersonal, standardized procedure that replaced actual suspensions with records of offenses.[87] Under his leadership, downward communication became a vehicle for creating an organizational memory independent of individuals.

Changes in Internal Correspondence

Finally, during the period from 1887 to 1900 the function and form of internal correspondence, along with methods and technologies for handling all types of internal communication, evolved to suit the changing needs of the company. The concept of using internal correspondence to document positions, not just to bridge distance, had became one of the tools in Fish's struggle to improve the upward reporting system. Correspondence among officers within the Chicago office also increased. For example, Second Vice-President Harahan wrote to General Superintendent Sullivan, located in the same building, about the poor performance of the line.[88] After delineating the rise in expenses in certain divisions during the previous month, he continued, "I have now talked all that I propose to, and some action must immediately be taken to bring these expenses into line." Harahan also sent a copy of the letter to Fish in New York, to let him know of both Sullivan's shortcomings and Harahan's own attempts to correct them. Clearly, using internal correspondence to document internal relations and to serve as weapons in bureaucratic struggles became important during Fish's presidency.

Patterns and functions of long-distance correspondence changed during Fish's presidency, as well. Because Fish remained in New York he needed constant nonroutine correspondence and special reports, in addition to the routine reports, to keep him fully informed about subjects that he would have known about first hand had he been in Chicago. General Manager Jeffery, his most frequent correspondent during the first years of his presidency, wrote to him almost every day and frequently more than once a day.[89] Many of these letters were status reports intended to keep Fish informed of progress on specific projects. For example, Jeffery sent Fish copies of special reports from two of his subordinates, with a cover letter that included the following explanation of his purpose in doing so:

> My object in sending these is not for the purpose of making recommendations, because I am not now prepared to make them. The subject covered by the correspondence, and illustrated by the map, may come up for discussion, and I thought it would be advisable to have in your hands in New York, a plat and copies of the reports to which you could refer in case I should write or telegraph you at some future time.
>
> This matter needs no special attention now, but at your leisure, some time in the next 30 or 60 days, it might be well to look over the papers and plat before filing them.[90]

These materials were not primarily documentary in the same sense that Harahan's letter was, but they had an important reference function in addition to the informative function.

The increase in amount and changes in function of correspondence between the New York and Chicago offices precipitated one minor change in form. Although all Chicago mail for a given day was sent to New York in a single special delivery envelope, Jeffery often wrote more than one letter in a day. Each one usually addressed a different subject, a fact differentiating these internal letters from earlier internal letters and from external letters. The documents themselves still looked like external letters, but the fact that they covered only one topic made it somewhat easier to locate information on a specific subject.

In conjunction with this change at the Chicago office, changes in handling and storage were instituted at the New York office. Very shortly after Fish took over as president, letters received by the New York office (whether from internal or external correspondents) began to be stamped and annotated.[91] The stamps designated what information needed to be included in the annotation (e.g., "From: ——— /Subject: ——— "), thus systematizing the process while reducing the amount of necessary handwriting. The annotations themselves helped Fish or his assistants in locating specific letters.[92] Even such minor office equipment as stamps helped impose system as efficiently as possible.

Although Fish's incoming correspondence continued to be stored in chronological order, changes in the indexing procedures also made specific items easier to find. Within six months of becoming president, Fish was filling up each volume of in-letters in about a month, with many of the letters from a small number of internal correspondents. One bound volume, for example, covered only twenty-four days, and the letters received included no fewer than thirty-six from Jeffery.[93] Clearly, indexing each volume separately and only by sender was less and less helpful in finding a given item. Although occasional subject entries had appeared in the index of Clarke's presidential in-letters at the end of his presidency, such entries grew more frequent in the index of Fish's in-letters. By 1889 the form of index entry had changed to include subject and sender of each item.[94] Beginning in 1895, a separate subject index was created to span volumes.[95] This loose-leaf index could be expanded at will to cover a period of years, thus eliminating the need to consult the subject index of each volume if the approximate date of a desired item was unknown.

Similar innovations were adopted to make locating documents in press books of out-letters easier. Each press book covered about six months at this time, with the time period decreasing to an average of four months for the next ten years.[96] Thus, indexing across volumes was somewhat less critical for out-letters than for in-letters, but indexing within volumes was much more critical. The individual press books were now indexed by subject as well as recipient. Initially the subject and recipient entries were mixed in a single index, though subject entries were written in red ink to make them stand out. Beginning in 1894, however, each volume had two separate indexes, one by subject and one by recipient.[97]

The fact that Fish remained in New York added another complication

to the system for storing correspondence. A "President's Office" continued to be maintained in Chicago. (In fact, the Code of Rules stated that
the President's Office was to be in Chicago.)[98] That office tried to maintain a complete set of Fish's outgoing New York correspondence in order
to have complete records that Fish could consult when he was in Chicago.[99] To create these second copies, the New York office adopted an
interesting combination of traditional and modern copying methods.
Carbon copies of Fish's out-letters were made when they were typed up,
and press copies of the original letters were made after typing. The press
books were retained in the New York office, while the carbon copies
were sent to Chicago for that office's records. The Chicago President's
Office did not take advantage of the fact that carbon copies were loose
by filing them by subject; instead, it simply stored them between pages
of the Chicago letter press books at the appropriate dates. This storage
method eventually caused a crisis because the carbons made the press
volumes too bulky to use in the letter press. Thus, in 1896, the policy of
forwarding carbons to Chicago was abandoned.[100]

Ironically, while the Chicago President's Office did not show any signs
of breaking out of the chronological tradition in storing carbon copies of
Fish's letters, indirect evidence suggests that around 1892 the General
Superintendent's Office in Chicago adopted some sort of subject-based
filing system (probably horizontal or box files, since vertical files were
only introduced in 1893) for certain types of documents. Starting at that
time, some circulars were stamped with a four-digit file number clearly
related to subject, not chronology.[101] (Circulars issued on different dates
and from different offices but on the same subject, for example, have the
same file number.)Unfortunately, no subject files have survived, and the
extant circulars are preserved in a chronologically organized bound
book. With this probable exception, however, bound chronological storage of correspondence evidently remained the norm in both the New
York and the Chicago offices.

All of the changes in methods of handling, storing, and indexing correspondence suggest that the Illinois Central officers were putting increased emphasis on systematic and efficient handling of the documents
at all stages, including retrieval. As the quantity of correspondence increased, locating items became more difficult. Moreover, as correspondence became more documentary, Fish and others were more likely to
want to consult an item later. Both these trends surely encouraged them
to develop more efficient ways of handling, storing, and retrieving the
documents. Of course, by making the documents more accessible, these
changes would in turn encourage increased future reference to correspondence. In either case, the evolution of function, form, and technology of internal correspondence is of a piece with the changes in upward
and downward communication discussed earlier. All reflect the shift
toward greater systematization set in motion by Fish.

Conclusion

By the end of the nineteenth century, the Illinois Central had grown considerably longer and had become much better managed than when Fish took over. A contemporary history of the railroad stated that before Fish took control, "The Illinois Central had the unpleasant reputation of being perhaps the worst regulated and slovenly of the large American railway companies."[102] In the past, the land grant and the lower level of competition had helped insulate it from extreme pressure to improve efficiency. Its managers before Fish were more interested in expanding south to increase business than in systematizing to improve efficiency and cut costs.

The same history credited Fish with changing that reputation. From the beginning of his involvement with the Illinois Central, Fish was intent on using statistical reports as the basis for monitoring and evaluating the company. Once he became president, he began a campaign to improve the reporting system. ICC reporting requirements aided his campaign, and falling freight rates (and profits) spurred him on. His residency in New York, rather than in Chicago, gave him additional incentive to make the reports more useful. Fish also systematized and depersonalized relations with employees. Rule books were introduced to provide a more complete and permanent organizational memory that transcended the individual. They standardized procedures, equipment, and rules throughout the organization. Finally, even internal correspondence took on a documentary function.

It is likely that Fish's ideas for reform were initially gained from other railroads, since he had been in the railroad business throughout his career and since he had connections with other lines through Harriman's vast holdings. And certainly all railroads were paying closer attention to managerial methods in the late century period of mergers and system building. Nevertheless, Fish's ideas may have been reinforced and clarified by the systematic management movement just gaining momentum in manufacturing. Certainly his reforms are consistent with that philosophy.

The results of his reforms are hard to pin down precisely, but evidence points to a relative improvement in the Illinois Central's performance. At first glance, the performance trend seems to have been downward rather than upward. In the ten-year report, we see that the dividend rate decreased from 7 percent to 5 percent between 1886 and 1896, a decrease of just under 30 percent.[103] In this period of falling freight rates, increasing expenses, and regulation, however, the financial performance of railroads in general was declining. The report also compared the Illinois Central's performance to that of a group of twelve other large railroads of various types. For this group, the average dividend rate decreased by more than 50 percent, from 5.38 percent to 2.61 percent. The Illinois

Central's higher initial and final dividend rate probably reflected the better financial and competitive position it had always benefited from, not greater efficiency of operations. The smaller drop in dividend rate, however, suggests that the Illinois Central under Fish did, in fact, improve its performance in comparison to other lines.

The many other factors make it difficult to isolate the effects of Fish's reforms. It is clear, however, that Fish transformed the management methods of the Illinois Central Railroad.

Chapter 6
Gradual Systematization at Scovill

The Scovill Manufacturing Company started out as a manufacturer of brass products in 1802—half a century before the Illinois Central Railroad was formed. In spite of its head start, Scovill, like manufacturing businesses in general, lagged behind the Illinois Central and other railroads in the development of its internal communication system. While in the Illinois Central a rudimentary communication system evolved early to assure safety, in Scovill systematic methods of management and formal internal communication evolved only slowly. Even though the firm integrated several distinct production processes in one facility from an early point in its history, it depended primarily on the traditional, oral methods of management for most of its first century of existence. Growth, a somewhat more progressive management team, and new communication technologies prompted initial changes toward the end of the nineteenth century. In the first two decades of its second century, the pace of change accelerated. In response to new ideas in management and rapid wartime growth, Scovill adopted systematic management methods that demanded extensive use of formal internal communication as a managerial tool.

Scovill's First Century: Slow Evolution

Although the company dates its origins to 1802, the Scovill Manufacturing Company did not acquire its corporate name until 1850.[1] It started as a series of family partnerships initially formed for the purpose of manufacturing brass buttons at a time when most buttons still had to be imported from England. In its mills on the Mad River in Waterbury, Connecticut, the family soon branched out into photographic plates, brass hinges (at another plant a few miles away from the main plant), and semifinished brass products.[2] In 1850 the three interlocking family partnerships engaged in these activities joined and incorporated as Scovill Manufacturing Company. Before its incorporation, management methods were primitive and communication was completely unsystematic; after incorporation, internal communication slowly began to evolve under the influence of new leaders and new managerial techniques.

The Early Years: Ad Hoc Management

Initially, management at the Waterbury plant was fairly ad hoc. According to Theodore Marburg, a chronicler of Scovill's early years, the partners attempted to divide duties only in a very general way: at any given time one managed and worked alongside the workers while the other was on the road, buying and selling.[3] Beyond this, division of management functions was almost nonexistent. There was, Marburg has shown, "an absence of delegation of tasks to subordinates, which led to the need for a great variety of personal activity on the part of the partners." The lack of even rudimentary formalization of policies and procedures was central to this problem.

By the 1830s, some delegation of tasks was attempted through such devices as setting up interlocking but separate partnerships to handle the new hinge business, hiring bookkeepers to maintain the account books and skilled workers to run certain parts of the manufacturing process, and occasionally paying piece rates to skilled workers who hired their own assistants. Initially the number of such quasi-managerial employees was small. In 1831, when the firm's principal product was still buttons, 3 men were on annual salaries, helping to manage a workforce of 69–75 other men.[4] By 1850, total employment was up to 157, working in essentially five departments: the rolling mill, the casting shop, the button shop, the butt and hinge department, and the photographic department. Each of these *departments* or *shops* or *rooms*, as they were variously called, was run by a skilled workman or foreman reporting directly to the owners.[5]

With this sort of structure, internal communication during the first half century was a simple matter. Beyond the account books, there is no evidence of written communication within the Waterbury plant during that period.[6] Notes may have been used, especially in communicating with the hinge mill a few miles away, but if so they were not considered important enough to save. Otherwise, management by word of mouth undoubtedly prevailed. Communication with the company's representatives outside of Waterbury—partners on trips and sales agents in other cities—depended on written correspondence until the telegraph was introduced into Waterbury in 1849.[7]

During this period Scovill's technology for producing, reproducing, and storing correspondence (whether internal or external) was simple but slow. Until at least 1830, letters were handwritten and then hand-copied into a bound blank book. Sometime between 1830 and 1854 (a period for which no letter books have survived), Scovill adopted the press book and letter press to speed up copying.[8] Incoming letters were kept folded, probably in pigeonholes.[9]

Postal service was frequently slow and always expensive at this time, especially for longer distances. Under good conditions, letters could get from Waterbury to New York (about one hundred miles) in one day and to Philadelphia in two.[10] In bad weather, however, transit time might stretch to almost a week.[11] From 1816 to 1845, posting a one-sheet letter from Waterbury to New York by U.S. Mail cost 12.5 cents and from Waterbury to Philadelphia, 18.5 cents.[12] The absence of postmarks on a few of these letters suggests that the high rates encouraged Scovill to have travelers hand-carry items when possible.[13] Before 1850, there is no evidence that Scovill carried on business with parties farther away than Philadelphia and Baltimore to the south and Boston to the northeast. This fact is not surprising, since a one-page letter to Chicago would have cost 25 cents and taken weeks to arrive, and transporting buttons would have involved even greater costs and difficulties.

Scovill's internal correspondence across distances was informal and unsystematic. The letters between partners when one was away from Waterbury were frequently long and full of information, but not at all regular in timing or organization. The frequency of the letters seemed to depend on the issues at hand and their urgency to either party.[14] Since postal rates favored length (up to one sheet) over frequency, the correspondents tended to fill a large sheet with writing before sending it. The letters might cover everything from issues of competitive strategy and requests to buy supplies to greetings from friends and family, often without even paragraph breaks to separate subjects.[15] This correspondence served to keep the partners in touch but showed little pattern and no concern for the reader's efficiency.

Up until 1846, when Scovill established its first New York store, commission agents in major cities such as New York, Boston, and Philadelphia were its primary mode of marketing its products.[16] The correspondence with them, though not strictly internal, set the stage for later internal correspondence with the company stores in those cities. Marburg has suggested that "one of the striking features in the relations between Scovills [sic] and their agents was the lack of standardization."[17] The correspondence through which, for the most part, these relations were maintained certainly was not very standardized. Twice a year the agents rendered their accounts to the firm in the typical account style of the period.[18] Beyond that, letters were reasonably frequent, although irregular in timing. In 1829, for example, Scovill received about seven letters a month from the New York agent and just under that from the Philadelphia agent, or on average a letter every four to five days from each, generally in response to letters recently received from Scovill.[19] The correspondence reflected the business of the moment, rather than any regular system of reporting.

The letters were no more standardized in form and content than in timing and revealed further lack of consistency in relations with agents. Like the letters between partners, they were disorganized and lacked any formatting to aid the reader. They covered a variety of subjects, including orders, prices, credit terms, complaints about money or other agents, special instructions for custom orders, possible new products, and miscellaneous errands the agent was asked to perform.[20] The letters reveal particular areas of friction. First, agents frequently complained about not receiving money owed them, with Scovill replying that its banks were too tight with money.[21] In addition, sales territories and authority for price setting were also subjects of dispute. One exchange concerned a complaint by the Philadelphia agents that the New York agent had come into their area and undersold them.[22] In response to a reprimand for this underselling incident, the New York agent complained, "If I have over looked any instructions please refer me to them. . . . I am bound to comply with your instructions but I do presume you do not wish me to adhere to a stipulated price when your competitors are selling for less."[23] While in theory the firm claimed sole power to set prices, the time lag in correspondence meant that curbing agents too much might lose sales. Consequently, the company continued to deal with agents on a case-by-case basis, a factor reflected in the correspondence.

In the company's first half century, oral methods of management dominated in the workplace, and the written communication demanded by distance was equally ad hoc. In the late 1840s, however, a series of changes occurred both in Scovill's structure and in the external environment that would eventually initiate gradual changes in the nature of internal communication at Scovill. As early as 1840, one Scovill partner wrote to the other that "our brass goes too much through agents hands."[24] In 1846 Scovill opened its own wholesale store in New York.[25] Around the same time, communication and travel over distances became cheaper and more rapid. The postal rate changes of 1845 and 1851 lowered Scovill's mail costs to only 3 cents for prepaid letters under one-half ounce to any U.S. destination.[26] In 1849, both the telegraph and the railroad came to Waterbury.[27] Perhaps made ambitious by the expanding opportunities afforded by these changes, in 1850 the various Scovill partnerships incorporated to form the Scovill Manufacturing Company. These events initiated some incremental changes in the communication system.

Communicating with the New York Store in the 1850s

Although the creation of the New York store to perform the functions previously performed by external agents transformed Scovill into a truly multifunctional firm, changes in its management methods and communication patterns came slowly at first. The letters sent by the New York store to Waterbury in the mid-1850s (the letters from Waterbury to New York during this period have not survived) suggest that the internalizing of sales had relatively little immediate effect on relations between New York and Waterbury; many of the duties and the points of friction remained the same. The New York store, like the agents before, solicited, relayed, and hurried orders for stock and specially made items. The New York store also complained frequently about its need for additional cash to carry on its end of the business.[28] Moreover, the manager of the New York store still occasionally (though less frequently) cut prices without consulting Waterbury when he felt it was necessary to avoid losing a customer to the competition.[29]

Yet the letters reflect a few changes in relations, seemingly tied to improvements in long-distance communication and transportation. Although the New York store performed some errands for Waterbury, that component of its duties, judging from the letters, was smaller than it had been for the agents earlier. Marburg has suggested that factors in this reduction of miscellaneous errands included "the improvements in communication such as the reduction of postal rates, which facilitated direct communication [with other firms] at a low cost, and the introduction of the telegraph in 1849."[30] Thus improved communication technology may have encouraged a very small step toward more regularity in relations and in communication between the two locations.

In addition, the decrease in postal rates and change in rate structure probably contributed to another small change in the communication patterns: an increase in the number and decrease in the length of letters. The frequency of letters from New York to the Waterbury headquarters increased from seven letters a month in 1829 to fifteen letters a month, on average a letter every other day, in 1854.[31] These letters still covered multiple subjects, but tended to be shorter than the earlier ones. The lowered mailing cost made it economic to transmit information and orders when they were received, instead of saving them for less frequent but longer letters.[32] The switch from hand copying to press copying sometime before 1854 may also have slightly encouraged the increased number of letters by making copying faster and easier. In everything but their shorter length, however, the letters resembled the earlier agents' letters. At this point no distinction in form and style was made between internal and external correspondence.

The telegraph may have been partially responsible for a third incremental change in internal communication: an apparent reduction in controversial price-cutting incidents. Unfortunately, the correspondence from the 1850s does not reveal much about the early impact of the telegraph, which had arrived in Waterbury in 1849. The telegrams were neither press copied into the regular press book nor confirmed in letters, though occasionally a letter referred to a telegram.[33] There is certainly no evidence that the telegraph was used in any systematic way at this time. The fact that price-cutting incidents were mentioned less frequently in the correspondence from this period, however, may suggest that the telegraph slightly increased Waterbury's control over the New York store. Of course, the shift from commission agents to salaried employees no doubt also played a role.

Changes in the communication system began, then, in the 1850s, but progressed only slowly. The 1850s also saw changes in leadership that would lead to more noticeable if still slow evolution in managerial methods and communication patterns starting in the 1860s. The first generation of owner/managers ended with the deaths of William H. Scovill in 1854 and James M. L. Scovill in 1857.[34] An interim period of rather desultory leadership was followed by a new generation of management.

Initial Signs of Change under Goss and Sperry

Although major changes in management methods and communication patterns awaited the turn of the century, lesser changes began much earlier. These changes were the product of the management team of C. P. Goss and M. L. Sperry, originally hired as bookkeepers around 1863 by the otherwise mediocre company president who followed the last of the Scovill family.[35] The team of Sperry and Goss, as secretary and treasurer respectively, proved to be the driving force in the company under F. J. Kingsbury, president of the firm from 1868 to the turn of the century. The company grew rapidly during this period: the 1850 workforce of 157 doubled to 314 by 1874, then almost tripled again to 1,157 workers by 1892.[36] In 1900, Kingsbury retired, and Goss and Sperry served consecutively as company presidents until 1920.[37]

Both Goss and Sperry respected and cultivated the written record. Early on, in fact, they increased office space to allow for the creation and storage of records.[38] The Board of Directors' Memorial Minute to Sperry after his death commented on his "remarkable clarity of composition" as well as his responsibility for the " 'law of the Company (like the laws of the Medes and Persians unchangeable) that no letters are to be mailed unless press copies are retained. In cases where you must either miss your train or not take the copy—*miss the train* and take the copy.' "[39] Building a dependable written record of correspondence as well as accounting data—a rudimentary corporate memory—was important to him. Moreover, both Goss and Sperry were remembered by those who worked with them for using written correspondence as a medium for admonition and exhortation.[40] Unlike the Scovill partners before them, they delegated responsibility to foremen and managers, then often controlled them from afar through written, not face-to-face, communication. These changes in management methods contributed to the slow evolution of the communication system.

Given their initial role in the company as bookkeepers, it is not surprising that one of the first management innovations Goss and Sperry undertook was to institute "a system of bookkeeping from which the profits of the various sections of the business—mill, . . . button, burner, aluminum departments, and so on—could be discovered."[41] This primitive attempt at accounts that functioned as a comparative measure for various units of the manufacturing operations, according to historian Philip W. Bishop, "survived as a tool of management until well after the turn of the century."

Such a bookkeeping system, of course, required more elaborate records than those previously kept in the ledgers and journals. In the 1870s, several printed forms were created to gather the necessary data.[42] Although most of these forms were basic transactional documents such as bill statements, order acknowledgments, and shipping forms, they helped track materials and finances more completely. Moreover, references to the "Monthly Statement" (no examples of which, unfortunately, have survived) suggest that rudimentary periodic reporting evolved during this period, as well. By this time the company had grown past the stage at which the president and officers could run it directly. For the most part, according to Sperry, they appointed foremen, then let them run the floor by whatever methods they chose.[43] The new system of accounts and the monthly reports provided a primitive method of evaluating the foremen and of encouraging and controlling firm growth. For the first time, face-to-face methods of management were supplemented by a small but systematic flow of data within the company.

This development coincided with some changes in correspondence, as well. Letters between Waterbury and its stores increased and began to show signs of change in the late 1870s and early 1880s. By this time, the New York store had been joined by a Chicago store and a Boston store. The trend toward more and shorter letters that began in the 1850s continued. The 1854 rate of about a letter every other day rose to an average of around a letter a day each direction in the 1870s.[44] While at least three different people in Waterbury wrote frequent letters to New York, a single one of them might write as many as four times on a given day.[45]

The increase in volume of internal correspondence probably played a role in several minor developments designed to make later reference to the letters or the press copies easier. First, these letters tended to cover fewer subjects than the rambling letters exchanged earlier, with most of them focusing on a single topic.[46] With the low postal rates, several single-subject letters on a given day were not prohibitively expensive to send, and they were easier to handle, store, and retrieve at both ends, an advantage that would grow as correspondence did. At this time, Waterbury also began to use subject headings to make the letters easier to locate. The subject lines were either order numbers written on the top of the page or phrases describing the subject, such as "Stair Rod Holder" or "Mr. Fuller's [letter] of the 6th."[47] Both of these changes slightly differentiated internal from external correspondence.

Such changes were only slowly adopted by the stores, whose managers guarded their autonomy from the greater control of standardization. Although Waterbury apparently did not demand that the stores use subject headings, by 1880 they did require the stores to write separate orders on separate sheets. The new procedure was adopted over the objections of some who preferred the more ad hoc ways of the past. The following letter from Goss to the head of the Chicago store reveals both the reasons for adopting such a system and the opposition encountered:

> On my return from N.Y. I find on my desk your letter of the 25th in reply to Mr. Hyde's letter of the 23rd. This letter was written by my

request as the request was made just as I was leaving for the cars [?].
We have at different times requested all the stores to make all orders as explicit as possible and *to write orders for buttons* on separate sheets of paper. Also to avoid as much as possible *interleafing*
[?] and writing across *the tops and Margins of the sheets*. These
were the points I intended to have Mr. Hyde convey in his letter.
And is there anything so very unreasonable in this request? N.Y.
Store and Boston Store saw the matter in proper light and instead of
"snapping us up" and telling us we are "wasting our time on that
subject" they at once complied with our request. In entering up
and keeping track of the large amount of orders which we receive[,]
some regular system and a division of labor and responsibility is
actually necessary. We certainly ought to be able to judge what are
the best methods for doing this. If you will glance through the letters you have written us this month and keep in mind what I have
said as to the disadvantages we labor under by the receiving of orders, [unreadable], and [unreadable] sheets you will readily appreciate what we are doing about it.[48]

Like many of his contemporaries during this period, Goss was responding to growth that threatened to overwhelm the company under the old,
haphazard methods by instituting "some regular system and a division
of labor and responsibility." Goss realized that he could only coordinate
and control the stores for maximum efficiency by regularizing the flow
of communication, necessarily restricting some autonomy previously
enjoyed by the stores.

Around this time the New York store adopted a measure of its own to
facilitate the handling of increasing amounts of correspondence: separate sets of press books for different types of correspondence. One series,
marked "Factory," was used to copy only correspondence addressed to
headquarters in Waterbury.[49] While the separate "Factory" press book
made a given letter to Waterbury somewhat easier to locate than before
by isolating it from external letters, the improvement was limited by
the lack of an index. Press books were traditionally indexed alphabetically by addressee. Since all letters to Waterbury were simply addressed
to the Scovill Company, such an index would have been useless.

In the late 1870s and early 1880s, stationery used for correspondence
between Waterbury and the stores was standardized in forms designed
for efficiency and ease of handling. One kind of letterhead provided the
return address, part of the date ("187–"), and the addressee (either "New
York Store" or "——— Store") printed on the top.[50] Another stationery
form, used for brief and very informal (sometimes penciled) notes from
New York to Waterbury, was printed on a half-sized sheet of paper
headed "Memorandum," a very early use of the term to refer to internal
correspondence.[51] These special letterheads were probably introduced
to replace more expensive stationery used for external correspondence
and to save the writer's time. They also made internal correspondence
more easily distinguishable from external.

Differences in style between internal and external correspondence were also emerging at this time. By the 1870s and 1880s, the wordier style of earlier times was sometimes, though certainly not always, replaced by a terser style for store letters. One letter to the New York store, for example, read as follows: "Kindly cancel order 0749 and return to us the sample Bale and Holder which we forwarded you a week ago. By doing this you will greatly oblige/Yours truly,/————"[52] This letter combined a terse, more compressed style with the conventional wordy close. Some notes written on the informal memorandum form omitted even the traditional complimentary close.[53] These changes in style probably reflected the increasing amount of internal correspondence. With several letters going each direction every day, the excess words of the conventional style added up. Letters to customers, as well as some internal letters, continued to be more traditional in style.

In the late nineteenth century, the adoption of several innovations in communication technology also began to affect communication patterns. In the late 1870s and early 1880s, the telegraph extended Scovill's market and, in a small and unsystematic way, its control over its own branches in different cities. The press books (which included copies of telegrams during this period) reveal that the New York store now frequently communicated with Waterbury over the telegraph. Almost one in ten communications copied in an 1877 press book, for example, was a telegram.[54] The following three messages about various orders, all sent on the same day, are typical: "Eighty five seventy four should have been spring metal How soon can you get out another lot"; "Eighty five seventy four send the spring metal"; and "Send Eighty-six one and eight six seven immediately answer."[55] The first two, with the intervening one from Waterbury (which did not survive), settled in a single day a mistake that would have taken three to six days by mail. By then the buyer might have turned to another source. The third rushed Scovill on two orders. These telegrams illustrate the value of the telegraph in helping Scovill serve customers from a wider market. The telegraph's role in internal control was less prominent, since it was not used for any systematic reporting. Nevertheless, pricing issues, once a point of considerable friction between the two locations, were increasingly negotiated quickly over the wires, lessening the New York store's autonomy at the same time that it reduced contention between the two locations.

The first internal telephone system to connect the office and the various buildings making up the plant was probably installed in 1881.[56] With that system and the improvements made to it over subsequent years, telephone conversations were possible between and perhaps within buildings of the plant. The telephone undoubtedly facilitated the oral management methods that were still predominant at Scovill at this time, but the typewriter was ultimately to be more important to the formal communication system.

In 1888 the first typewritten letters appeared in Scovill's press books, interspersed with the still more common handwritten ones.[57] The typewriter arrived just in time to help out with the ever growing but still primarily (with the exception of correspondence with the New York store) external correspondence of the company. The Waterbury office had filled an average of over five one-thousand-page press books a year from 1880 to 1885, but for the next five years it used up just under nine books a year.[58] The first typed letters appeared near the end of that second five-year period, indicating that they certainly did not cause the increase but did help handle it. By 1889 Scovill had hired a female typist, or "typewriter."[59] In 1893 there were at least two typists, along with a system of initials indicating who typed each letter, suggesting the company's increasing belief that standardization and accountability were desirable.[60] The typewriter and its operators reduced the cost of correspondence and lessened the burden on Sperry, Goss, and other high-level managers.

By the turn of the century, the Scovill office operated very differently than it had in the middle of the nineteenth century. Goss and Sperry had introduced slightly more systematic accounting methods, had regularized minor aspects of relations with the stores, and had adopted new communication technologies. Incremental changes in communication had resulted. Yet in the factory the company still depended on ad hoc management. The foremen still ruled on the shop floor, with virtually no intervention from above. They communicated with the workers orally, establishing their own rules. Scovill did not even post a set of general rules. Given the foreman system of governance, Sperry explained in 1887, such rules seemed unnecessary and useless:

> We have never had any shop rules printed. There is a general understanding that ten hours constitute a day's work and that the hands are expected to do a day's work if they get a day's pay. Each department is under the direction of a foreman, in whom we trust and who sees that the hands are industrious and attend to their business. If they do not do it, he sends them off and gets others. . . . We do not think printed rules amount to anything unless there is somebody around constantly to enforce them and if such a person is around printed forms can be dispensed with.[61]

Systematic management, and the formal internal communication system so necessary to it, had not made much progress at Scovill. That awaited the twentieth century.

Before the War: The Advent of Systematic Management at Scovill

The pace of change accelerated after 1900 as Scovill responded to the demands of continued growth by adopting the values of systematic management. The firm grew from just over one thousand employees in the 1890s to four thousand by 1914.[62] With this growth and the addition of more brass products to its line came inevitable changes in structure. The managerial hierarchy expanded, with department superintendents supervising an increasing number of foremen and subforemen.[63] In the early twentieth century, the division of basic manufacturing into Button, Burner, and Rolling Mill departments began to break down and various other configurations were tried. By 1911, production had been consolidated into a single department that was divided into seven divisions known as classifications. These numbered classifications had a total of fifty-four subdivisions under them.[64] This complicated structure covered only manufacturing; other structures evolved for marketing, supply, accounting, and other departments.[65]

In this climate of flux, a new generation of upper management led by John H. Goss (see fig. 6.1), son of the long-time treasurer and now president and general manager, C. P. Goss, saw the increasing drawbacks of the old methods of management. In a 1905 "Report Made to the General Manager on Timekeeping in the Departments," J. H. Goss listed some of the causes of inconsistency in timekeeping for payroll accounts:

> 1. Rapid growth;
> 2. The consolidation of departments that were originally separate;
> 3. The classification and concentration of machinery and operations;
> 4. The separation of original departments into subdivisions or new departments;
> 5. The inevitable increase and subdivisions of the clerical work and system due to these changes.[66]

These changes had caused problems throughout the company, not just in timekeeping. Initially as superintendent of the Burner Department, then as superintendent of manufacturing in 1908 and as the first general superintendent of manufacturing by 1909, J. H. Goss adopted and actively pursued the goals of systematic management to provide better managerial control.[67]

The analysis following the above list in his report is worth quoting at length, because in it Goss clearly stated the management problems and solutions as he saw them:

> These changes have been slow and almost unnoticeable in their consequent effects until now there are in some respects wide divergence in both methods and opinions regarding methods, and surely the time is ripe for those in authority over *all* the departments to get together with the data and decide upon a plan which shall be simple and at the same time applicable to all alike.
>
> As it is now, while the honesty of none is brought into question, there is undoubtedly not only an opportunity for dishonesty with

6.1.   John H. Goss, who introduced the ideals of systematic management into Scovill. *(Courtesy of the Baker Library, Harvard Business School.)*

little chance of its being discovered until it is far reaching, but also every likelihood of money being paid out by the Company unnecessarily and therefore at a loss, because the responsibility unchecked rests entirely upon the department superintendents and their subordinates to systematize or leave unsystematized those details which are vital to the payroll scheme and many things are allowed to pass in a department, which in themselves at the time may not seem important, but which unchecked have become precedents in that department and at present are unquestionably the cause of large losses to the Company.

The main trouble at present is due to the fact that the force on the payroll in the main office have apparently no authority to stop a method that is wrong and that has crept into a department system, but seem as much controlled by the precedent in each department as the clerks in that department itself are. They at least consider that they have no authority to compel the superintendents of departments to stop such a practice.

The best way to stop this difficulty would seem to be to allow the superintendents to make no special agreements with any of their help that are not recorded in the main office and that have not been approved by the supreme authorities; and then have the difference from the regulation adjusted in the main office for each case and for each pay day, including extras for overtime or full time service.

Perhaps a suggestion as to a set of regulations to govern the time records out in the factory would not be out of place at this point. These could be printed in the front of each payroll book issued and then there could be no chance for misunderstanding due to changes of one sort or another.[68]

This report reads almost like a primer of the theory and methods of systematic management. Goss objected to the inefficiencies and losses inherent in an organization where power was completely decentralized in the hands of the superintendents and foremen, who might or might not "systematize" details. The answer to this problem, he said, was "for those in authority over *all* the departments" to create a simple but consistent system for all parts of the company, independent of individual superintendents and foremen.

The methods by which Goss suggested introducing system were heavily dependent on documentation. He had already mentioned the increase in routine clerical work such as payroll accounting as one of the causes of chaos. More and different types of written communication were needed, he indicated directly and indirectly in his carefully reasoned report, to achieve an efficient system. He wanted the system to be based on data that had been collected in written records. For dissemination and implementation, he wanted the new regulations printed up so that "there could be no chance for misunderstanding due to changes of one sort or another." Written regulations would create a corporate memory of the system and a deterrent to accidental or intentional misunderstanding of the centrally mandated rules. Finally, any exceptions, he argued, had to be recorded and approved in writing.

In the decade following this report, Goss was to make considerable progress in systematizing management and reducing the heretofore almost unlimited power of the foremen. Bishop credits him with creating the first formalized labor policies and with "open[ing] up to the light of day some of the more obscure corners of the plant until now dominated by foremen who, in their way, guarded secrets as jealously as the boss-[brass]casters, also on the way out at this time."[69] Following the blueprint he had laid out in the 1905 report, Goss developed the communication system as his principal method of achieving better control over the company in general and the foremen in particular. Consequently, the early years of the century saw a flowering of downward and upward written communication at Scovill.

Regaining Executive Control through Downward Communication

The philosophical objection to downward written communication voiced by Sperry in 1887 gave way in the early years of the century under the pressure of the growing workforce and J. H. Goss's desire to control that growth by shifting the balance of power from the foremen to upper management. By at least 1905, Goss was issuing written notices, orders, and announcements to the foremen and workers in the Burner Department, of which he was then superintendent.[70]

In doing so, Goss had to address some practical problems of dissemination. Printing was far too expensive and slow for the small scale of his initial efforts. Early forms of the mimeograph and other duplicators were available by the turn of the century, but there is no evidence in the records that duplicators were used at Scovill before the second decade of the twentieth century. The company had, however, discovered the advantages of carbon paper used with the typewriter. As superintendent of the Burner Department, Goss made two or more copies of each notice, keeping one and circulating the other from foreman to foreman, each one reading it and checking off his room on a list rubber-stamped on the back of the notice.[71] As general superintendent after 1909, he had enough carbon copies of notices made to send one to each department or classification head for circulation among the foremen when applicable. Although this procedure sometimes required more than one typing with the maximum number of carbons, it was nevertheless workable as a mode of dissemination.

Storage also posed problems. Press books were used for copying and storing Scovill's outgoing correspondence to external parties or to company stores, but no system existed for the still-rare written communications within the Waterbury plant. When he was head of the Burner Department, Goss stored his own copies of notices on a Shannon file.[72] This system gave him a complete record, but it did not allow the foremen easy access to the notices at later times. At least one foreman, E. G. Main, retyped all internal notices and notes he received and stored them in chronological order on his own Shannon file, indexed alphabetically by subject.[73] This personal file solved the problem of accessible storage for him, but at the cost of retyping all of the notices, a time-consuming and therefore expensive process. Whether others did so as well is impossible to tell, since only Main's set of notices has survived. Not until the next decade is there evidence of duplicating machines that allowed each of the dozens and dozens of foremen, subforemen, and superintendents to keep a copy.

Despite only partially solved problems of dissemination and storage, downward communication flourished to fulfill Goss's goal of increased central control. The mere issuance of all these directives reinforced the trend away from foreman autonomy and toward executive control and companywide system. In the past, foremen answered to the central office only in terms of their output; now Goss and other executives also dictated methods. One of Goss's announcements, for example, reduced the power of foremen by subjecting them to the same rules their workers were subject to: "Please note that hereafter actual time in attendance during working hours must be reported by your time-keepers for everybody, including foremen."[74] The recently added timekeepers had already reduced the foremen's control over their own workers. By making the foremen subject to the timekeepers, as well, Goss was further reducing the foremen's autonomy.

Goss also used downward communication to curb the power of foremen and other lower-level managers in more subtle ways. In a note to E. G. Main, foreman of the Store Room, Goss wrote, "Perhaps it would be well for you to issue written instructions to each man, covering, in a general way, his specific duties as elevator operator. Before issuing the same to them, submit it to me for my approval."[75] By retaining right of approval, Goss limited Main's power over his workers. Moreover, the written instructions would create an organizational memory of that particular job independent of Main and of the individual operators. Even simple announcements of new managerial appointments could carry a veiled threat to the autonomy of lower-level managers. In 1911, George A. Goss was appointed superintendent of the manufacturing departments under J. H. Goss, now general superintendent.[76] When J. H. Goss announced the appointment, he reinforced George Goss's power over department heads, superintendents, and foremen by stating that he would be answerable "only to the General Executive Officers of the Company"—that is, to J. H. Goss and those above him—and that he "will be obeyed and respected accordingly."

All of J. H. Goss's orders and instructions were not aimed at reducing the power and autonomy of the foremen, for these low-level managers continued to play an important role within the more systematized framework. Sometimes the notices alerted the foremen to a problem but told them to find ways to solve it. In one such communication about workers using carelessly issued passes to sneak out early, he said, "All of these things are entirely under the control of the foreman or superintendent who has charge of the man to whom the passes are issued, and in order to keep the matter under control we must be careful in making out passes and issuing them."[77] He went on to enlist their cooperation by appealing to their moral leadership, saying, "Foremen who persist in going out early cannot expect to have as good control over their help on this point as those who themselves wait until the quitting hour. The moral influence, both on the help and on the management, is more strongly in favor of the man who lives up to the requirements of the Company as to the working hours than the man who does not."

In addressing another problem with worker behavior on the job, Goss also put responsibility on the foremen: "There is evidence that in different places throughout the shop some of the help are selling candy, tobacco, fruit and such articles to other help in the factory. This, of course, cannot be allowed, and I wish every foreman would make a rigid investigation of the conditions in his room, and be sure that no such thing exists in any territory under his jurisdiction. Please act immediately upon this."[78] Thus, Goss left some responsibility and moral authority in the foremen's hands to maintain their commitment, while limiting their responsibility in other areas to improve efficiency.

Although many policies designed to systematize operations and curb lower management's power were initiated and communicated by J. H. Goss, his father C. P. Goss (now president and general manager) and even the company's directors occasionally demonstrated that they, too, appreciated both the value of systematizing and the power of written communication in doing so. In a 1908 notice to foremen and superintendents, for example, C. P. Goss set up a specific procedure for fulfilling employees' purchases of labor or materials from the company, a function previously carried out through the foremen in a haphazard manner.[79] The directors of Scovill occasionally issued policies intended for the foremen or even the workers during this period, although they were generally passed through one or more mediating levels. For example, a 1905 announcement designated "Per order of Directors" is quoted in full (and commented on) in a communication to the Burner Department foremen from J. H. Goss, then superintendent of the Burner Department.[80] Directors' announcements posted for workers were generally signed by intervening managers (starting with the foreman of the room in which each copy was posted) to avoid completely undercutting them.[81] Still, the power of foremen over workers was increasingly limited by the establishment of consistent, companywide labor policies implemented through notices.

As the organizational hierarchy grew in size and complexity early in the second decade of the century, written communications were issued at more levels and for more reasons. With the reorganization into several classifications around 1911, the classification heads began to issue notices, which carried designations such as "Classification #1 Instruction" or "North Mill/Instructions."[82] Whatever level they came from, these communications established methods and policies to systematize an increasing number of procedures. By 1914, even brief instructions on specific jobs were written, using a printed form entitled "Instruction

Sheet."[83] According to a notation on one such surviving form, the instructions were initially given orally; however, the written instructions provided documentation so that responsibility could be assigned for any failures to live up to the orders given.

The written word was a powerful, but not foolproof, tool in creating efficient and systematic procedures and in asserting managerial control. Skill at employing this tool was, in J. H. Goss's view, critical to his success as general superintendent. According to a memoir written much later by a company historian,

> John Goss had said, in discussing the great speeding up of time calling for frequent and important decisions . . . that his chief difficulty was not in arriving at a conclusion but in finding a way to put that into effect. It called for the shaping up and dissemination in new instructions. It was easy, he said, to express clearly what he had in mind. The difficulty was in getting just that thought into the reader's mind, clearly and unequivocally. Every foreman, he said, had his own ideas and methods, and was, however unconsciously, reluctant to change them. Thus he would tend to interpret the new instruction so as to make the least alteration in his existing pattern. So, he said, he had to take much time to read, re-read, alter and re-phrase his message before its issue. And it usually had to be followed up to check the result, so it might be depended on. "It is my hardest task."[84]

The style of these communications, then, had to be perfectly clear and unambiguous to reduce the possibilities for interpretation.

In order to achieve that goal, Goss (and others) abandoned many conventions of traditional letter-writing style in favor of a more direct though still formal and even legalistic style. For example, one communication from Goss to foremen and superintendents read as follows: "No one is to operate the elevators unless they have been duly authorized by me as qualified to operate them; or by Mr. Main who has charge of the elevators in both the Burner Department and the Gilt Button Department. These instructions are to supersede any instructions or any custom which has been followed to the contrary heretofore."[85] The second, rather legalistic sentence suggests that Goss had already encountered resistance to changes in custom and wanted to leave no loopholes. He also used other rhetorical strategies to gain compliance to new rules. Many notices explained the rationale behind an order, to enlist reasoned rather than blind obedience. Many also enumerated the exact procedures to be followed, to avoid any misunderstandings. In the following notice he used both of those tactics:

Superintendent's Order #137
All nickel-plated scrap before being sent to the Mill should have
the nickel-plate stripped off, as it materially increases the value.
If in addition to the nickel-plating, the scrap has solder attached

to it in any way, it is a question for discussion in each case as to whether it would pay to strip the nickel off.

Please therefore observe that hereafter—

1. All nickel-plated scrap that does not have solder attached to it shall be sent to the Dip Room to have the nickel stripped off before sending to the Mill.

2. All this kind of scrap which has solder attached to it, when accumulated in sufficient quantities to warrant sending it to the Mill, before sending it out the quantity and character should be reported to the Superintendent for his decision as to whether it should have the nickel stripped off or not.[86]

Some notices even had "THIS IS VERY IMPORTANT" typed at the bottom.[87]

These downward directives throughout the company took on a different appearance from traditional letters to suit their use and to facilitate later reference and retrieval. The salutation "Dear Sirs" was abandoned for "Instruction #87" or a subject heading such as "How to State the Value of Express Shipments."[88] Although most notices were still signed, they omitted the complimentary close of the letter form. Now they differed clearly from letters, both in form and in style.

The emergence of downward written communication at Scovill was a major step in the evolution of the communication system. These notices, orders, and instructions both systematized procedures and provided a corporate memory independent of individuals.

Systematizing through Upward Reporting

The goals of systematic management could not be achieved by downward communication alone; J. H. Goss and other executives needed upward flows of data and analysis to serve as the basis of their efforts to make operations more systematic and efficient. During the years preceding World War I, records and reports designed to fulfill this function began to proliferate at Scovill. Forms, tables, graphs, and prose reports all contributed to this development, which continued at an even faster pace during and after the war. Furthermore, specialized types of routine and nonroutine reports grew up to support the research function. All of these reports aided executive control.

In instituting new records and reports, Scovill executives justified the additional paperwork by the increased managerial control and consequent improved efficiency. For example, when C. P. Goss established the system for filling employee orders of Scovill goods and services, he distributed pads of newly created forms ("Employee's Order Sheets") to be filled out by the employee, signed by both the foreman and the department superintendent, and sent to the main office for recording. After explaining this process he pointed out, "This will of necessity make more delay in delivery than has been formerly the case, but it will also make for more accuracy and less loss, and also be less expensive."[89] He was stating one of the basic assumptions of systematic management, that system, with its consequent bureaucracy, pays. Individual convenience and efficiency had to give way to standardized procedures designed to provide maximum efficiency across the company. Of course, in any given case the costs and benefits to the organization as a whole would have to be assessed. Frequently, however, Scovill's executives of this period seemed simply to assume that increasing control automatically increased efficiency.

In addition to their role in systematizing a procedure, sets of forms, tickets, and other primary records also made possible better descriptive and analytic reports based on them. In a 1909 communication, J. H. Goss made this connection clear: "It is desirable to get systematized as soon as possible the method of receiving goods on our various orders from outside sources of supply, so that we may be sure of getting a reasonable report from the proper source of information upon the quality of each lot of material that arrives."[90] Only with such "reasonable reports" could this aspect of the supply function be evaluated and improved. Thus, the paperwork created to systematize a process contributed to the efficiency of that process indirectly, by providing data sources for further analysis, as well as directly, by institutionalizing complete and consistent procedures.

Accurate records also helped to establish managerial accountability, necessary in evaluating the many lower-level managers in the growing firm. In a 1912 notice, George A. Goss referred to a system of receipt tickets that had evidently been in place for awhile but was not working correctly:

> In many instances the clerks seem to have lost sight of the fact that the original plan for use of receipt tickets was to hold the man responsible who sends out the goods until the next man had receipted for them. It seems now that many clerks are using their own copies for checking up their records and not obtaining the original receipt. This is the cause of much work being lost when it is routed to the wrong room and left by the truckers by mistake in the wrong room. There are no receipts to check up these mistakes and in these particular instances the records are practically worthless.[91]

When the system of receipt tickets was used correctly, it clearly defined responsibilities at any given time, creating a paper trail to the source of any problems. Moreover, such written records also promoted accountability by providing the data needed to compare the performances of individual foremen and superintendents and to monitor their performance over time.

The data collected through the new systems had to be compiled, analyzed, and communicated before being used to monitor the efficiency of the company and of specific managers. In the early years of the century, report formats developed in ways to facilitate comparison. Starting in 1907, J. H. Goss had hand-drawn tabular Cost Analysis Sheets filled out for the Burner Department at monthly intervals.[92] These analytic tables assigned costs to maintenance and repairs, supplies, interdepartmental credits, and direct and indirect payroll. Each of these four categories was further broken down and recorded for each of the rooms of the Burner Department. A yearly summary, which included a column for the previous year's figures, was also compiled. Such tables facilitated comparisons across time and between rooms.

When J. H. Goss was promoted, this technique spread. The 1908 monthly and yearly tabular report forms were printed up for the newly created Manufacturing Department (composed of the old Burner and Button departments), and the comparison column filled in with the figures for the previous year's Burner Department.[93] A similar set of printed tabular forms was created for the Rolling Mill. Moreover, by 1910 all of the rooms within each department were using tabular forms to submit the periodic reports on which the monthly and yearly compilations were based.[94] These periodic reports, continued under the supervision of the company auditor until 1920, provided data on the basis of which various comparisons and analyses could be carried out.

These reports and analyses also resulted in some of Scovill's first graphs.[95] In addition to having tabular comparisons of the 1907 Burner Department and the 1908 Manufacturing Department made up, J. H. Goss also had some graphic comparisons of costs drawn up for the two years (see fig. 6.2). These divided bar charts made visual the relative magnitudes of various elements of costs. By 1911 and 1912, more graphs were being used to display the statistical data collected internally. These included line graphs and layered line graphs with labels such as "Rolling Mill (and Casting Shops) Costs and Materials, 1908–1912."[96] Increasing analysis of data thus encouraged the adoption of new methods of presentation to facilitate rapid understanding. Scovill would continue to develop and use graphs as managerial tools during and after the war.

Prose reports also seem to have increased in quantity during this prewar period. Those that have survived were special rather than routine reports, although references attest to the earlier existence of some regular prose inspection reports.[97] The surviving special reports had developed more in function than in form at this time. They were frequently agents of systematization. For example, J. H. Goss's report on timekeeping, quoted above, advocated establishing a stricter timekeeping system with no individual exceptions. He also occasionally used reports to enforce compliance with an order or policy by announcing a new policy in a downward communication, then requesting reports from foremen and department heads explaining how they would implement the policy.[98] In other cases, a middle or upper manager might request a report to be used as the basis for systematizing some process.[99] In form, however, this and other prose reports still resembled correspondence.

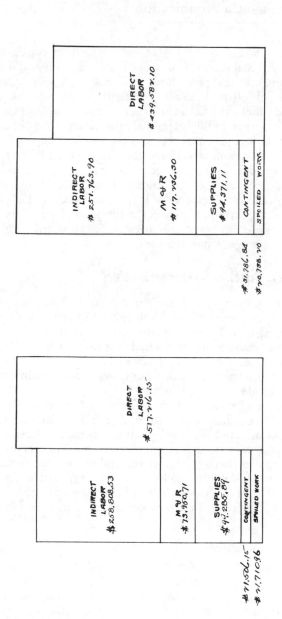

1908

MANUFACTURING DEPARTMENT
[OLD BURNER AND BUTTON DEPTS]

INDIRECT LABOR
251,763.90

M & R
117,736.30

SUPPLIES
94,371.11

CONTINGENT

SPOILED WORK

31,786.84
70,738.70

DIRECT LABOR
439,587.10

1907

BURNER DEPARTMENT

INDIRECT LABOR
258,808.53

M & R
73,980.71

SUPPLIES
40,285.84

CONTINGENT

SPOILED WORK

71,506.15
71,710.96

DIRECT LABOR
517,716.15

6.2. Early Scovill graph comparing costs for two different years. (Courtesy of the Baker Library, Harvard Business School.)

By the beginning of World War I, J. H. Goss's efforts to systematize management at Scovill had resulted in considerable upward and downward communication. Clerical procedures were also being developed at this point to keep track of the increasing paperwork. As early as 1907, notices, reports, and other internal documents bore the marks of rubber stamps. Some were stamped "File" to indicate the destination of a copy. Carbon copies often bore a stamped set of initials to replace the signature of executives. Soon communications were stamped with the room receiving them and the date of receipt. Increasingly, the typist's and the writer's initials appeared on a communication, and sometimes even more information, such as "Dictated by J. H. Goss to C. L. D." Such stamped and typed designations helped make the flow of written communication within the company more efficient.[100] Thus, the tools of system were themselves being systematized.

The War and After: Further Systematization

During the war years, enormous growth and the need for even greater efficiency were forces in the continued evolution of internal communication as a managerial tool. By 1918, the amount of written communication within the Waterbury plant demanded four internal mail deliveries a day, although only one external delivery was received.[101] By the end of the war, Scovill's internal communication system bore almost no resemblance to that of 1880.

The developments of the war and immediate postwar period began with a major change in methods of storing and retrieving documents, a change that may have affected the subsequent emergence of lateral correspondence. Then the extensive development of upward reporting led to attempts to rationalize that system. By the end of the war, the more formal labor policy that had evolved led to some new types of downward communication. All of these changes helped Scovill complete the transition from a typical nineteenth-century communication system to a modern one.

Vertical Filing for Improved Access

Although Scovill did not adopt vertical filing until 1914, the pressures promoting this change had been building for awhile, and the technological adaptation allowing it had already begun.[102] Incoming correspondence had outgrown pigeonholes and graduated to box files well before the turn of the century.[103] The problem of guaranteeing accessible storage for the growing bulk of Scovill's written communication had become more and more urgent, however, in the early years of the century. There were two key sources of pressure: the growth of outgoing correspondence (including correspondence with Scovill's stores in New York and other cities) and the evolution of internal written communication within the Waterbury plant. Both pressures made the inadequacies of the old system increasingly apparent and ultimately led to the adoption of vertical filing.

The rapid growth of Scovill's outgoing correspondence, which had begun in the 1880s and 1890s, continued into the new century. By the early twentieth century, Waterbury had established a separate series of press books for letters to the New York store, probably to reduce the bulk of the main series.[104] This New York store series averaged two to three volumes a year from the turn of the century to 1910. At this point, the amount of correspondence with the New York store soared, with seven volumes for 1911, and twelve and thirteen each for the next two years.[105] This increased amount of long-distance internal correspondence, probably accompanied by a similar increase in external correspondence to customers and suppliers (the external press books from this period have not survived), posed a huge storage and handling problem for Scovill.

The correspondence contains evidence of some attempts to deal with this problem. In 1910, for example, a letter from Waterbury to New York reiterated as policy earlier requests that store correspondents limit each letter to a single subject to make internal routing in Waterbury easier.[106] By 1911, the delivery of correspondence to various rooms and offices within the Waterbury facility had also been systematized.[107]

Whatever the attempts to impose order on growing correspondence with Scovill stores and with external parties, its volume made accessible storage under the old system of press books and letter boxes inadequate. Locating all of the correspondence to and from a single company on a specific subject might require searching many press books as well as several box files. A letter from Waterbury to New York only a week before the new vertical filing system was introduced dramatized the problem: "Replying to yours of the 24th regarding terms to Jos. L. Porter & Co., we are sorry that our record for 1908 is quite as inaccessible as yours seem to be, and, unless you consider the matter of enough importance, you will let the matter pass."[108] The volume of correspondence had apparently outgrown the company's ability to deal with it under the old storage system.

The problem created by increased correspondence with the world outside the Waterbury plant was compounded by the development of written communication within the Waterbury site. Since this was neither outgoing correspondence to be copied in the letterbooks nor incoming correspondence to be filed in the letter boxes, it was copied and stored haphazardly, if at all. E. G. Main, for example, stored his notices and instructions on a Shannon file. Another manager saved some of his own incoming and outgoing internal correspondence, probably in letter boxes, and later added them to his vertical files.[109] Notices and reports from this period reveal that various other makeshift filing systems existed for special forms and for card records.[110] While ways were found of handling specific types of communication, lack of consistency from office to office remained a problem, as did the absence of a system for storing general internal correspondence. Furthermore, internal correspondence about a specific customer or order was not filed with the related external correspondence, complicating any search for all correspondence on a specific order or customer. Clearly, the company needed a better system of storage.

By 1911, Scovill took the essential first step of introducing unbound rolling press and carbon copies for its regular correspondence.[111] Creating loose copies was necessary but not sufficient to precipitate the more critical change in organizing papers. Instead of merging outgoing, incoming, and internal documents to make access to information easier, Scovill initially bound the loose copies of outgoing correspondence into chronological volumes.

By the end of 1912, however, Scovill was seriously investigating vertical filing systems at other companies. An internal report entitled "Vertical Letter Filing; as practiced by the Bridgeport Brass Co." closely examined the details of the copying, indexing, and filing apparatus and procedures at a rival company.[112] The report's opening discussion of Bridgeport's basic filing principles suggests the issues considered important at Scovill:

> 1. About two years ago they discarded entirely the numerical system of filing letters and since that time they have followed the strictly alphabetical method.
>   2. Copies Of Out-Going Letters
> They make carbon copies of all out-going letters. No press copies whatever are made. They have never had any trouble on account of changes in original letters not appearing on the carbon copies. The carbon copy is not attached to the original letter, but each is filed strictly under its respective date.
>   3. Store Letters
> For stores like New York, Chicago, W. P. Swoyer, etc., they proceed as follows:
>   a. If the letter refers to a customer, use the customer's name at the top of the letter as a subject and file all such letters in the general alphabetical file under the name of such customer, no matter where such letter originated, whether with a store or with Swoyer or with the customer directly. This keeps all correspondence from and about a given customer in one folder.
>   b. In addition to the above, for customers like Swoyer they use a short, simple subdivision of subjects. They will send us a typical list of such subjects as used for Swoyer.
>   For customers like General Electric Co. they have a new folder for each month, but no attempt is made to subdivide by headings correspondence from such people. They only subdivide by headings where they control the correspondent, as in the case of their own agents or stores.

The report went on to give details about the cabinets and file folders used, as well as the procedures for sorting and filing the letters.

This study (and perhaps others that did not survive) led Scovill to adopt vertical files one year later. In a letter written to the New York store less than a week before the end of 1913, Waterbury announced:

> On January 1st we shall start a new system of filing our letters and we shall use the vertical system of filing. By this system we shall distribute the store letters through the general file under the heading of the subject discussed in the letter. . . . The subject of the letter will be the name of the customer to whom the letter refers.
>
> We shall then file through our general file under the name of the customer all your letters pertaining to said customer and all our replies about said customer will be headed and filed in the same manner.
>
> In the cases of letters about matters that do not relate to particular customers, such as financial matters and other general subjects, these will be filed in the general store folder, same as we have been doing in the past. We hope that by filing in the general file all letters pertaining to or about customers we shall be able to reduce to a small number the letters that must necessarily be filed in a miscellaneous folder.[113]

This explanation, because tailored to the New York store's needs, did not indicate how communication internal to Waterbury would be filed. Most likely, upward and downward communication concerning customers or suppliers would be filed in the alphabetical files, while other communications between the main office and various departments would be filed in miscellaneous files for each department or even each room.

Vertical filing was Scovill's answer to the growth of correspondence. Now the company could build up an organizational memory about each customer or supplier, and about each department or room. Access to records was by subject rather than just by origin and chronology. Individual items were organized chronologically within the folders, a chronology much more useful than that cutting across subjects as in the press book. Thus, the entire story of relations with a particular company could be retrieved in a single folder.

The influence of vertical filing went beyond improving accessibility of documents: it affected their form as well. In the letter to the New York store announcing the upcoming conversion to vertical filing, Waterbury instructed New York on the format and content of letters:

> It will, therefore, be necessary after January 1st in writing us to observe the following requirements:
> 1. Write about only one subject on one sheet.
> 2. Place at the top of the sheet the name of the subject about which the letter is written.

3. The subject of the letter will be the name of the customer to whom the letter refers.

Although Waterbury had requested single-subject letters and subject headings before, the new filing system provided another compelling reason for the demand. Only days after the letter just quoted, and one day before the vertical filing system was to be instituted, E. O. Goss himself reiterated the connection between the filing system and the format:

> It is my impression that you have been requested to head all of your letters with the names of the customers referred to therein, and in case an order is referred to to use the order number as well as the name.
>
> Your letters of December 29th some of them bear the order number but only three bear the customer's name, as a heading.
>
> We are changing our system of filing, and we must INSIST that you pay particular attention to this matter.[114]

The admonitions clearly indicate that the new filing system promoted standardization of the format of internal correspondence and its differentiation from external letters. Although the communications still resembled letters in their salutation and complimentary close (which were not replaced with modern memo headings until 1920),[115] the subject heading differentiated them from external letters. E. O. Goss's letter itself had a subject heading, *"Headings of Letters,"* suggesting that Waterbury was also concerned with helpful labeling of letters that would be classified as miscellaneous correspondence with the New York store.

The company's central file was not the only application of vertical filing in the firm. Over the next few years, department heads and even foremen set up their own vertical files.[116] These files were organized alphabetically by the names of internal correspondents or, occasionally, by the name of a department or committee. At lower levels, too, information was becoming more accessible, and organizational memory being created. The evidence is suggestive, although by no means conclusive, that this proliferation of files encouraged the growth of internal lateral correspondence, the next form of internal written communication to emerge in Scovill.

The Evolution of Lateral Communication

Reports designed to carry information up to the head office and notices designed to carry orders, policies, and announcements down were in evidence a decade before the filing system was introduced. Yet virtually no lateral correspondence dated earlier than 1914 has survived. During the war years, however, lateral communication virtually exploded at Scovill. In the one apparently complete set of files from this period (those of William H. Monagan, foreman of the Casting Shop), by far the largest component is inter- and intradepartmental correspondence.[117] The sudden appearance of lateral communication in the records at this time indicates that the new vertical filing system and its decentralized offshoots were responsible at least for its survival. Moreover, as in the case of the Pennsylvania Railroad, discussed in Chapter 2, the existence of files at the departmental level and lower probably encouraged an increase in lateral correspondence.

Scovill's interdepartmental communication apparently grew partially as a function of growth and partially in response to the values of systematic management. Within the Scovill plant of earlier years, bridging distance was never a real problem. As Scovill's Waterbury complex grew, however, foremen and department heads could no longer talk to each other by walking across a hall. During World War I alone, plant area grew two and a half times.[118] Moreover, between 1914 and 1918 Scovill's employment rolls tripled from about four thousand employees to almost twelve thousand, of whom slightly over one thousand were executives and clerks, and the rest factory workers.[119] Thus, coordination was no longer a simple matter. Although Scovill had a factory telephone system, individuals were not always near their telephones. As one manager suggested, "Time is lost, confusion results and money is spent in endeavoring to locate executives [for telephone calls] when they are out in the plant."[120] Written messages served as a method for bridging physical distance in cases in which immediate contact and the give and take of conversation were not essential. Moreover, writing was better than conversation for accurately conveying complex and precise information such as a formula.[121]

Systematization may also have encouraged written documentation of critical relationships and transactions. Systematic management's principle of documenting procedures and results to create an organizational memory could readily be extended by analogy from vertical relations to horizontal ones. Moreover, as the firm grew and the number of internal units increased, managers and employees began to identify more with their own units and to view other units almost as other companies, thus reinforcing the impulse toward documentation of relations. Phrases such as "Confirming our discussion" or "Confirming our telephone conversations" indicate that many lateral communications were written less to convey information or opinions than to document positions for future reference.[122]

The desire to document lateral relations was by no means universal; when growth and shifting of responsibilities led to tensions, however, documentary correspondence seemed to increase radically. By far the biggest component of lateral communication in the files of Foreman Monagan was with various representatives of the Research Department and the Research Committee that oversaw it.[123] As the foreman in charge of the Casting Shop, Monagan occupied a position that had lost considerable power to the Research Department since the beginning of the century. Although the casting bosses used to be the sole repositories of the secrets of casting, now alloys and casting methods were studied by the Research Department. The tensions between the two units were predictably great, and interactions were scrupulously documented in written communication. For example, Monagan sent the following note to a manager of comparable level in the Research Department:

> The second paragraph of your letter of March 14th reads as follows:
> "It is my understanding that you will suspend weekly reports, and that you will not require analyses of the metal, at least, for some time."
> I would like to go down on record as saying that if at any time I wish to have any part of the program, which has been agreed upon by the Research Committee and myself, suspended I will notify the Research Committee. This to my mind is a serious subject and unless you feel that it is impossible for you to give us the required information, in so far as analyses are concerned, I will expect that the program agreed upon by the Research Committee and myself, be adopted and not the program outlined in your letters of above dates.
> I will see to it that reports, agreed upon in the original program, shall be passed on to you. Trusting that there is no misunderstanding and if there is I stand ready and willing to make all necessary explanations.[124]

Both Monagan and his correspondent seem to have been less concerned with agreeing on a strategy than with making their positions clear to each other and to others in the organization. Similarly, a conflict in responsibility between Monagan's Casting Shop and a segment of the Manufacturing Department over the correct labeling of scrap material led to an extended exchange of written messages in order to document each side's views of what had caused the problems.[125] Copy lists including other managers as well as "file" suggest that these communications were intended to document the conflict, not just for those directly involved, but for others with authority as well. Sometimes such exchanges ended with documentation of a resolution to the conflict that had been negotiated face to face.[126]

As the company grew and became more systematic in its management, at least some of its managers adopted policies of documenting their relations with other units in a manner similar to, though not as extensive as, the documentation of policies, procedures, and operating data in upward and downward communication.[127] It is in this documentary function that the advent of accessible filing surely made the biggest difference. Only with adequate storage and retrieval systems would such communication have much use. The many references to this documentary function in the lateral messages of the period suggest that vertical filing encouraged the development of lateral correspondence within the firm. Thus, lateral correspondence joined vertical written communication to complete the network of communication flows in Scovill. Meanwhile, under pressure from the growth precipitated by World War I, upward and downward communication continued to evolve.

Rationalizing the Reporting System

By the end of the war, a vast web of reports had grown up at Scovill. In 1919, at least two hundred different reports—including inspection reports, schedule reports, production reports, employment reports, maintenance reports, cost reports, electrical power reports, and others—were routinely compiled.[128] The entire range of report formats was represented, from forms and tables to graphs and prose. Most of the reports ultimately made their way to the general superintendent's office.

These reports had grown up rapidly and haphazardly as a response to wartime demands, and eventually it became clear to Scovill management that such reports needed to be rationalized themselves in order to make them effective agents of system in the company. In 1918, a statistician named E. H. Davis (later to become an amateur historian of the firm) was brought into the company to create a Statistics Office in the Research Department. From the start, Davis saw clearly that his first and major task was not to perform statistical analyses of the data but to gain control over what data were collected, how, and why. In an August 1918 special report to the head of the Research Department, he presented his recommendations on the role of the soon-to-be-formed Statistics Office.[129]

Davis's initial goals, as stated in this report, all centered on the records (by which he meant primary records such as job tickets as well as routine numerical reports):

> X. A general survey of the existing statistical or record situation in the plant as a whole through a study of printed forms now in use.
> Y. *Repair* work in department record systems, by filling gaps, remedying present deficiencies and inaccuracies, etc.
> Z. *Standardization* of department records.
> W. *Anticipating* and providing for changes of record necessitated by future changes in the work of production departments.

Such goals required a thorough understanding of how records were created and used at this time, as well as a clear picture of what needed to be changed and why. He demonstrated his understanding of the process two pages later in the report:

> In this plant as in all plants, notation is made locally covering productive progress. In the beginning certain basic records—such as memoranda [i.e. a foreman's notes to himself], job tickets, etc.—were essential for the consistent performance of the work itself. From this, gradually, more permanent records were organized—either as files in which the working records were preserved or in the form of summarized rescripts [i.e., reports] from them—and thus a sort of numerical history was developed as a background in the more important production offices. Subsequently they became the basis for comparative estimates, contract making, etc.

Although this general movement of information from original records to more highly summarized and analyzed reports was appropriate, the existing procedures had evolved haphazardly and needed to be rationalized to serve the current needs of the company: "The problem of the present day is, briefly, for a more reliable and comprehensive application of this traditional process."

Davis went on to state succinctly and elegantly the credo of systematic management as it applied to numerical records: "The greater responsibilities which are now involved demand that such basic data shall be accurate in *fact*, certain (and therefore complete) in *significance*, and comparable (and therefore standard) in *form*." His plan for standardizing records and reports in content and form was as follows:

> Primarily, as I understand the local situation here, there is necessary a general *survey* of the existing record situation, as a fundamental to any extensive constructive organization. This survey . . . appears to be most directly achieved by a study of the *record forms* or *blanks* now in use at the plant. I propose to continue and complete my collection of the blanks now in use for record or report on plant operation, to classify these by offices and functions (and in special cases, by materials dealt with) and then to prosecute a series of departmental enquiries along the lines so indicated. Thus the statistical office will become a clearing house of reference as to *what* current records actually exist, and *where* they are available. Eventually, this office may become the regular depository of carbon copies of many such reports or records, as they are made; and thus a general body of statistical data covering the entire plant will be accumulated, by a sort of evolution rather than by radical change effected by executive order.

So important had records and reports grown in managing Scovill that a whole office had been created to serve as a "clearing house" for them. It was to be "a *service* organization" that he compared to "a supply office, a traffic office, or a pay office."

Davis's efforts were soon felt throughout the organization. Three months later in a November report he indicated that he was beginning to standardize records and reports in six areas: "Production Statistics," "Process Statistics," "Inventory of Stores Statistics," "Industrial or Employment Statistics," "Financial or Monetary Statistics," and "General and Economic Statistics."[130] Initially he established an "intermediate step," in which "the receipt of original reports is now, in many cases, followed by a consistent tabulation of the information, and then by charting the results." Eventually he intended to standardize all of the reporting forms themselves, but this intermediate step enabled Davis to begin bringing statistical data under control.

He turned to new office technologies to help him in his task. Between his August report and the November one, he ordered a Powers tabulating machine and a reflecting lantern. The former was a mechanical tabulating machine using punched cards, which had appeared on the market in 1913.[131] It would help Davis's staff handle the increasing amount of data that he was trying to process. The latter was that day's equivalent of an overhead projector. He was concerned not just with receiving and processing the reports from elsewhere in the organization but with creating his own reports, to be presented in "illustrative conferences" to senior management. In addition, he used a great deal of vertical filing equipment to store and make accessible the growing number of reports and records he was accumulating in his attempt to compile a companywide statistical data base.[132]

Early on, Davis also attempted to standardize report formats, beginning with the graph or chart. Attached to his first report was a note to him from a subordinate asking how the Statistics Office should sign its charts. At the bottom of this note are some scribbled notes in different handwriting, presumably Davis's own:

> Standard Chart
> Standard Scale
> Standard Form of Title
> Standard Form of Legend
> Standard Form of Signature
> Standard Form of Data
> List of regular recipients—on chart.

From the start, Davis saw the need to standardize both the data and the format of graphs.

One possible influence on his desire to create graphical standards was a report found in his files, apparently obtained from an external source, on "Graphs in the Presentation of Business Statistics and Reports."[133] Although the report is undated and unsigned, internal evidence suggests that it was issued on a subscription basis sometime within a few years after 1915. The report covered some of the same material presented in Brinton's book on graphics (which it cited), as well as suggesting some specific graphs of use to a superintendent of manufacture, a sales manager, and a purchasing agent. The section entitled "The Standardization of Reports" could easily have served as Davis's inspiration:

> The form of statistical or graphic reports made in any business
> house should be carefully worked out and standardized, so that re-
> ports may be compiled from month to month as a matter of clerical
> routine rather than as a matter of special investigation. By thus
> standardizing the reports the data likely to be required can be on
> hand promptly when needed by the executives and can be compiled
> at small cost, since such compilation is a matter of routine and
> likewise is apt to be more free from error than a special investiga-
> tion, which also requires expert work.

Whether or not this treatment of graphs was his initial impetus, Davis moved quickly on this matter. Two months after his first report, he

issued a "Scale Selection Chart and Plotting Guide" to standardize "ac-
cumulative" charts.[134] Given a yearly actual or estimated accumulation
figure for the subject being graphed, this table indicated the correct chart
paper and scale to use for various types of graphic curves. He also promoted standard chart forms for other types of graphs.[135] By his November report, Davis could list as one of his accomplishments that "an effort
has been made to establish a standard scale table for the chart sheets
now in use. A standard practice has been adopted governing all chart
work as to legend, signature, etc." In the years following that report,
graphs played a prominent role in analytic reporting at Scovill. Metal
mixtures used in casting were the subject of daily graphic analysis.[136] A
major study comparing Scovill's statistics in 1914 and in 1920 was conducted primarily in graphic form, with tables as back-up.[137]

If graphs were relatively easy to standardize, the forms and tables used
to report numerical data in the first place were harder to get under
control. He attempted, for example, to establish a system for collecting
comparable data from all the production departments, to be compiled
regularly into comprehensive tables. Progress was painfully slow. Even
getting the production reports routed to his office for analysis, much
less modifying them for consistency, was no easy task.[138] More than six
months later, Davis described to his superior the remaining inconsistencies in one unit's reports and what would be needed to bring them into
line with the others, concluding: "Unless this should be desired by the
officers who now receive Mr. Roper's report, I would advise leaving the
matter alone, for the time being at least. The present work of the sort
which we are doing occupies the whole time of one clerk, or about one-
fifth of our present clerical force."[139] Davis did not give up his crusade,
although he did have to defer parts of it: "The general question of production reports, their form, content and analytical detail, is one which
I think may properly be raised for consideration at some appropriate
time." In spite of difficulties, he stuck by his credo: "The uniformity
and comparability of reports of this sort are as important as their completeness and almost as important as their accuracy."

Davis's efforts to rationalize the reporting system had a seemingly
unexpected side effect. In his attempts to make the Statistics Office a
"clearing house of reference," he uncovered and probably accelerated
the dismantling of certain unnecessary routine reports. From 1919
through the mid-1920s, Davis compiled voluminous lists of the periodic
reports that the general superintendent's office was supposed to receive.[140] When he found gaps in the sets of reports received, he sent
inquiries to the person involved, announcing, "Routine reports to the
General Superintendent are filed in my office and I have the responsibility of following them up when they do not issue according to schedule."[141]

These inquiries frequently precipitated the official elimination of reports that were already being silently eliminated. The person receiving

Davis's inquiry often wrote a note to General Superintendent J. H. Goss asking whether the reports were really needed. In many cases, including the one quoted below, Goss replied in the negative:

> With reference to the weekly fire inspection report and the report of indicator post gate shut-offs made by Mr. Barker of the Plant Protection Department, I do not think it necessary to send these reports to me in the future unless to draw attention to some peculiar or abnormal condition. I shall depend upon you to keep the inspections going, but the clerical work of making out the reports can be saved. If there are any other similar reports which you think can be cut out please give me an expression of your opinion with reference to the same.[142]

By this time, the clerical work involved in routinely compiling some reports was seen as a burden not always compensated for by the value of the report itself, especially when it was purely descriptive. As Goss put it in another such case, "I am giving up certain reports that do not seem to me worth while to continue further in view of the labor required to compile them."[143]

In officially eliminating these reports Goss was doing more than reducing clerical costs; he was also reacting to information overload. In the note concerning the fire inspection reports, he delegated responsibility for monitoring the inspections to the foreman, who was to let Goss know when anything interesting developed. He used this same strategy in eliminating other inspection reports, as well.[144] By this postwar period, presumably Goss could afford to loosen control in some areas in order to protect his own time and to reduce clerical costs; however, he was by no means giving up the whole concept of routine reporting to gain information and control. In response to a misunderstanding in one such case, Goss's secretary explained: "Probably I was at fault in stating to Mr. Davis that *all* Bowen's periodical reports would be discontinued. I had in mind all inspection reports. I think the reports as to compressed air used by different departments, distribution of cost of gas, steam used by departments, and distribution of charges for City water should be continued, of course, as in the past, and know you will agree with me."[145] Goss wanted to harness the power of routine reporting without letting it reduce efficiency.

Judging by Davis's lists, the net effect of his work both in combining some reports and eliminating others was an initial reduction of reports to the general superintendent from over two hundred in 1919 to around one hundred fifty in 1920 and around fifty in 1921 and 1922.[146] Thus, the initial proliferation of routine reports was curbed in the name of efficiency. Those needed to pull critical information up the hierarchy were standardized and often combined to facilitate comparison, while others were dropped or handled at lower levels. Both of these developments helped to rationalize the routine reporting system that had grown up so rapidly and haphazardly.

The routine operating reports were not the only reports to grow up during this period. Another set of reports evolved to serve the needs of the emerging research function. Bishop asserts that Scovill's first laboratory research to define alloys took place around the turn of the century, and that by 1909 research into the technical processes involved in brass casting was intensive.[147] Sometime before 1913, the Research Department had been formally created. This establishment of research expertise outside the casting and rolling rooms themselves was, of course, a serious infringement on the long-established powers and privileges of the foremen, but it was very much in line with the firm's efforts to systematize. Creating a corporate memory of such research was an important part of the department's function.[148]

Documenting research work involved at least four different types of reports: the proposed research report, the preliminary report, the weekly progress report, and the summing up report.[149] Special reports on important experiments might also be issued.[150] These research reports had developed far from the letterlike reports of Scovill's earlier days. All of them had prescribed formats designed for easy reading. They began with standard pieces of identifying information, had one or more prose sections of variable length (which might include tables and graphs), and ended with a space for comments by the Research Committee overseeing the research efforts. The sections of the summing up report, the only one of the four prescribed report forms to have more than one prose section, also demonstrated the department's preoccupation with creating reports that could be read or skimmed rapidly. Unlike the chronological reports of earlier days, this form prescribed the more modern, conclusion-first order. A "Pamphlet of Instructions for Members of the Research Department" explained the format in some detail.[151]

Although the instructions were occasionally circumvented,[152] the norm was clearly laid out, and most of the reports adhered to it. Like other types of reports, research reports had been standardized in a format designed for efficiency.

Controlling Labor: New Forms of Downward Communication

During the war and after, the stream of circular letters, bulletins, and notices continued at a faster rate than ever. They were issued by every level of management and by both line and staff units to announce changes in personnel and to standardize procedures.[153] At the same time that this routine communication continued to systematize and depersonalize the firm, the wartime increase of employees spurred the adoption of two new genres of downward communication: the in-house magazine and the employee manual. These channels of mass employee communication served to convey, respectively, general norms and specific responsibilities for all employees; the former also helped counter depersonalizing effects of other forms of downward communication.

Scovill's in-house magazine started out as the organ of the Scovill

Foremen's Association. According to Bishop, that organization was formed in 1913 "to create 'closer personal relations among foremen[, to promote] their collective advancement in technical and supervisory leadership,' and 'to encourage a broader interchange of ideas among its members.' "[154] By providing facilities and supporting the association, Scovill's upper management encouraged these foremen to identify with each other, rather than with the workers they supervised. This club could personalize certain aspects of the workplace and reduce friction among foremen, thus indirectly increasing company control over them. While this organization apparently did not transcend its social agenda to become a form of managerial committee such as those discussed in Chapter 3, it did spawn another new form of communication. In 1915, *The Scovill Foremen's Association News,* renamed *The Scovill Manufacturing Company Bulletin* starting with the second volume, first appeared on the scene.[155]

This in-house magazine, which gradually broadened its audience to include workers as well as foremen, became an indirect channel for downward communication from (and occasionally upward communication to) upper and middle management at Scovill. For the first few years it was edited by C. H. Stilson, superintendent of the Cost Office. As the head of a small staff unit, he was a good link between the foremen and upper management. In the opening issue, Stilson explained its purpose as follows:

> The Foreman's Association has been formed with certain worthy objects, one of which is to encourage a broader interchange of ideas among its members. It is hoped that this little publication will become a factor in furthering this object. We shall try to put into it every month a little news, a little knowledge, something of inspiration and perhaps a laugh or two.[156]

A close look at the first volume of the *S.F.A. News* in 1915–16 shows the extent to which the newsletter lived up to that goal.[157]

Table 1.   Space in *S.F.A. News* Devoted to Various Purposes

Purpose	% of Space Devoted
Education	43%
S.F.A. news	22%
Inspiration	19%
Humor and misc.	15%
Specific work advice & orders	2%

As table 1 indicates, the largest amount of space was devoted to educating the foremen on such matters as technical processes (e.g., new metal casting techniques); managerial principles (e.g., safety and systematic management); and the work of various Scovill departments, especially the newer ones concerned with employee welfare (e.g., the hospital). The articles on technical processes and on various departments broadened the foremen's understanding of the company's business beyond their own departments, thus indirectly increasing their identification with the company as a whole and reducing the isolation of various shops. Articles on managerial principles had a more overt purpose of promoting desirable principles in the foremen.

Of these educational articles, safety was the most pervasive single topic, accounting for fully 17 percent of the first volume. Safety had become a key issue for manufacturing companies all over the country in the early twentieth century, for both monetary and humanitarian reasons. The welfare movement, as one of the educational articles explained, had made safety one of its causes:

> The Safety First Movement began in the United States about six years ago. The cause of the widespread campaign in this country may be due to the growing socialistic spirit, it may be due to the efforts of various groups of charity workers who have taken the cause of the dependent working man's family to the legislature, or it may be due to a wider humanitarianism in the people at large; at any rate in the effective prosecution of safety work there are two main things to be accomplished:
>
> First—The plant must be put in good physical condition throughout and must be maintained in equally good condition thereafter without intermission.
>
> Second—The human element must be educated, trained and disciplined.[158]

The articles on safety addressed issues involving both plant and labor, attempting to inculcate principles and general procedures. One article, for example, reported on the previous year's safety record and exhorted the foremen to try to improve it for the following year. In this matter management used the newsletter to promote the welfare of employees by promoting safety as a value; at the same time, the company communicated to readers that it was interested in their personal welfare, thus in part countering the depersonalizing influence of much other downward communication.

The other major focus of educational articles in this first volume was the systematic management philosophy itself. These articles addressed both the general philosophy and some of the mechanisms for realizing it. An article entitled "Systems" attempted to justify a key aspect of systematic management—increased recording and reporting—to foremen who saw it as a waste of their time or usurpation of their autonomy:

> There has always been opposition to system, and a feeling that much of it is red-tape. The opposition occurs because men dislike to do things in a certain pre-determined way—they want to do it their own way. The accusation that it is red-tape is made because,

usually, men do not know why system is important. From time to time, the connection between system and manufacturing will be touched upon in these pages. In the meantime, everyone concerned can feel certain that the Scovill Mfg. Co. seeks at all opportunities to avoid unnecessary clerical work.[159]

In this passage the editor, while acknowledging the foremen's opposition to systematization, communicated and defended management's values to them. An article entitled "Indispensability" defended another key principle of systematic management—the interchangeability of personnel—by arguing against the old-fashioned practice of making oneself indispensable in a job, in favor of the "newer practice" of "transform[ing] the room or office into a 'self-starter.' "[160]

After educational articles, items about activities of the Scovill Foremen's Association constituted the next largest component of the magazine (see fig. 6.3). These articles were clearly aimed at repersonalizing relations in the company by adding a social dimension to the workplace. But even these items sometimes carried a message from upper management. One article about an association banquet, for example, included direct exhortations to the foremen to participate in the organization's activities:

> It is to be sincerely regretted that a greater percentage of the club members were not in attendance, first, because a club is a club and requires the support of its members; second, because a member's absence always gives the wrong impression. The direct benefit of social intercourse between fellow-workers is not appreciated as keenly as might be; in reality, there is nothing more valuable than the intimate acquaintance and good-will of those about one.[161]

Such encouragement to be part of the foremen's association promoted the broadening identification of foremen with each other and the company, and not simply with their own fiefdoms.

The next largest components of the newsletter were inspirational items, then humor, recreation, human interest, and other miscellaneous items. The inspirational items were brief pronouncements on such subjects as teamwork and success. These, too, could be seen as inculcating values, but not in the sustained manner of the longer educational articles. The jokes and humorous stories made up a smaller proportion than might be expected, only about 4 percent. These items plus short items on recreation (e.g., a pool tournament) and on specific employees together made up 15 percent of the magazine. They injected a personal note into the magazine and provided the pure entertainment value necessary to encourage the readers "to look forward with interest to the appearance of these pages," as the editor said in his opening statement.

Finally, direct instructions or orders, which made up the bulk of downward communication in other media, were only 2 percent of the first volume of the *S.F.A. News.* Even these items were clearly not in-

THE
S. F. A. NEWS

ISSUED MONTHLY BY THE SCOVILL FOREMEN'S ASSOCIATION

VOLUME I.   WATERBURY, CONN , AUGUST, 1915   NUMBER 4

CROOKED WORK.   ALL TOGETHER!   ANCHORED.

THE OUTING.

They all say it was the best outing the Club has had. Promptly at 9.00 a. m. July 3rd a car full of stalwarts left the club house, being delivered f. o. b. cars the Shoreham in undamaged condition at 10.45. Immediately on reaching the hotel, Fred Stone, the photographer, herded all into a group and took our picture. He fretted a great deal and lost several pounds trying to place Dan Page so the picture wouldn't be out of balance. The laughing commenced then and did not let up until some time after dinner. After the agony was over (it is a good picture, remarkably few of us looking like undertakers), the "stunts" started.

Event number one was the 12-pound shot-put, won by Wallace of the Mills Department. Littlejohn excelled in this event, as he did in most of the others, but was unanimously disqualified as being, under the circumstances, rankly professional.

The second event, Hop, Step and Jump, proved to be one of unusual interest to the grand-stand for a number of reasons. This event was also won by Wallace, an indication that the Mills Department expects to keep in the front rank. Several members of other departments took part in this event, one of whom promised to come a close second to Wallace. In fact the only thing Dan Allman needed was a little practise. As it turned out, it was a beautiful exhibition of the Graeco-Roman Hop, Step and Jump, except that something went wrong with the timing of the leg-action between the step and the jump. There was a heated discussion as to whether it was between the hop and the step, or just after the step—at any rate he saved his hands by lighting on his face. Taken any way you like, the event stands out prominently from the morning's fun.

Event number three proved to be a dressing contest for the club dudes. Among those to face the starter were Messrs. Dunnigan, Ganley, Hickox, C. Snowman, Bennett and Adrian Wolff, Jr. Mr. Wolff crossed the line first, and, although a protest was made that his cravat was not adjusted in the latest Fifth Avenue style, the judges decided it good enough for an outing, and awarded him first prize.

Next came the Egg Carrying Race. The idea being to see who could navigate a specified distance (with an egg in a spoon as a companion) in the shortest space of time. The six contestants got away to a good start.

6.3.   Article in *S. F. A. News* aimed at humanizing the workplace. (S.F.A. News, *August 1915. Courtesy of the Baker Library, Harvard Business School.)*

tended to be a primary source of the instructions; rather, they reminded foremen of existing orders and exhorted them to action:

> Do you know what to do in case a sprinkler-head goes off on the ceiling of your room and begins to flood everything? There is a framed set of instructions hanging in your room giving the location of the shut-off and drawing-off valves. Have you seen these valves since the first of the year? Could you find them now? Have they been moved? Better drop this paper and read the instructions—then see if you can find the valves. DO IT NOW.[162]

Clearly, to the extent that this newsletter was intended to convey something from upper management to lower management (in addition to simply personalizing work life), it was intended to communicate general values, not specific orders. By this time the company had grown so large that the values of upper management could not be communicated by example—the upper levels were simply too far away from the foremen. This in-house magazine provided a medium through which general values could be communicated in a form made palatable by its entertainment value.

The *Scovill Bulletin*, as the magazine was renamed in its second year, soon broadened its audience to include all employees of the firm. By 1920 the masthead read, "Published Monthly in the Interest of the Employees by the Scovill Foremen's Assoc."[163] With this expansion of audience and with the enormous growth in workforce brought by the war, the contents of the *Bulletin* evolved, as well. Educational articles continued (see fig. 6.4), though they became less predominant. The humanizing appeal was increased and broadened with items about activities of groups other than the foreman's association, such as the newly formed Scovill Girl's Club.[164] The influx of immigrant labor brought stories on English classes and other forms of welfare work.[165] An increasing amount of space was also devoted to the war effort, with short pieces on Scovill employees serving overseas, victory gardens, war bonds, and so on.[166] It even took on a two-way communication function. In 1917, suggestion boxes were set up to solicit employee input on improving safety. Some suggestions were printed and answered in the *Bulletin*.[167] In general, the *Bulletin* tried to maintain a personal element that would encourage employee loyalty in a period of rapid growth and systematization. As the editor stated the *Bulletin*'s scope and purpose in 1920: "The Bulletin's field is as big as the factory; it is just that. . . . Whatever concerns Scovill's and Scovill folks is our news and we want it. . . . It is this sense of solidarity, of mutual ownership, of a big family feeling, of the Scovill spirit, that the Bulletin exists to cultivate."[168]

This attempt at personalizing the workplace was also a needed and, ultimately, relatively successful attempt to counter labor unrest. The labor agitation of the early twentieth century peaked in the postwar contraction. In 1920, part of Scovill's workforce joined other workers in

ONE END OF THE SCALE DEPARTMENT.

KEEPING UP THE STANDARD

One of the wonders of modern manufacturing (and it is something developed by Americans mainly) is what is called intensive production; instead of making a few hundreds of this, or several thousands of that, we make a hundred thousand or half a million. And then turn around and make a half million more—all without batting an eyelash. More than that, you cannot tell the millionth article made from the very first one; shake them all up in a hat and you cannot tell one from another.

How is this done?

By setting standards and by keeping up to these standards. All this means many gauges, well designed jigs, and carefully adjusted measuring and recording devices.

Not the least important among such devices are weighing scales; there are in this plant over 637 scales of capacity to weigh articles from one ten-thousandth of a pound to several tons. All these are in charge of the Scale Department whose duty it is to keep a record of the history of each scale and its whereabouts, also to see that they are kept in adjustment. Scales on which goods are weighed before shipment from the factory, are tested daily.

To properly test and readjust scales quite an equipment is necessary, as well as men trained in delicacy of measurement, and nicety of adjustment. This equipment as indicated in the pictures, includes snap gauges, range sticks, leveling devices, etc. The large picture shows a devise for sealing beams and levers by their correct multiplication that will take any lever from 2 up to 15 feet in length, sealing them with any weight from 1 to 2000 pounds.

The scales shown in the small picture were made in this factory and are sensitive to $\frac{1}{10}$ grain. Another scale built here, but not shown in any of the pictures, is larger and is sensitive to $\frac{1}{2}$ grain.

Keeping things up to standard is a requisite of success in manufacturing. The Scovill Standard has been placed high and must be kept there.

A SET OF STANDARDS.

6.4.   Article in *Scovill Bulletin* intended to educate readers. (Scovill Bulletin, *February 1917. Courtesy of the Baker Library, Harvard Business School.*)

the Naugatuck Valley in a strike. During the strike the *Bulletin* served as a communication link between the company and the workers. In one issue, for example, the *Bulletin* printed letters to President E. O. Goss from workers, along with his responses.[169] When the strike ended soon after, the *Bulletin* presented "The Company's Attitude toward Returning Employees."[170] Soon after the strike was over, however, the magazine returned to its more indirect humanizing function. Thus, in its first five years, the in-house magazine shifted from a primarily educational role to a primarily humanizing role to an active two-way communication role during a time of tension. In all these roles, however, it indirectly reinforced managerial control in the growing firm.

The in-house magazine was not the only new form of downward communication to emerge during the war years; it was soon joined by manuals for employees and foremen, a medium of downward communication explicitly devoted to increasing control. As Scovill grew during the war, the problems of organizing and communicating with the enormous workforce grew with it. External imperatives such as growing labor unrest in the country exerted additional pressures. In 1918, an employment office was established in the Industrial Service Department.[171] In the same year, that office issued the first employees' manual.[172] As stated in the introduction to this first edition, "These instructions are issued by the company for the guidance of its employees. . . . Any person who becomes an employee of the company should endeavor to follow the instructions of the booklet."

The foremen's manuals, which were introduced during the same period, contained much of the same material present in the employees' manuals, but they also included additional sections pertaining to supervisory duties. A memo concerning the 1922 revision of the foremen's manual explained its role and contents:

> The purpose of the Foremen's Manual is to gather into one place the various systems and regulations of the factory in order that those in positions of authority may be familiar with the various forms of procedure which relate to their duties. It will thus include a good deal of information not comprised in the Employees' Manual. On the other hand, everything that is in the Employees' Manual is of interest to Foremen and others in authority, not only because of their relation as employees but also because they are expected to explain and administer the regulations to their subordinates.
>
> It therefore appears that it would be more convenient to Foremen if the entire substance of the Employees' Manual is included in the Foremen's Manual, together with any additional material applying *only* to Foremen and others in authority.[173]

These foremen's manuals were in loose-leaf form, allowing upper management to update them during the year by adding or deleting pages. Moreover, the foremen's manuals for various departments could be made up separately so they would only include the foremen's sections relevant to that department.

Employees' and foremen's manuals carried the documentation of rules and responsibilities inherent in the systematic management movement one step further than the downward instructions and notices. In the past, new or changed rules, or rules that were being violated, were communicated and documented as needed in the fragmentary notices and bulletins. The manuals took on the more ambitious task of standardizing and documenting *all* rules and procedures. They tried to create a comprehensive organizational memory of rules, one that was independent of individuals. They transcended personal management in favor of personnel management.

The manuals completed the spectrum of downward communication at Scovill. By 1920, management had an array of modes of downward communication to use as managerial tools. Notices and manuals functioned directly to gain control over actions at lower levels. The *Scovill Bulletin* functioned indirectly through its education and its humanizing role to maintain control in the face of growth and depersonalization.

Conclusion

From its beginnings as a button factory to its postwar expansion and diversification, Scovill moved gradually from ad hoc and personal management methods to an impersonal management system. In the process, it developed a vast web of formal internal communication. By the immediate postwar period, Scovill had matured into a large firm in which the communication system was a major agent of control.

Chapter 7
Du Pont's First Century
Conservatism in Family and Firm

E. I. du Pont de Nemours and Company was founded in 1802, the year of Scovill's origin, on the Brandywine River outside Wilmington, Delaware. Although Du Pont[1] has long been a diversified chemical giant, it started out with a single product: black powder for guns and for blasting. The powder came in various grades, but the ingredients were basically the same. Unlike Scovill, which offered a variety of both semifinished and finished products within its first half century, Du Pont added only one, closely related product in its entire first century—military smokeless powder in 1891. Although internal written communication appeared early in Du Pont, its development of formal communication as a managerial tool was no more rapid than Scovill's in its first century. While Du Pont grew steadily (and in fact much more rapidly than Scovill), its management methods changed slowly. Only in Repauno Chemical Company, a spin-off from Du Pont, did systematic management make significant inroads in the first century.

Beginning in 1902, however, Du Pont was transformed almost overnight. A new generation of the du Pont family took over the business and within two years had consolidated much of the American explosives industry into a single operating company. At the same time, the management of the newly expanded firm was rapidly systematized. A complex and extensive formal internal communication system grew up practically overnight. Thus, the changes that came gradually in Scovill came suddenly, rapidly, and completely at Du Pont.

During its first century, Du Pont was profoundly conservative in managerial methods. Both the nature of the powder business and a family predilection for documenting events encouraged the development and survival of a certain amount of internal written communication earlier than in Scovill. Relatively early use of telegraph and telephone also facilitated communication among the partners and their agents. Yet this communication, like the management methods of the company, was ad hoc and unstructured. During the last two decades of the nineteenth century, Du Pont lagged behind other companies that were beginning to adopt systematic techniques of management and the new technologies of written communication.

In these same last two decades of the century, however, the Repauno Chemical Company, a dynamite firm established by a member of the du Pont family and partially owned by the Du Pont Company, was at the forefront of the managerial developments of the times.[2] The techniques of systematic management and communication that Repauno developed would serve as the nucleus around which many of the early twentieth-century developments at Du Pont would take place.

1802–1880: Conservative Management and Communication at Du Pont

In the second decade of the company's existence, French exile baroness Hyde de Neuville painted Eleuthère Irénée du Pont, its founder, up on the balcony of the first du Pont family house overlooking the Brandywine River (see fig. 7.1). He is speaking through a speaking trumpet to someone out of the scope of the picture, presumably directing manufacturing operations at the powder plants a hundred yards below the house. The scene, assuming that he is indeed overseeing workers at the mills, suggests several important characteristics of management and communication at Du Pont in its early days.

Most obviously, it illustrates the personal involvement of the du Pont family in the business. A close relationship existed between work life and family life on the Brandywine. The company was an extension of the large du Pont family, and family and business matters were closely interwoven.[3] Unlike at Scovill, where the founding family was displaced by managerial employees by the second half of the nineteenth century, the founder of Du Pont was succeeded by more family members, both as heads of the firm and as partners, for several generations. Although the company would ultimately expand far beyond the possibility of personal superintendence by the owners, in 1850 the firm turned down an offer to buy another powder mill on the grounds that "it would not be proper to own powder mills we could not personally superintend."[4] As the company began to set up subsidiaries in other locations, family members were supplemented by some nonfamily managers. Even these, however, frequently married into the family.[5] This family dominance contributed to the company's conservatism in its first century. Ironically, the radical transformation of the company at the beginning of the twentieth century was also a product of family dominance.

The painting also depicts the word-of-mouth style in which workers were managed. Although written communication was common among the du Pont partners themselves, as well as between the partners and their agents in other locations, management of the company's workers at the mills was primarily oral. The speaking trumpet in the picture simply extended the first owner's ability to give orders to foremen or

7.1. Baroness Hyde de Neuville's sketch of Eleuthère Irénée du Pont speaking through a trumpet, presumably to workers at the powder mill below. *(Courtesy of Hagley Museum and Library.)*

workers orally, allowing him to work from his residence as well as from the mills themselves.[6] This oral management tradition was to continue virtually unchanged until the turn-of-the-century upheavals in the company.

Finally, the picture hints at yet another theme: the need to establish communication between facilities located close to one another, but not at the same site. Because of the dangers inherent in the production of explosives, facilities were built in small clumps separated by buffer zones.[7] The three main sets of mills, built between 1802 and 1836, were spread over a three-mile stretch of the Brandywine. The distances were too great for the speaking trumpet pictured in the painting, but not great enough to require use of the mails. Patterns of written and oral communication at Du Pont reflected the need to bridge these distances. Moreover, communication technologies that supported that need were sometimes adopted more readily than other communication technologies.

The picture, then, suggests several of the key themes that characterized early management methods and communication patterns at Du Pont. It does not, however, suggest another central theme: a family tendency to record things for posterity.

Early Unsystematic Documentation

Although foremen and workers were managed orally, a significant tradition of internal written communication grew up early in the firm's history. Perhaps in part because of the complicated relationships between company and family and in part out of a sense of history brought with them from Europe, the du Ponts frequently recorded and preserved facts and opinions on issues, documenting their relations with each other and with those outside the family. Yet during the company's first century, this extensive documentation was unsystematic and unaccompanied by attempts to use the records as a managerial tool.

In the various family and company collections of documents, a surprising amount of written communication among the partners has survived, much of it apparently written not when one of the partners was on the road, as at Scovill, but when both parties were at their home base.[8] Some of these letters concern financial affairs of the du Pont family (and, necessarily, the firm); others were analyses of business issues or responses to inquiries or proposals. There were even isolated technical papers and reports, some of them more modern in format and organization than was characteristic of this period in general.[9] Tables of experimental data, for example, appeared as early as 1820.

Although the use of local written communication at Du Pont is not terribly surprising, given the small-scale dispersal of the powder mills, both the nature and the survival of these communications suggest a strong desire for internal documentation. With the several partners overseeing operations at three different sites on the Brandywine and transacting financial business in Wilmington, notes or letters carried by messengers or left to be picked up later would have been needed between face-to-face encounters. Many of the letters, however, were more extensive and formal than the distance alone would demand. For example, in one 1853 letter that appears to have been local, Henry du Pont apologized to his brother Alfred for the time it had taken him to reply to a letter dated about two weeks earlier.[10] The letter concerned some complex intrafamily financial questions about mortgages and interest. Since they probably would have seen each other during that period, the information may have been written to convey it precisely and to document the issue, rather than simply to communicate it. In addition, the technical papers and reports were clearly intended for the record, not for immediate communication. Moreover, the fact that these internal documents were folded and annotated for pigeonhole storage also suggests that the du Ponts had an exceptionally strong impulse toward record keeping and preservation.

In spite of this early and evident desire for documentation, the company did not show any similarly precocious desire to systematize such communication or to use it to improve management practices. The letters, reports, and other papers seemed to respond to specific immediate needs, or to a general feeling that documentation was important for historical reasons, rather than to serve as part of a routine flow of information and communication. Even the technical data, though sometimes displayed in tables, were not consistently or routinely collected, and technical reports were isolated rather than part of a system of regular reports. In all of this early material, then, the partners used written communication to document issues for posterity but did not attempt to establish a coherent system of upward or downward communication to use for purposes of control. The company simply preserved written documentation without putting it to any real use.

Du Pont's communication with agents, like Scovill's, was also irregular and unsystematic. Correspondence with William Kemble, Du Pont's principal New York commission agent up to the 1850s, was typical.[11] Kemble, like most commission agents at that time, served several companies. Nevertheless, in addition to his sales activities he was expected to perform a range of extraneous services for Du Pont, including arranging many financial matters, procuring supplies, and researching technical issues. He was even asked to see to the outfitting of two company workmen going to Yucatán to work on a company project.[12] Internal references indicate that letters generally took only one day between New York and Wilmington, although the schedule of the two daily mail deliveries affected that timing.[13] In the early and mid-1840s, before the reductions in postal rates, the company and Kemble each sent letters roughly every other day, about the rate Scovill maintained with its store in the 1850s after both reductions.[14] After the first drop in rates, Kemble's exchange of letters with Du Pont increased significantly, to one letter every one and one half days.[15] In spite of the greater frequency of this correspondence, however, it was as haphazard as that between Scovill and its agents. Except for rendering quarterly accounts of his receipts and sales, Kemble had no regular reporting requirements. He wrote only in response to specific needs.

Richard J. Ruth's detailed study of the correspondence of F. L. Kneeland, Du Pont's principal agent in New York from about 1852 to 1884, shows increasing quantity but still no system.[16] By the later years of Kneeland's association with the company, when he worked as a sole and eventually salaried agent, Kneeland wrote to the head of Du Pont as often as two or three times a day to transmit orders, answer queries, and relay critical market information. But, according to Ruth, while "he was frequently asked to investigate and report specific matters of interest to the firm," he was "not required to submit regular formal reports" beyond his quarterly accounts.[17]

Thus, communication with agents followed the pattern for communication among partners: extensive but unsystematic. The partners' penchant for documentation showed up in the fact that they occasionally saved even drafts of letters to agents.[18] One long anecdotal letter to an agent in Philadelphia was labeled "Copy of letter to John J. Twells— original not sent as the substance is to be given in conversation."[19] In effect, this letter documented an oral conversation.

Yet despite these early manifestations of one principle of systematic management, the du Ponts' motives for documenting internal relations and transactions seem to have been more conservative than progressive; that is, they wrote and preserved documents for their own sake, rather than as tools for improving the firm's efficiency or effectiveness.

Technological Conservatism at Midcentury

The Du Pont Company's conservatism also showed up in its adoption and use of various communication technologies available in the mid-nineteenth century. General Henry du Pont, head of the company from 1850 to 1889, delayed in adopting innovations to speed up writing or copying letters. He insisted, for example, on using quill pens, rather than the more practical steel ones that gained popularity in the midcentury period, until virtually the end of his tenure.[20] And while Kneeland, the New York agent, began using press copying in 1854 (and Scovill was using press copying by at least that year), the Du Pont Company continued to use hand copy books through at least 1857.[21]

Moreover, in spite of the importance the family accorded to documentation, the firm did not store its internal or external correspondence consistently. An office had been built in 1837 to replace the president's home as the location from which the company's affairs were managed.[22] Although the firm's copy books for outgoing and pigeonholes for incoming and internal correspondence were centrally maintained there, the scattering of mills and houses along the Brandywine dictated that correspondence was sometimes written or stored elsewhere. Rather than delaying the dispatch of a letter, a partner sometimes sent it and later added a brief "memorandum" to the copy book summarizing its contents.[23] Thus, while in some cases even drafts were saved, in others no complete copies were retained.

A similar inconsistency appeared in the storage of incoming and internal letters and documents. General Henry's desk had sufficient pigeonholes to hold folded and annotated papers for the previous year, and a large pigeonhole cabinet held those from several more years.[24] Correspondence that was older still was retired to the attic in boxes. Yet the appearance of company correspondence in the personal collections of various members of the du Pont family suggests that it was not consistently kept in the central office files.[25] In spite of these problems with centralized storage, the conservative company did not adopt more suitable methods until the end of the century.

With the telegraph, another new communication technology of the midcentury period, the Du Pont Company managed to be progressive and conservative at the same time. The company first began using the telegraph in 1847, three years after its introduction in the United States.[26] From that year through 1855, Du Pont used the telegraph only sparingly, generally to communicate with agents in more distant cities over urgent matters. For example, a telegram to its Boston agent was used to agree to a saltpeter deal the agent had negotiated.[27] Ironically, these telegrams to distant locations were sometimes written out and mailed to the Philadelphia agent, requesting that he send the message on by "magnetic telegraph."[28] Such incidents probably reveal either problems stemming from competing telegraph companies or an unwillingness to make an extra trip to the Wilmington telegraph office several miles away. Up to 1855, the company sent only about a dozen telegrams in a year, though references in letters indicate that it received more. As with other changes, it dragged its heels in adopting this technology.

By 1855, competitive pressures to use this fast new communication medium had apparently increased enough to modify the company's inherent conservatism. To make use of the telegraph, however, the company needed better access to the telegraph office in Wilmington. Thus, it was driven to install a private line connecting the Brandywine offices to the Wilmington telegraph office.[29] The company announced the connection to some regular correspondents, noting that when they telegraphed the company they should "use the Morse line *exclusively*" (a necessary explanation in those days of multiple telegraph companies). This early installation of a private line, well before the 1870s, when the printing telegraph helped popularize such lines, was quite progressive.[30]

The private telegraph line allowed Du Pont to communicate instantaneously with the Wilmington telegraph office and, through it, with the rest of the telegraphic world. The connection had an immediate impact on Du Pont's long-distance communication. The wires were installed at the end of December in 1855. In the three-day period beginning on January 2, 1856, the company sent telegrams to agents in New York, Cincinnati, and Philadelphia confirming orders and arranging rapid transportation for powder.[31] In each case the telegram was confirmed by letter. In addition, a letter to another Philadelphia agent requested that he keep Du Pont posted by telegraph on the sailing of a vessel.[32]

Du Pont installed the private line to take advantage of the telegraph's competitive value in serving distant customers. Locally, it took advantage of the private line in external communication but ignored possible internal uses. According to John W. Macklem, a long-time company employee whose recollections are the source of much information about the company in the latter part of the nineteenth century, the hardware stores in Wilmington that marketed the powder locally could send their orders from the Wilmington telegraph office directly to the Du Pont Company office at no cost.[33] Yet, the company did not extend the line to connect its several facilities to allow internal use of the telegraph. With the mills that performed various parts of the powder-making process spread along the Brandywine, the company clearly needed rapid methods of communicating among them. Nevertheless, as late as the 1880s, messenger boys still carried written and oral messages back and forth.[34] In this matter, Du Pont once again displayed its inherent conservatism. Throughout the period, it failed to use the technologies of communication in ways that could have facilitated internal communication and promoted efficiency.

1880–1902: Conservatism at the Core, Innovation at the Periphery

In the last two decades of the nineteenth century, Du Pont grew significantly, both by expansion and by acquisition. Some expansion beyond the main works had occurred previously. In 1859 the company had established a powder mill in Wapwallopen, Pennsylvania, run through the subsidiary Du Pont Company of Pennsylvania, and in 1876 it had purchased the Sycamore Mills in Nashville, Tennessee, and the Hazard Powder Company in Hazardville, Connecticut.[35] At the end of the 1870s, in an effort to control competition, Du Pont had acquired major shares in several other powder companies, including Lake Superior Powder Company in 1876 and Oriental Powder Mills in 1880.[36] In the 1880s and 1890s the company was to build two more major facilities: a huge black powder plant in Mooar, Iowa, in 1888; and an experimental smokeless powder plant across the Brandywine at Carney's Point, New Jersey, in 1891. This latter plant, built and operated by partner Francis G. du Pont, served as a primitive research and development center as well as producing the new smokeless powder that he had developed.

This growth came during the same period that saw the introduction of new communication technologies and a change in top management. The telephone and typewriter made their debuts on the Brandywine in the 1880s, and frequent use of the telegraph led to the development of a specialized telegraph code. In 1889, the profoundly conservative General Henry du Pont died and was succeeded as head of the firm by Eugene du Pont. These changes, however, had remarkably little effect on the management of the company. This was the period during which the first signs of system in management and communication began to appear at Scovill and other companies, but changes in business methods and in the communication system at Du Pont came unevenly and often slowly.

The Introduction of New Communication Technologies

A series of changes in communication technology occurred during the 1880s and 1890s at Du Pont, some eagerly embraced and others resisted as long as possible. In general, innovations tended to be adopted and championed by the younger generation and by peripheral parts of the company, such as the Carney's Point facility and the Iowa Powder Mills, rather than the main office and plant. When they were adopted by the main office, they were typically used in ways that reinforced ad hoc methods of management.

At the younger partners' instigation, Du Pont acquired an internal telephone exchange quite early. In 1877, only one year after the telephone's invention, the young and progressive Lammot du Pont had the new device demonstrated for him over Du Pont's private telegraph wire.[37] Impressed with the demonstration, he had a telephone installed in his home the following year, connecting him to the Delaware Bell Telephone Exchange.[38] When Lammot left the Brandywine to found Repauno Chemical Company in 1880, young inventor and chemist Francis G. du Pont championed the technology. That year, at a time when there were fewer than five hundred public exchanges and only a handful of private ones, a very small telephone exchange was established at Du Pont's Brandywine works.[39] The central station of the exchange was at the Hagley Mills, the middle of the three mill complexes on the Brandywine, with telephones at only four other locations: the Lower Yard, the Refinery (or Upper Yard), the office, and Francis G. du Pont's house (see fig. 7.2). In 1883, the Du Pont exchange was connected to Wilmington's public exchange.[40] This early telephone exchange was replaced at least twice and perhaps three times before the end of the century.[41]

The adoption of the telephone illustrates a generational pattern that was to recur in Du Pont. The early experimentation, initial installation, and later updates of the telephone system were promoted and handled first by Lammot du Pont and then by Francis G. du Pont, the younger, more progressive members of the family. General Henry, a member of the previous generation, refused to use the telephone at all.[42] Eugene du Pont, Francis G.'s older brother who ran the company from 1889 (when General Henry died) to 1902, used the telephone but did not champion it.

Although Du Pont's telephone system was installed early, thanks to the efforts of the younger generation, its effect on the communication system was limited to supporting ad hoc oral communication. The scattering of Du Pont facilities along the Brandywine made this function a valuable one. Nevertheless, the telephone exchange was not initially used to its full capabilities. In the early years it often broke down, according to Macklem, forcing the company "to resort to the old way of transmitting our messages afoot."[43] Even when the telephone system was working, much communication that might appropriately have occurred over it did not because of the older generation's resistance to the

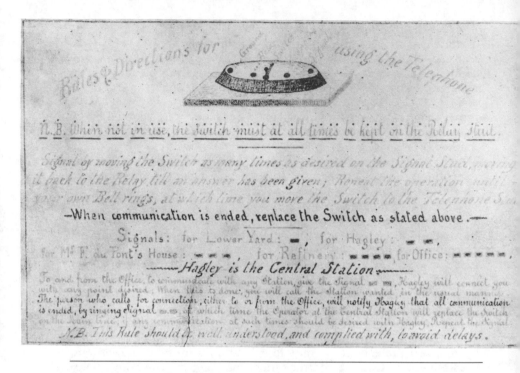

7.2.   Instructions for using the small telephone exchange established in Du Pont's office around 1880. (*Du Pont Magazine, January 1928, courtesy of Hagley Museum and Library.*)

new technology. Messages to and from General Henry, for example, continued to travel by messenger.[44] In addition, as long as General Henry headed the company, various of the partners gathered with him in the office meeting room "every afternoon for their after dinner talk and smoke."[45] These informal gatherings, at which General Henry questioned the other partners about operations in various parts of the mills, obviated the need for many telephone calls. When Eugene du Pont took over after General Henry's death, the telephone probably played a more important role in the firm's informal communication.

Potentially more important than the telephone to the evolution of formal communication as a managerial tool were the technologies of written communication. These innovations, however, were introduced slowly and unevenly across the company, frequently reaching the main office last, if at all.

Even though Du Pont had been early to adopt a private telephone exchange, it was not on the cutting edge with the typewriter. Although typewriters had become relatively common in business by the 1880s, the main office waited until 1888 to purchase one, doing so only after agents and younger partners had proved the invention's worth. Chicago agent E. S. Rice started using a typewriter in his correspondence with the company in 1886.[46] A few days after he began using the machine, he informed the company that he would be hiring a typist: "Owing to an increase of business and our recent troubles I have found it necessary to employ a stenographer and type-writer [i.e. typist]. . . . This, of course, shortens my work considerably and gives me more time for outside work. Answers to all letters received by the office are dictated by me, taken in shorthand and transcribed on typewriter."[47] From this point on, his letters were almost always typed, and he was later to comment to Francis G. du Pont that "when used to using this little machine you will find it as much a labor-saving machine as anything you have ever adopted in the manufacture of Gunpowder."[48] Rice, then, immediately saw the time-saving potential of the typewriter, especially when it was operated by a skilled typist.

General Henry, however, was not initially interested in trading off tradition for efficiency. In January 1888 Rice began a letter to Francis G. du Pont by commenting, "I have your valued favor of the 13th and am pleased to note first of all that the typewriter has found its way on-to the Brandywine in spite of the quill pen, etc."[49] In fact, further correspondence reveals that the typewriter had only found its way to Francis G.'s home, where he kept his own fully equipped office, rather than to the company office.[50] Once again the younger generation was more open to new ways of doing things. Initially, Francis G. did his own typing.[51] He found the typewriter a wonderful, time-saving invention, commenting that he now hated writing with a pen.[52] In 1890, he bought a Graphophone so he could dictate some of his letters for later transcription, though he apparently continued to type on the machine himself much of the time.[53]

With both an agent and one of the younger partners leading the way, the typewriter finally made its way into the main Du Pont office in March 1888, three months after Francis G. du Pont acquired his own first typewriter.[54] Even then, the innovation was almost forced on General Henry by his office staff. According to Macklem (then a junior clerk), he and the head correspondence clerk, Mr. Haley, rented a typewriter for a month, practicing on it at home. "Before the month was up, Mr. Haley prepared some of the company letters on the new device and placed them on Mr. [Henry] du Pont's desk for his signature. He instantly became interested and issued orders that we purchase one for office use."[55] Ironically, while General Henry was the last one to adopt the machine, he immediately saw its value as a mechanism for shifting the labor of producing letters to his clerks. Unlike Rice and Francis G. du Pont, he was not interested in using it himself. "It is needless to say," Macklem continues, "that thereafter Mr. du Pont ceased to write his letters in longhand, but scribbled them off in the rough for the boys to type."[56]

Changes in duplicating and filing technology came to the Brandywine in an even more piecemeal fashion, reaching peripheral outposts but apparently not the main office. By the 1880s, according to Macklem's accounts, the company used a fairly traditional system of press books and pigeonholes.[57] The letter press and the press books for more recent years were kept in the office of the correspondence clerk, who handled some letter writing and all press copying. Older press books were retired to the attic. Incoming correspondence was kept in pigeonholes in General Henry's office, with current correspondence in the pigeonholes on his desk and correspondence for the previous few years in a pigeonhole closet. Another pigeonhole closet in the bookkeeper's office contained agents' quarterly accounts and perhaps letters.

Two years after General Henry died, in 1889, and Eugene du Pont took over as head of the firm, the company offices were moved to a new, larger building on the Brandywine. With this move came some minor changes in office staff and equipment, but no major changes. The office force grew from five to eight, including the company's first stenographer.[58] By at least 1896, the company had acquired some sort of stencil duplicating apparatus used with handwriting.[59] In spite of these changes, available evidence suggests that press books and pigeonholes were still the major storage devices used for handling mail. No outgoing correspondence for this period has survived, though incoming and a little internal correspondence has, indicating the existence of two separate systems, with outgoing correspondence almost surely still in press books. The evidence for continuing use of pigeonholes for incoming and internal documents is fairly conclusive. The surviving correspondence itself bears the fold marks and annotations of papers stored in pigeonholes.[60] Moreover, a photograph of the second office, shown in figure 7.3, reveals a large pigeonhole cabinet.[61] Finally, an 1897 internal note refers to "pigeonholing" a report, suggesting that the firm used pigeonholes for storing internal as well as incoming documents.[62] Although this pigeonholed correspondence was sorted (either originally or later) by sender, it was still isolated from related outgoing correspondence.[63]

In contrast to these traditional methods of copying and storing correspondence in the main company office, some members of the firm were beginning to adopt new methods in their own personal and company paperwork.[64] In 1888, Francis G. du Pont ceased to fold and abstract his incoming correspondence for storage in pigeonholes, and instead began using Shannon files, the horizontal files that held papers in place with an arched wire.[65] He clearly found the Shannon files preferable to pigeonholes for maintaining accessible storage, for in 1889 he recommended them to one of his correspondents for keeping letters "ready for frequent reference."[66] By 1890 he was considering upgrading from box to cabinet Shannon files.[67] At this same time, he also purchased a rolling press copier that produced loose, rather than bound, copies.[68] A few days later, he began combining loose press copies of his letters with incoming

7.3.   Pigeonhole cabinet in the second Du Pont Company office. *(Photograph S20-11, courtesy of Hagley Museum and Library.)*

responses from correspondents, both in the Shannon files.[69] The disappearance of Shannon file holes in his correspondence from 1895 on suggests that at this point he adopted vertical filing, which had been introduced in 1893. Thus, Francis G. was keeping up with modern technologies in duplicating and filing, even if the company as a whole was not.

Similarly, agents and other members of the younger generation of du Ponts adopted new technologies for producing and reproducing documents. F. G. Thomas, manager of Du Pont's Iowa Powder Mills, had a typewriter by 1888 (although he used it only infrequently until he had a clerk to type) and Shannon files by 1889.[70] In 1890, his most frequent company correspondent, Francis G. du Pont, suggested that he get a Graphophone to allow him to "speak your letters into it, and get Mr. Simons to transcribe them" on the typewriter.[71] In 1892, E. S. Rice in Chicago began using carbon paper, either in addition to or in place of press copying.[72] By 1893 Pierre S. du Pont, then a young man working for Francis G. du Pont, had also begun to use carbon copies instead of press copies and to combine related incoming and outgoing correspondence, probably in vertical files.[73]

By the late nineteenth century, then, several innovations in the technology of written communication had reached the periphery of the growing company, but only the typewriter had reached the main offices. In fact, of all the new technologies of the 1880s and 1890s, only the telephone was adopted early in the main office, and it merely extended informal oral communication. The main office lagged behind peripheral locations and the rest of the business world in adopting the technologies most likely to facilitate formal and systematic internal communication.

Communication on the Brandywine: Old Patterns

Communication patterns in the facilities along the Brandywine changed very little during this period. Written communication within the main Du Pont powder mills still seems to have been used primarily for the sake of immediate convenience or unsystematic documentation rather than for use as a managerial tool.[74] The internal telephone system obviated the need for some of the notes used earlier to bridge the distance separating the scattered mills, though brief notes were still used to convey precise details such as mixes of ingredients in powder.[75]

Upward reporting had developed slightly in quantity and form, but not at all in essential nature. Lower Yard lists of materials on hand, for example, had been made out annually since 1853.[76] By the 1890s, the report was issued weekly as well as annually.[77] Moreover, it was now submitted on a form, which was hand drawn but stencil duplicated. The forms made compiling the report more efficient; nevertheless, the simple descriptive list of materials had apparently not changed much in nature since the 1850s. During the 1880s and 1890s, two other descriptive reports were filled in on duplicated forms each week: packing lists and

---

NOTICE!

Owing to the great danger of loss of life and property, which might be caused by PARLOR and BLUE HEAD MATCHES. The use of them on this place is STRICTLY PROHIBITED.

☞ SAFETY MATCHES MUST ONLY be used.

The loss to the Company would be only in Dollars.

The loss to the People would be the LIVES of husbands, family, and friends!

The Company to protect the lives of its men and the property, will DISCHARGE on the spot, any man found in the Yard with a Parlor Match on his person.

Do the families think it worth while to help protect the lives of their friends, — by stopping the use of these Matches on the place, where ever they can.

7.4.   Posted notice on matches. *(Acc. 504, courtesy of Hagley Museum and Library.)*

lists of patterns to be made up.[78] The amount of descriptive information collected may have increased somewhat by this time, but its use had not changed fundamentally. The weekly lists did not collect any additional information for analyzing or interpreting the information, nor is there evidence of any systematic analysis.

Even though the duplicated forms demonstrate that Du Pont owned some sort of stencil copying device by the 1890s, the firm apparently did not use it for duplicating mass circulation directives such as circular letters. The only written employee communication to survive from this period is a printed and posted notice prohibiting the possession of any matches other than safety matches in the powder yard (see fig. 7.4).[79] Because of the dangers of powder production—certainly comparable to dangers of railroads—they adopted this minimal form of downward mass communication earlier than Scovill did. They did not, however, set up more routine channels of downward communication.

Ironically, a mechanism typically used to repersonalize the workplace was instituted before it had been significantly depersonalized. Francis G. du Pont, as usual slightly more progressive than the older partners, started a club for workers in the Wilmington area plants in 1892, after a period of some labor unrest.[80] Although his action could have been used to initiate more regular labor relations, there is no evidence that it played such a role. Workers were apparently still individually managed by the foremen and individual partners, with no company-mandated standard procedures beyond the rule about matches.

Minor Progress at Peripheral Locations

Although the managerial methods and communication patterns of the Du Pont main office did not change significantly in the 1880s and 1890s, there were preliminary signs of change in some peripheral sites run by partner Francis G. du Pont. In 1891, he established the Carney's Point factory just across the Brandywine in New Jersey to develop his new invention, smokeless powder.[81] As a scientist and an inventor, Francis G. saw himself as progressive and open to new, more scientific methods of doing things. In the 1890s, he adopted some of the trappings but little of the reality of systematic management and communication at Carney's Point.

Francis G. was interested in new communication equipment and supplies. In addition to his early involvement with Du Pont's telephone exchange and his adoption of typewriter, flat files, and rolling copier, he experimented with other office supplies. He had several types of special stationery printed up primarily for internal purposes. By at least 1888 he had some stationery with a memo-like heading, "From Francis G. du Pont"/"To ———."[82] He also ordered special imprinted pads of paper "for recording experiments and for making reports that are to be sent to this side of the river."[83] Finally, in 1889 he obtained dating stamps for stamping incoming notes with the date they were received and the date they were answered.[84]

Although Francis G. clearly was aware of new trends in management and readily adopted the physical accouterments of system, his actual management practices at Carney's Point, as reflected in part in his internal communication, did not demonstrate significant systematization. The experimental records from his laboratory at Carney's Point, for example, show little consistency in form or content, ranging from miscellaneous figures and calculations to tables and even very rough attempts at graphs, all mixed together with no discernible pattern or progression in form.[85] There is no evidence that he wrote any systematic reports of results to headquarters, either. His management practices at Carney's Point were apparently even more ad hoc than his scientific documentation. His young cousin Pierre S. du Pont, who was to play a leading role in the post-1902 reforms, left the company in great part out of frustration at the disorganization and lack of planning in Francis G.'s running of the plant.[86]

A similar pattern emerged in Francis G.'s management from afar of the construction of the Iowa Powder Mills. In 1888 he began to direct construction of the world's largest black powder mills in a location later to be designated Mooar, Iowa.[87] The extensive and detailed correspondence he carried on with F. G. Thomas, the distant cousin who oversaw the construction and early operation of the mills on site, reveals that Francis G. generally managed the construction of the plant in the old-fashioned way, trying to be involved with everything.[88] He did, however, introduce modern communication technology and even some minimal system into the new enterprise, in part to retain some control over the plant when it began operations.

Francis G. du Pont felt a personal involvement in the Iowa Powder Mills from the very start as he designed and oversaw the facility from a distance. He demanded frequent letters from Thomas on all aspects of its construction, complaining, "At the distance that you are from me, it will be necessary for me to have frequent communications from you, and I have a special interest in the new plant, and I need to know all that is transpiring, and yet I do not."[89] While Thomas tried to write daily letters, the press of things to be done around the plant continually interfered.[90] Even so, during the year of construction Francis G. wrote more than every other day, while Thomas wrote only slightly less frequently.[91]

These letters were supplemented by frequent telegrams to give or request instructions on specific problems. The interlocking sequence of telegrams and letters was frequently quite confusing, since telegrams were received the day they were sent and letters took as long as four days.[92] Thus, a telegram countermanding or clarifying a letter might arrive before the letter itself. At certain critical points in the construction, almost as many telegrams as letters passed between Wilmington and Iowa.

Although the correspondence was extensive during the years of construction and start-up, it was generally not systematic in content or organization. During the construction period, Thomas sent Francis G. monthly cash accounts and requisitions, but no other regular reports.[93] The letters were long and packed with information on various subjects, rather than short and limited to a single subject. Francis G. praised Thomas when he wrote long, chronological accounts of events, informing him that "you need not fear writing too long letters, because your letters are all I have to keep me advised of what is going on."[94] Efficiency in communication was not yet valued at Du Pont. The two correspondents introduced only one change in format to facilitate this frequent internal correspondence: they numbered each letter to clarify sequence and to alert each other to any letters that might go astray.

Francis G. and Thomas were more concerned with methods for creating, duplicating, storing, and transmitting the letters than with their form. Thomas had a typewriter but never typed well enough to use it routinely.[95] In an attempt to enable him to write more frequently during busy times, Francis G. suggested that he buy a Graphophone for dictation.[96] Francis G. even made the innovative—though ultimately impractical—suggestion that they use the same type of dictation machine, so they could send cylinders back and forth and hear each other's voices.[97]

Storing the correspondence accessibly was important to Thomas because of the multitude of construction details transmitted in Francis G.'s letters. Francis G. recommended that Thomas used a Shannon file to "keep my letters in such a shape as to be ready for frequent reference."[98] As the return letter made clear, Thomas had already adopted such a file for incoming correspondence, though his outgoing correspondence was still copied and stored in a press book. He did not reap the advantages of merged files as Francis G. did by using a rolling press and interfiling the copies on his Shannon files.[99]

Finally, rapid transmission of messages was important in their correspondence. Early in the project Francis G. expressed dissatisfaction with the slow speed of the mails.[100] Soon they discovered that special delivery stamps could reduce the normal four days to as little as two.[101] Then, at Francis G.'s suggestion, Thomas further improved mail delivery by getting a U.S. Post Office established at the train station near the mills and convincing the Post Office to use freight trains as well as passenger trains so that the plant could get two mail deliveries a day.[102] The two correspondents also agreed that Thomas needed the telephone to facilitate rapid communication within the Iowa site and with the nearby railroad station and other local firms.[103]

As the plant shifted from construction to operation in May 1890, Thomas began corresponding with the main Du Pont office as well. Francis G. announced that while he wanted to be kept informed about everything concerning the mill's operations and sales, he was turning over money matters to the company.[104] In addition, once the plant began production, several regular reports were instituted, though they were still basically descriptive rather than analytic.[105] These new reports included various weekly lists, as well as a monthly statement on a form that seems to have been developed for Wapwallopen, Du Pont's Pennsylvania mill.[106] These monthly report forms caused Thomas some problems. As he put it, "We do not seem able to make head or tail of them according to the way we have the charges made on our books[.] [T]hey may fit Wap[wallopen] all right but we have to do a lot of fitting to work it in with our books."[107] Thus, even these limited attempts at instituting consistent reporting were achieved only with difficulty.

Based on Francis G.'s disorganized management of the Carney's Point and Iowa facilities, it seems clear that his interest in systematizing was not matched by his managerial ability. He was familiar with some of the changes going on in the business world, and he readily adopted technologies and devices for systematizing. But he either did not fully understand the philosophy behind it or found it incompatible with his own abilities and preferences. In either case, Francis G. introduced only the most limited elements of system into Du Pont's operations.

Managing Sales Agents in the Field: Superficial Change

The number of agents was reduced after General Henry's death in 1889, and sales were consolidated into about ten "branch offices" or agencies generally run by salaried agents.[108] In practice, however, these agents still had a great deal of autonomy. No significant system was introduced when the principal sales agents became salaried employees. Each major agent contracted with independent commissioned subagents in as many towns as possible. The subagents agreed in their contracts to sell at the prices established by Du Pont, to be communicated by the company through the agent.[109] Although this additional level of hierarchy created an opportunity for mass downward communication on two levels, none developed. The agents and subagents were kept advised of the current prices through individual letters and telegrams.[110] In fact, other than the printed contracts that established the general terms of the agent relationship, the only rules widely disseminated to agents came not from Du Pont but from the Gunpowder Trade Association, the association established by the major powder manufacturers to control competition.[111] Except for rendering their quarterly accounts, as required by contract, most agents and subagents were not required to submit regular reports.[112]

The essentially unchanged nature of the correspondence with agents suggests that the consolidation of sales offices had very little effect on management methods. In a highly competitive environment, and in spite of the Gunpowder Trade Association's efforts, subagents were in perpetual warfare with agents and agents with the company over prices and competition. This warfare dominated the correspondence, as it did the correspondence between Scovill and its stores during this period. Price discounting and other competitive tactics among the various powder companies created, as one salaried agent commented, the "difficulties of men like myself, who are expected to control the actions of your Agents."[113] Even in straightforward matters where the company's instructions, as conveyed by the agent to the subagent, were quite clear, individual subagents did not feel bound to obey.[114] Although the company was constantly involved in acrimonious correspondence over breaches of policy, it did not institute any systematic procedures to prevent them. Neither mass downward communication nor a system of upward reports was established to provide better control.

The company's correspondence with Elliot S. Rice, the Chicago agent, illustrates the nature of relations between the company and its principal agents during the 1880s and 1890s.[115] Although Rice was on salary, he still viewed himself as an agent more than as an employee. When an explosion at the Chicago wharf forced Du Pont and several other powder companies to find new locations for their powder depots, he complained, "Other companies have representatives here who are taking the responsibility; and while I am disposed to obey all orders and do the best I can, I should feel much better if relieved from the responsibility attending the purchase of new site and expenditure of a large sum of money which must follow."[116] In spite of this reluctance to see himself as a representative of Du Pont, he certainly felt that he was an important (if not the most important) agent.[117] He carried on an extensive but unsystematic correspondence with both the company itself (i.e., General Henry and later Eugene du Pont) on sales matters and Francis G. du Pont on certain special projects.

In his letters to the company, Rice generally limited himself to one issue per letter but wrote frequently. He averaged almost two letters a day to company headquarters, covering a broad range of subjects that included prices, infractions of Gunpowder Trade Association agreements by competition (and occasionally his own subagents), orders and urgent requests for their shipment, payments needed from customers or the company, and proposals for new subagencies.[118] As the head of a major agency, Rice submitted monthly rather than quarterly reports of sales in his district. Though his reports came more frequently, they still provided only the most basic listing of transactions. The company had no system for collecting and analyzing data to assess the relative efficiency of agents.[119] His letters were generally formal and respectful, though he took offense at criticism rather easily.[120]

His less official letters to Francis G. du Pont, in contrast to those to the company, were long, rambling narratives, each of which covered a wide variety of subjects.[121] In one exceptionally long letter about the establishment of the Iowa Powder Mills, he excused himself as follows: "My anxiety to give you as much information as possible, must in a measure explain and excuse these long letters, as a matter of fact you have probably decided that my disposition to long letter writing is at fault as well."[122]

In addition to his many letters, Rice also sent frequent telegrams to the company and to Francis G. du Pont. In 1888, Rice developed a company telegraphic code that replaced frequently used names and phrases with random words such as "wax" (for a person's name) and "weasel" (for "result of examination satisfactory").[123] The code was intended to secure accuracy, economy, and privacy. In the first place, Rice pointed out, "When [the code is] completed, our correspondence had better be carried on with the aid of this, which will very materially lessen the danger and annoyance from mistakes by operators," presumably by making operators pay closer attention to the words.[124] While few of the actual telegrams have survived, correspondence between F. G. Thomas at the Iowa Powder Mills and Francis G. du Pont during the period right after the code was introduced suggests that the code certainly did not eliminate errors, and may have increased them. One telegram, for example, came through with "thill" instead of "chill" and "picker" instead of "bicker," resulting in a major misunderstanding and the loss of a day's work.[125] Eventually Francis G. decided that "it is a good plan to always verify the telegrams by letter."[126] In its first goal, then, the code was not particularly successful.

Its second, and perhaps most important, goal was to reduce telegraphing costs. Before even sending the new code book to the company, Rice announced to Francis G. that he had tested the code on a real telegram, and "it works like a charm, cutting down number of words nearly one half and simplifying very materially."[127] After the code had been put into use by the company, he commented, "I am pleased to note that the code is working to your satisfaction, and it seems to me that it is saving a great expense."[128] The amount saved depended, of course, on how often it was used in telegrams. Because the actual telegrams from this period have not survived, the extent of its use is not clear. The correspondence between Francis G. du Pont and F. G. Thomas in Iowa suggests that the code was not well suited to their communication about constructing the Iowa Powder Mills.[129] However, the code may have been used more extensively by Rice and other agents, for whose purposes it was designed.[130] If so, it could have generated considerable savings.

Finally, the code was intended to guarantee privacy and security of messages on confidential matters. Rice went so far as to create two extra pages of code especially relating to the Iowa mills project that was just being negotiated at the time he created the code.[131] Only two copies of these pages were created, one for Francis G. du Pont and one for Rice himself. In a postscript to the letter transmitting the code, Rice added, "Kindly acknowledge receipt of code by wire so that I may know when to 'pull the curtain.' " The curtain of secrecy was obviously important to Rice, and presumably to the company as well.

Such a code presented an opportunity for encouraging or at least facilitating standardization of agents' reports. The code's potential in this regard seems to have remained unexploited, however. In the case of Rice, the code probably simply reduced the price of ad hoc communication. In spite of the company's consolidation of sales offices and creation of salaried sales agents, very little progress was made toward systematizing their relations with the company.

Du Pont finished the century clinging to its old-fashioned ways. The few advances in management methods, communication mechanisms, and communication technology—initiated by the younger generation— came from (and often remained in) peripheral plants or facilities. Even Francis G. du Pont, a scientist who was more open to new technologies than General Henry or Eugene du Pont, had not fully internalized the lessons of systematic management, though he adopted some of its outward manifestations. This conservatism in business methods in a period of new ideas elsewhere in the business world frustrated some of the younger members of the du Pont family. Lammot du Pont left the company in 1880 out of frustration with its resistance to new ideas, and his son Pierre did so in the 1890s.[132] Meanwhile, however, the company Lammot founded, Repauno Chemical Company, nurtured the early stages of systematic management; the techniques it developed would come back to the Brandywine at the beginning of the twentieth century.

The Rise of System at Repauno Chemical Company

The Repauno Chemical Company was formed in part as a progressive reaction to the Du Pont Company's conservatism. Lammot du Pont, General Henry's nephew, had been a key figure in Du Pont since the late 1850s, handling most of its business away from the headquarters on the Brandywine. In the late 1870s, he became fed up with General Henry's autocratic and conservative control of the company, as well as his refusal to get directly involved with the new explosive, dynamite.[133] After protracted negotiations, Lammot left the parent company to start a new dynamite firm, funded by the Du Pont Company, Hazard Powder Company (wholly but secretly owned by Du Pont), and the Lafflin and Rand Powder Company (a cooperative rival firm). From the start, Lammot du Pont was concerned with developing more systematic and effective forms of management for the company, and after his death in an explosion in 1884 his successors continued to pursue this goal.

Lammot du Pont's Legacy: A Belief in Systematic Methods

While Lammot was still part of the Du Pont Company, he showed an interest in analyzing issues more systematically than did General Henry or the other partners. Although he shared the family penchant for documenting events and issues, he went beyond documentation for its own sake. He also believed in analyzing the past and the present so they could provide guidance in the future, thus anticipating a basic principle of the systematic management movement as it began to emerge in the 1880s.

He applied some of these beliefs in projects undertaken while he was still at Du Pont, though his efforts seem to have been ignored. In the 1850s, for example, he brought together drawings of the company's machines and facilities over the years and wrote a historical note on them.[134] His motive for this, according to his biographer, was "a sense of history" and "an archival impulse," combined with a realization that if he did not save these documents, an important source of data on the company would be lost forever.[135] Similarly, in 1871 he wrote a long memorandum compiling all known information on every explosion at the Du Pont mills from the company's founding to that year, analyzing their causes, and, when possible, describing how he believed they might have been avoided.[136] In this latter study, especially, he showed his interest not just in posterity but in understanding what the past could teach the present.

His analyses did not always focus on the past, however. He was far ahead of most of his colleagues in the company and in manufacturing companies in general in analyzing the present as the basis for improving the future. Perhaps his most revealing analysis was of the 1870 census statistics for industry, from which he discovered that although Du Pont was the largest producer of black powder in the country, its labor productivity and its return on investment were the lowest.[137] This analysis confirmed him in his belief that new managerial methods were needed at Du Pont. He performed an analysis of the work flow through the Hagley powder mills, showing the inefficiency of both the layout and the workers. In this study, as Norman Wilkinson points out, he "was anticipating by two decades the spirit and aims of Frederick W. Taylor."[138] There is, however, no evidence that General Henry ever paid any attention to this analysis.

Written analysis was only one of Lammot's vehicles for improving Du Pont's management. He also advocated regular meetings in which the partners could discuss new ideas and proposals. In an undated memorandum, he proposed that the partners meet once a month to hear one of them present a paper on some new idea.[139] The group would then discuss the idea and either subject it to further study or make a decision based on it. Nothing in the records indicates that his proposal was accepted. In fact, the appearance of a similar recommendation for monthly meetings in a proposal Lammot wrote shortly before he left the company suggests that his earlier suggestion had not been heeded.[140] General Henry preferred to get his information from the other partners at the afternoon smoking sessions in the office, then to make his own decisions privately.[141] He did not want any formal meetings that might dilute his decision-making powers. Lammot's proposal of a committee management system was to resurface in Repauno and then in Du Pont after the 1902 change in management.

Lammot du Pont had seen the need for more modern and systematic methods of management even before he left the Du Pont Company to found Repauno, and he himself had used or recommended various methods of communication to support them. Consequently, it is not surprising that, from the start, Repauno was more modern than Du Pont in this regard. For example, Lammot's clerk and stenographer was using a typewriter by 1882, six years before such a machine appeared on the Brandywine.[142] Although Lammot ran Repauno for only four years before he was killed by an explosion, his desire to modernize the company and improve management methods survived in his successors. By the 1890s, a series of changes were occurring at Repauno to put it in the front ranks of systematic management and communication. Routine reporting and downward communication were instituted to control the far-flung sales organization, and committee management methods were adopted at headquarters. The format and technology of internal communication evolved to fit the new functions. All these changes put Repauno well ahead of Du Pont.

Systematic Communication to Control the Sales Force

In 1892, when Du Pont sales representatives and agents were still acting independently, even though many were on salary, Repauno reorganized its sales function, put all agents on salary, and instituted a system of sales reports.[143] Certain information useful in dealing with customers and in monitoring and evaluating the various sales agencies was collected regularly by means of two types of reports. First, "every office is required to make out a slow report on the 15th of each month on all customers whose accounts were due the previous month and up to the 15th . . . remained unpaid."[144] This report enabled the main office to monitor financial affairs of the sales offices much more closely and to rein in agencies that were overextending credit.

Trade reports were required even more frequently: "They shall be made out each day when trade is visited, or when anything is heard of interest to report regarding present or future customers."[145] Sales representatives wrote the reports on a form with blanks for certain basic pieces of information such as the name of the customer visited, the date of the visit, and the prospect for future sales.[146] Thus, the content and format were being standardized across the various sales offices. Moreover, as President J. Amory Haskell explained in several letters, these trade reports were the company's main source of information in evaluating the representatives: "He cannot afford to neglect making out trade reports, as, if things should not turn out as well as he expects, he could, at least, put himself on record as having tried to secure trade, and exercising energy in looking after business."[147] For the first time, through these two new types of reports, the sales representatives were being held responsible for the *process* as well as the *products* of their efforts. This shift in emphasis was very much in line with the goals and methods of systematic management emerging at that time.

If these reports established a flow of information upward (and inward) from the agencies to the main office of Repauno, circular letters established a flow in the other direction. In order to systematize the operations of the agencies, Repauno executives issued circular letters at least a decade before either Du Pont or Scovill did.[148] While no examples of these circular letters have survived to tell us how they duplicated them, references in letters indicate that the company occasionally used carbon paper. Both stencil and gel duplicating devices were also available at this time. These circular letters helped establish the uniformity that Repauno management desired, though the many letters reminding agents of missing reports suggest that the concepts of systematic management were not easily accepted by the agents. They may well have resented the increased control being exerted by the company. Evidence suggests, however, that the company achieved a certain amount of consistency.[149]

The telegraph supplemented the reports, circular letters, and individual letters in maintaining communication between Repauno's home office and the various agencies. As early as 1884, four years before Rice introduced a private telegraph code into Du Pont, Repauno had developed its own private code suited to its business.[150] As Repauno's auditor pointed out to a company trying to sell them a code more than a decade later, "We have a most complete code which was gotten up years ago, and is adapted to our business; consequently no matter how good another code may be, generally unless the technical terms peculiar to our business were incorporated, we could not use it."[151] The letter went on to explain that the code was periodically revised "for the purpose of placing therein sentences applicable to our own business furnished by our agents from time to time." The periodic updates are evidence that it was actively used.

The code was intended to promote both secrecy and brevity. Its author refused to show it to an outsider, asserting that it was "for private use."[152] Moreover, the code assigned ciphers even to one-word names, a substitution clearly intended to maintain confidentiality rather than to save money.[153] The code also achieved the savings of brevity by substituting single, ten-letter combinations (each of which was charged as one word) for many phrases commonly used in the business. Such a code was particularly effective in a systematically run sales organization, in which communication was already partially standardized.

Other Developments in Internal Communication

Other new modes of communication evolved within the Repauno plant and between it and Repauno's headquarters in Wilmington. The regular meetings that Lammot had tried to institute at Du Pont were instituted, though in a slightly different form, at Repauno. Under the direction of General Manager Hamilton Barksdale, department heads met regularly, presenting written papers concerning managerial issues and problems.[154] The papers were then the subject of group discussion and problem solving. These meetings formalized and documented the give and take of face-to-face problem solving.

Repauno's management also created methods for collecting operating information and sending it up the hierarchy. Letters refer to "the various weekly reports" that the superintendent of the plant was responsible for getting to the company offices in Wilmington.[155] The reports, probably tabular compilations of operating data, provided the data needed by the general office to evaluate operations, just as the sales reports allowed the evaluation of sales offices.

By the mid-1890s, some changes in the direction of greater efficiency and consistency were also taking place in Repauno's general correspondence, whether to sales offices, within the headquarters, or to external parties. All letters (internal or external) were still copied in the company letter press book, but carbon copies were made as well when extra loose copies were needed.[156] Thus, the traditional comprehensiveness of the bound press book was being supplemented by the newer, more flexible copying technology. Moreover, certain conventions of format, designed to make letters easier to read and quicker to refer to, were being adopted. In both internal and external correspondence, Barksdale and Haskell used subject lines and subheadings to indicate the subject or subjects covered in a letter.[157] These modifications to traditional letter format were probably designed both to highlight the subjects for the reader and to make the letters easier to locate later for reference. These changes in format were an initial step in making correspondence function as an accessible corporate memory, a step that had not been taken at Du Pont.

By the end of the century, major changes in the methods of handling correspondence were occurring at Repauno, as the traditional press book and pigeonhole system failed to keep up with the needs of the growing company and with available technologies. By 1895, Repauno had begun to maintain some sort of flat file, either vertical or horizontal, of carbon copies of letters written by company officers when they were away from the Repauno office in Wilmington. Notations were added to the company press book to indicate the existence, location, and subjects of letters not copied in the press book.[158] The centralization of all correspondence, required by a press book system, did not accord with the realities of company needs, but Repauno, unlike Du Pont itself, devised ways of dealing with that problem.

Up until May 1898, the company had copied internal correspondence into the press book. At that point, the main office began instead to maintain a list designated "Letters to various Departments/Building."[159] For each item the list included the date, sender, receiver, and subject. A notation on the front of the list indicated that this list was the only record of such letters sent, although the recipients kept the originals of the letters. While the reason for this separation of internal from external correspondence was not stated, the increasing number of internal letters (as many as seven a day sometimes) took time and space to copy in the press book but were not very accessible in press books indexed only by the recipient's name.

By 1901, vertical files had transformed Repauno's system for handling correspondence.[160] Incoming correspondence, carbon copies of outgoing correspondence, and internal correspondence were all merged for the first time in these files, arranged numerically, with an alphabetical card file to provide subject access. At the beginning of 1901, a press book of outgoing correspondence was still maintained, though carbon copies of letters were also made for the vertical files. Then, in May 1901, the press books were abandoned forever.

As Scovill would discover when it changed to vertical files more than a decade later, Repauno found that subject-based vertical filing demanded some changes in letter format. Barksdale wrote to one of his associates, "In order that the new system of filing our correspondence, which I have adopted in my office here, may be effective, it is essential that each subject corresponded about should be included in a separate letter, as when 2 or more subjects are written about in the same letter, it is necessary for extracts of each to be made for the different files."[161] Barksdale followed his own prescription, writing separate letters for each separate subject, sometimes several to the same person on the same day, with the subject typed in all capitals to highlight it. Moreover, Barksdale's outgoing letters began to include the file number in the upper corner, as well. This new system for handling correspondence was designed to be much more functional, and to create a subject-based corporate memory accessible at any time.

Thus, Repauno, under the initial inspiration of Lammot du Pont and the later guidance of other far-seeing executives, jumped far ahead of the Du Pont Company in modernizing its management methods and, necessarily, its communication system. It set up systems of upward and downward communication to standardize, control, and evaluate its sales force and to collect operational data. It saw the full potential for various communication technologies, from telegraph to carbon paper, and made use of them. It was not afraid to abandon old methods for new as the parent company was. Early proponents of the principles of systematic management, the executives of Repauno built up an efficient company with the communication system necessary to support it. It was to serve as the nucleus around which the post-1902 Du Pont was re-created.

Chapter 8

Du Pont, 1902–1920
Radical Change from a New Generation

In 1902 Eugene du Pont died, leaving the company without an obvious successor. At this time of crisis, a new generation of du Ponts emerged to take control.[1] Alfred I. du Pont, a junior partner in the old partnership, banded together with his cousins Pierre S. du Pont and Thomas Coleman du Pont to reconstitute the company. Of the three, Alfred was the only one at that time working in the company. Pierre, Lammot du Pont's oldest son, had left the company in frustration at its old-fashioned ways three years earlier; T. C. du Pont, who became president of the new company, had never worked there, having gained his business experience in coal, steel, and street railways. The three young men took over first the family company and then, within two years, much of the explosives industry in America. By buying up all of the companies in which the Du Pont Company held stock, purchasing its longtime friendly rival Laflin and Rand, and taking over a number of other smaller companies, Du Pont brought under its financial umbrella well over half the explosives production (powder and dynamite) in the country.

The three partners did not, however, stop with financial consolidation; they also wanted to reap the benefits of operational consolidation. They set themselves the task of creating a single, centrally administered company out of the many small companies and plants that they had absorbed. Their engineering education (all three had attended the Massachusetts Institute of Technology, though only Pierre had completed a degree) and T. C.'s and Pierre's exposure to the business world beyond the Brandywine made them very much aware of the current ferment in managerial philosophy and technique. Moreover, they brought into the company from Repauno and other firms several executives who shared their modern perspective. The cousins and their associates intended to introduce the principles of systematic management into the newly enlarged company to enable it to realize its full potential. As T. C. du Pont put it in 1903, "The essential point is to bring these various departments together with one system and a common interest."[2] Another executive of the new company spoke of the Executive Committee's "earnest effort to introduce system and a general plan into what we are doing."[3] System was indeed to be one of their watchwords as they transformed the stodgy old Du Pont Company into a modern corporation.

To achieve this transformation, they created a new communication system. First, they developed new patterns of communication at the executive level, establishing themes that would also penetrate lower levels. Then, as the many dynamite plants were consolidated into a single High Explosives Operating Department, managerial practices developed at Repauno were introduced in order to unify and systematize them. At the same time, the special needs of research and development at the Experimental Station drove the evolution of somewhat different communication patterns there. Other significant developments occurred in other departments and in the company as a whole.

The Transition: Management by Committee

In the years immediately following 1902, even as they progressed on the financial side of the consolidation, the executives of the new Du Pont struggled with the problems of bringing the many diverse firms and plants into a single operating entity with functionally defined departments. Changes in the communication system both reflected and addressed these problems. A system of executive management by committee was quickly established, reporting was developed as an analytic tool, and internal correspondence evolved to suit new demands on it.

In the late 1870s, Lammot du Pont had proposed to modify the traditional Du Pont system of one-man rule by instituting monthly meetings of all executives and by vesting this group with decision-making power. His proposal died for lack of support from the head of the company, and autocratic rule continued to prevail. Lammot's judgment was vindicated and one-man rule finally ended at the beginning of 1903 when a seven-member Executive Committee was created to oversee and direct the new operating company.[4] The executives of the committee (each of whom except T. C. du Pont also headed a major functional department) were to make broad company policy, allocate financial resources, and coordinate the departments.[5] This committee was made up of the three cousins (with Alfred heading the Black Powder Operating Department and Pierre the Treasurer's Department); Hamilton Barksdale, previously president of Repauno (heading the High Explosives Operating Department); J. Amory Haskell, president of Repauno before Barksdale but most recently president of now-absorbed rival firm Laflin and Rand (heading the Sales Department); A. J. Moxham, a close associate of T. C. and Pierre in the steel and transportation businesses (heading the Development Department); and Francis I. du Pont, later to be replaced by T. C.'s brother-in-law Henry F. Baldwin (heading the Smokeless Powder Operating Department).[6] With some minor changes in membership and responsibilities, this committee would run the company for the next decade, and, with periodic major changes in form and mandate, succeeding committees have run it up to the present.

The committee as a mechanism for executive control was thus introduced to Du Pont at least a quarter of a century after Lammot first suggested it. The committee members had to learn to work together effectively and efficiently in meetings and to recognize what work was better left to individuals, as this letter from A. J. Moxham to T. C. du Pont during the early months of the committee's existence reveals: "Executive committee meetings are doing great good and I think in every way broadening the control of the general situation. We have to learn to leave details to Heads of Departments instead of trying to make them subjects of Executive action, and to leave to the Executive Committee only the big questions. I think we are improving upon this at each meeting."[7]

Initially, the committee held regular meetings once a month, plus extra meetings as required.[8] In the next few years, the expansion of the company outstripped the improvement in committee processes Moxham had referred to, leading to two- and three-day monthly meetings. Consequently, in 1905 two shorter meetings with different agendas replaced the one long one, and an Operative Committee of seconds-in-command in each department was established to handle many of the details that were bogging down the Executive Committee.[9] In 1911 and again in 1914, the membership and duties of the Executive Committee were changed significantly in response to problems that had emerged. These changes, which have been described and analyzed in detail by Chandler and Salsbury, generally attempted to segregate the day-to-day operation of the departments from the long-range oversight and planning function of the committee. In spite of the need for such periodic adjustments, Executive Committee meetings continued to be the medium for making major decisions.

Two types of written documents played important roles in the committee's work: minutes and reports. The minutes recorded what went on at the meetings, including reports presented, issues discussed, and decisions reached. These minutes continued the company tradition of documentation, but made that documentation both more systematic and more accessible. The minutes were meticulously recorded by the secretary of the Executive Committee after each meeting and approved at the beginning of the next.[10] Moreover, they were kept in accessible files near the Executive Committee meeting room and the executives' offices in the new Du Pont Office Building that had been built in Wilmington.[11] These minutes were periodically referred to or quoted in correspondence, suggesting that they formed an active corporate memory.[12]

The Executive Committee also set up a system of reports to provide the information and analysis needed in their decision making. Each department submitted regular monthly reports as well as an annual report to the committee. As Chandler and Salsbury have described them, "Each report included statistical information appraising the current work of a department, described problems or issues, and made suggestions for proposals or actions."[13] These dense and frequently long reports (the 1911 annual report of the Experimental Station, for example, was 147 pages long)[14] were distributed to members a few days ahead of the meeting at which they would be discussed, to give committee members time to read them. Even with this extra time, however, information overload quickly became a problem as the number of departments increased and the business each handled grew.

Over the years, the committee made many attempts to reduce the burden of report reading. As early as 1904, some members suggested presenting the complex statistical data of the Sales Department report in graphic form, but this approach was ultimately rejected on the grounds that it oversimplified the data.[15] In 1905 and 1906, Executive Committee members were asked to categorize issues raised in their departmental reports according to importance, and to try to eliminate extraneous materials.[16] When exhortations failed to solve the problem, a Reports Committee was formed with the following duties:

> 1st. To examine all reports of the several departments submitted to the Executive Committee, including the regular reports from the Manufacturing, Sales, Treasurer's, Purchasing, Development, Real Estate and Legal Departments, with a view to transmitting to the Executive Committee a digest of important matters, together with recommendations as to their treatment.
>
> 2nd. To recommend the form in which reports should be made, with a view to giving the greatest amount of information with the least work.
>
> 3rd. To call for special reports, from or on the regular reports, as may be shown necessary from time to time.[17]

This committee, which functioned from 1911 to 1914, developed an accessible format within which it provided summaries of the reports and suggested points for discussion, keying them to pages in the original reports. When the 1914 reorganization apparently eliminated this Reports Committee, the departments were asked to provide their own synopses of their reports.[18] The demands of efficiency were beginning to shape these reports.

The need for consistency also affected Executive Committee reports. Systematic monitoring of operating statistics over time, one of the key tools of systematic management, required comparable data. In a report from the Executive Committee to the board of directors, written after Du Pont was successfully prosecuted for antitrust violations and required to spin off parts of the company, the committee showed its concern for maintaining comparability:

> Owing to the dissolution under the Government Decree, our various reports for this year will not be comparable with previous re-

ports, and therefore, the benefit of comparisons with past records will be lost, unless our previous data can be rearranged to correspond with the present conditions. The Accounting Committee has, therefore, been requested to give this matter consideration with a view to seeing whether some method can be devised for securing the benefit of past comparison in our reports this year.[19]

This concern for statistical integrity shaped records and reports over the years in the Executive Committee and elsewhere in the company.

In addition to the meetings, minutes, and reports, the members of the Executive Committee communicated with each other extensively through internal correspondence. Although the executives were grouped together in the new Wilmington office building, rather than scattered along the Brandywine as in the past, correspondence among them increased rather than diminished. As the Executive Committee members worked out an administrative structure for the company, many points of friction emerged. While the men undoubtedly discussed these conflicts orally, they also frequently stated their positions and explored resolutions in written correspondence. For example, a dispute about the boundaries between the Sales Department and the Development Department generated a series of letters among various executives in the Wilmington office.[20] In these letters, the parties to the dispute tried to delineate precisely where one department's duties left off and the other's began, frequently defining terms and presenting various scenarios to clarify the boundaries. Such precision was more readily attained in written than in spoken communication.

Moreover, the letters documented their positions and agreements for future reference. In one letter to Moxham, J. A. Haskell stated his documentary purpose in the opening sentence: "Regarding our conference yesterday with the President and the understanding arrived at outlining the respective functions of the Development and Sales Departments, [I] would say that my understanding is approximately as follows . . ."[21] In another, Moxham wrote to T. C. du Pont, "It has been agreed that Mssrs. Barksdale, Haskell, you and myself should discuss this situation, but it seems to me best that the matter should be made one of record, hence this letter."[22] In one of his several attempts to settle this dispute, T. C. du Pont cited an earlier letter and made clear his documentary purpose: "I request in case of any uncertainty the dividing line be drawn as follows . . ."[23] Furthermore, he ended the letter by stating, "I am sending copies of this letter together with copy of your letter to all members of the Executive Committee so that they may be entirely familiar with the position I take." The letter would communicate his decision to the other executives and would serve as a permanent record to be referred to later. Documentation was less for posterity, as it often had been in the nineteenth century, than for constant reference and use.[24]

To serve that reference function, the correspondence had to be readily accessible. The modern developments in duplicating and filing, earlier adopted at Repauno, were used to make it so. Immediately after the three cousins took over, the old press books were abandoned in favor of carbon paper and vertical files.[25] Multiple carbon copies could be made at the same time with the thin onionskin paper provided for copying, thus allowing T. C. du Pont, in the example cited above, to send copies to all the members of the Executive Committee. The various sets of files that have survived show that some and perhaps all of the members of the committee kept their own vertical files, in addition to the central Executive Committee file of minutes and official reports.[26] Since in the early years the members of the Executive Committee were also department heads, they had departmental files, as well. In some cases, they may have merged executive communication with departmental files, but in general they kept separate sets.

The various sets of files were organized differently, but always in ways intended to facilitate use. They were generally grouped numerically by subject, probably with card indexes to match numbers to subjects. Although internal and external correspondence were merged in these files, by the second decade of the twentieth century internal letters were easily distinguishable by the blue or green stationery used for originals and the distinctively marked onionskin paper used for carbon copies.[27] Between the committee's files and the personal sets of files maintained by the executives, correspondence among committee members was made readily accessible for future reference.

As the importance of written communication to the workings of the Executive Committee became clear, the executives paid more attention to the documents, frequently rewriting them more than once. In fact, sometimes the process of drafting and redrafting served important communication functions itself. In a transmittal note to his two cousins, T. C. du Pont referred to one of his several drafts of the document recommending the 1914 reorganization as "the boiling down of my thoughts on the subject for the past sixty days" and requested comments on it.[28] By sending the drafts to other members of the committee for comments, he was also trying to garner support for his proposals, a process that was particularly important when most decisions were made by group. In such cases, members of the Executive Committee sometimes commented on matters of form as well as content. When T. C. circulated a later draft of the recommendations for the 1914 reorganization, for example, Moxham suggested several modifications designed to make the document easier to read.[29]

Although the members of the Executive Committee spent a lot of time and effort on establishing communication flows at the executive level, they occasionally dealt with communication at lower levels, as well. During the attempt to define boundaries between the Sales Department and the Development Department, one of the major issues was the propriety of communication between lower-level employees of the two departments about confidential information. In a series of letters, all of the parties agreed that employees of the two departments should follow strict vertical communication channels, passing the information across department lines only at the level of the department heads.[30]

In its first decade, then, the Executive Committee had to set up a series of communication mechanisms. These methods of communication contrasted to a considerable extent with earlier communication patterns at Du Pont. The work of absorbing more than half of the country's explosives capacity into a single, efficient operating company demanded effective executive communication. The necessary communication systems to bring together the new functional departments were even more extensive, but they exhibited many of the characteristics established at the executive level.

Systematic Management and Communication in the High Explosives Operating Department

The newly consolidated company had three production departments: the High Explosives Operating Department (HEOD), which produced dynamite; the Black Powder Operating Department; and the Smokeless Powder Operating Department. The HEOD, with twenty-five hundred employees in a dozen plants, was the largest of the three (and by itself about as large as all of Scovill in this period).[31] Initially headed by Hamilton Barksdale and directed by Harry G. Haskell (J. A. Haskell's younger brother), it was built upon the systems these men had earlier established at Repauno. Barksdale and Harry Haskell sought to systematize procedures and standardize materials and equipment throughout the geographically dispersed and previously independent high explosives plants. To achieve control over these plants, they created an extensive communication system. Immediately after the massive acquisitions, regular meetings of the plant superintendents played a major role in the consolidation of plants into a department. A system of downward communication was developed to standardize procedures, and an extensive system of reports was created to pull information of all types up the hierarchy for analysis and evaluation. Internal correspondence evolved toward more efficient forms. By World War I, the HEOD had become the epitome of the systematically managed organization with its extensive internal communication systems.

The Plant Superintendents' Meetings: Unity and System

In 1904, Barksdale and Harry Haskell were faced with the task of absorbing into a single efficient operating department the many dynamite plants that had been acquired by Du Pont since 1902. Drawing on their experience with meetings as a management tool both at Repauno and in the Executive Committee, the two men adopted the plant superintendents' meeting as a principal means of unifying and systematizing the department.[32] These meetings began in November 1904, when the bulk of the financial consolidation had been completed but operational consolidation was just beginning. The first gatherings, initially held monthly, brought together Barksdale, Haskell, and eight or nine plant superintendents to discuss matters of common interest. By 1907, the meetings had evolved to be less frequent but larger, and included super-intendents and assistant superintendents from all the plants and the home office of the HEOD. The meetings were held twice a year through 1909, then once a year after that.

The early superintendents' meetings were a mechanism for broaden-ing the participants' allegiance from their own plants to the department as a whole while exposing them to new managerial ideas and tech-niques. In 1904, most of the superintendents of dynamite plants were accustomed to working almost entirely autonomously. The early meet-ings, as Haskell described them retrospectively in his opening address to the 1911 meeting, tried to initiate cooperation and the exchange of ideas among them: "These monthly meetings were held with the object of getting the Superintendents together to discuss matters of interest, in an attempt to get co-operation through the consolidation of the West Coast Companies, which had been taken into the duPont Company. Everybody had been going along more or less on their own hook, and it was thought desirable to hold meetings and exchange ideas."[33] Within this cooperative context, the meetings also provided a mechanism for spreading the managerial and technical knowledge gained at Repauno to other plants.

In addition, the meetings provided a chance for Barksdale and his colleagues in the HEOD headquarters to learn about the various plants and to monitor and adjust the department's introduction of new rules, procedures, and reports. The early assemblies were quite informal, al-lowing plenty of opportunity for two-way communication between Barksdale and Haskell, on the one hand, and the plant superintendents, on the other. In effect, the department's executives used the meetings to humanize managerial relations at the same time that they were impos-ing impersonal systems. Thus, problems on both sides—the main off-ice's problems in getting superintendents to follow new procedures and the superintendents' problems with the new procedures—could be ad-dressed as they arose.

Over the years, Haskell continued in his 1911 address, the meeting procedures and preparations grew more formal and elaborate: "At first we listed subjects for discussion. Gradually we got into the habit of preparing papers, and later on in order to permit discussion of the papers they were written ahead of time and sent around so that the Superinten-dents and others could discuss them more readily." Beginning with Meeting 31 in October 1909, the minutes of each meeting—including the papers presented as well as a transcription of the group discussion—were printed or duplicated, bound, and issued to all attendees to serve as permanent records. As the program became more formal, two-way communication was maintained by building it into the schedule. Starting at least in 1907, the program always included a session devoted to discussing problems between plant superintendents and the main office.[34] Barksdale and Haskell thus created an ongoing mechanism for dealing with any resentment created by the increased control.

Moreover, these sessions served as another opportunity for the departmental executives to convey their managerial values to the superintendents. For example, during one such session, Barksdale made a personal pitch for cooperation between the superintendents and headquarters, saying, "I know of no point at which there is failure [by the superintendents], except to appreciate the full importance of cooperation, and unless that cooperation is achieved—unless we secure that—the organization as a whole can't reach its full measure of success, and I, consequently, as an individual, fail."[35]

He went on to explain why reporting tasks that the superintendents labeled as red tape were necessary to headquarters, then ended on a personal note: "I am going to make a personal plea to you gentlemen to try and make it a point to take the broad view, not to look upon your own plant as the only thing to which you can render assistance, but to place yourself more in the position, as I stated before, of one of several who are working for a common end." These personal pitches reinforced principles of systematic management while humanizing the upper levels of departmental management.

The HEOD superintendents' meetings continued to be an important element of the formal communication system in the department well after the initial integration of the various plants, encouraging and spreading managerial principles and innovations as well as creating esprit de corps among the superintendents. The opening addresses by Barksdale or Haskell were used as vehicles for communicating their views of what the department had achieved and for reinforcing their priorities. This personal persuasion could be quite effective. In the opening address to the 1911 meeting, for example, Harry Haskell reminded the group of Barksdale's 1910 address on "the desirability of paying enough wages to get the very best kind of workmen that we need in our factories."[36] Haskell noted that most of the plants had increased their wages in response to the address and stated that the net effect had been to reduce the cost of producing explosives.

Many of the special reports delivered at the meetings by members of headquarters staff (of HEOD or of Du Pont as a whole) were intended to educate the superintendents in technical and managerial issues. These papers ranged from general interest features like those in the *Scovill Bulletin* to detailed studies of specific common problems. A relatively small percentage of the papers presented at the meetings concerned purely technical subjects, such as "Concentration of Weak Nitric Acid" and "Improvement in Chloroform Treatment."[37] More often, the focus of educational papers was on managerial issues such as safety, costs, efficiency, standardization, and labor. Safety was a major topic in the HEOD superintendents' meetings, with members of the Safety Division at headquarters presenting papers on topics such as "Danger Features of the Powder Line" and "The Personal Side of Work Injuries."[38] Other educational papers discussed aspects of efficiency and systematic management. For example, the head of Du Pont's Accounting Department was invited to give a guest paper on the "Object of Accounting," in which he explained the principles behind cost accounting.[39] In it he discussed one of the basic principles of systematic management: collecting data not just for the record but for analyzing and monitoring operations.

Not all papers came from headquarters staff. In his 1911 opening address, Harry Haskell noted a progression from main office domination of the program to increasing participation by plant superintendents. In fact, he commented humorously, while initially he had experienced difficulties in getting the superintendents to participate, "the problem now is to stop the Superintendents talking." In some cases, special committees composed of plant superintendents and main office personnel were established to report to the meetings on various subjects. For example, a Nitroglycerin Stability Commission was established at one meeting to look into various technical problems relating to the unstable chemical. That superintendents' committee continued to meet separately and to report regularly to the superintendents' meetings.[40] Other committees were established to set standards for equipment and procedures in various parts of the high explosives plants and to report their detailed findings and recommendations to the superintendents' meetings.[41] In still other cases, individual superintendents presented reports sharing their insights and techniques with each other. These contributions covered a variety of topics. For example, new and systematic methods of labor relations, tempered by the concern for worker welfare of the corporate welfare movement, were shared and discussed by superintendents in programs such as "Training Staff Men and Handling Plant Labor," "Physical Fitness in Employees," and "Treatment of Slums (discussion)."[42]

In addition to their educational functions, the meetings also had an important social function, creating a group spirit and loyalty to the department among the superintendents. One manifestation of this spirit was the *H.E.O.D. Knocker*, a humorous newsletter put together by the superintendents in conjunction with one meeting.[43] In satirizing both superintendents and headquarters executives, it defused hostility and provided a channel by which superintendents could give headquarters their perspective. In general, the meetings promoted a healthy relationship between the plants and department headquarters.

*Downward Communication: Consistency, Comprehensiveness,
and Control*

Although the HEOD superintendents' meetings established a favorable climate for systematizing the department, downward communication provided the direct mechanism for instituting new policies. The system of downward communication in the HEOD derived from that of the Repauno Chemical Company but quickly went beyond it. From a very early date, downward communication in the HEOD had two main components: circular letters and similar downward communications, issued as needed; and a comprehensive manual of rules, regulations, and procedures entitled "HOW." The circular letters, modeled directly on those used at Repauno, established rules and procedures for consistency and control. HOW, an innovation that seems to have gone beyond anything at Repauno, offered the comprehensiveness lacking in the more fragmentary circular letters.

Although comparable directives probably appeared earlier, beginning in 1906 HEOD headquarters issued a continuous series of numbered bulletins, and in 1907, a similar series of circular letters to its plants throughout the country.[44] These communications showed the benefit of Repauno's previous experience in their design and use. They were reproduced by a duplicator (either hectograph or mimeograph) onto special printed forms. These forms had a title and a space for the number at the top (e.g., "Circular Letter #———") so that they could easily be referred to when desired. Typed in below the heading was HEOD's file number (e.g., "OUR FILE OH-25-Y."), as well as the place of origin and the date. The file numbers indicate that the location of each circular or bulletin in the subject-based filing system was determined when it was written, demonstrating that the function of the files as an organizational memory was taken quite seriously. At the bottom, the forms had a perforated tear-off receipt with printed instructions: "This receipt must be detached and RETURNED PROMPTLY." On the receipt were the circular letter or bulletin number and the file number (filled in by duplicator) and spaces for the date received and signature of the recipient. Thus, headquarters had built in a mechanism for checking that each person who should have received a copy did, and that no one could deny having received notification of any new rule or instruction.

The distinctive printed forms clearly differentiated bulletins and circular letters from correspondence. Initially the documents duplicated onto the forms were formatted like letters, beginning with "Dear Sirs:" and ending "Yours very truly." By 1911, however, many of them had subject lines, and gradually they ceased to have letter salutations and closings.[45] The genre was now completely distinct from letters.

HEOD headquarters used circular letters in its efforts to standardize and systematize as many materials, pieces of equipment, and procedures as possible across the previously independent plants. Circular letters would periodically announce that certain office items had been chosen as standard for the HEOD or for the company as a whole.[46] Other circulars standardized procedures for activities such as replying to inquiries from other departments.[47] Still other circular letters reinforced instructions or procedures already established, working against the inevitable inertia of plants not used to such systematization. One letter, for example, tactfully explained the importance of accurate records and reminded superintendents that although they had the authority to make small changes in construction and operation of plants, they were required to report those changes to the HEOD main office so that the records would be accurate.[48] Another circular, using a technique also discovered by Scovill's J. H. Goss, pointed out a common lapse, then requested "a statement of just what steps you will take to see that the instructions in HOW are carried out."[49]

While the majority of circular letters established or reinforced standards, bulletins and some circulars achieved other miscellaneous purposes such as transmitting reports and relaying general information (such as changes in railroad regulations) that might affect the shipping of supplies to or finished products from the plants. These communications were less critical to systematizing procedures, but they still helped to create common knowledge and bonds among the disparate plants that came together to form the HEOD.

To perform their important standardizing function, the circular letters had to be accessible for reference, not just in the headquarters, but in the plants as well. The file designations on circular letters reveal that headquarters had files with subject-based numerical indexing. They were almost certainly vertical files based on the system established at Repauno in 1901. The plants, however, evidently had different systems. At one of the 1907 superintendents' meetings, according to the minutes, Harry Haskell suggested that inaccessibility of documents in the plants' filing systems might be reducing the effectiveness of circulars: "Mr. Haskell said that one general trouble we had with the Works, due probably to the filing system, is that there seems to be a tendency to forget Circular Letters or instructions and that Wilmington is constantly having to write about small things, which might just as well be done in the first place from instructions already issued."[50]

Resistance to control was probably at least as responsible for these failures as were inaccessible files; nevertheless, by 1910 headquarters had mandated a special storage system to assure that each plant had a complete set of HEOD directives:

> In order that we may be sure that all plants have a complete and
> permanent file of circular letters and that all are filed in a corre-
> sponding way, it is requested that these be kept in the regular circu-
> lar letter binders, known as Shipman's Common Sense Binder, and
> that they be filed in numerical order. Those plants not keeping
> their circular letters in this way, are requested to do so, sending in

stationery requisition for a binder, specifying thereon the kind just mentioned.

Any plant that desires to keep circular letters in correspondence files, can make copies of the ones that are sent out from time to time, for their files but it is particularly desired that a complete set of circular letters be kept in the way above mentioned.

The binder should be provided with an index, according to subjects, with proper reference thereon to the circular letter numbers, which should allow easy and quick reference to any particular letter that is desired.[51]

This system guaranteed that each plant had a complete set of HEOD instructions but isolated them in a separate chronological file. Circular letters continued to include appropriate file numbers for the files at headquarters, suggesting that HEOD headquarters continued to file copies of the letters in its subject file as well as in a separate set. Only the separate set, however, has survived.

HEOD headquarters found the bulletins and circular letters so useful in systematizing the department that they also advocated (though did not require) a similar type of downward communication, called office orders, *within* the plants: "The office orders provide a means for transmitting to interested individuals on the works, written instructions or advice originating with the superintendent, and are also used to convey to them, information, etc. contained in Circular Letters, Bulletins, and other letters from the Wilmington office."[52] These office orders were to be handled like the circular letters, in that the recipients as well as the issuing superintendent were to keep complete sets of them in binders. The circular letter suggesting this system went on to discuss its advantages:

This system has many good features, the principal one being that it provides a superintendent with a record of instructions issued to his various assistants. In addition, in some cases in the past, circular letters received at plants have simply been referred to the party or parties interested and they have initialled the letters which are afterwards passed to the file. With this system of office orders each person would be provided with a record of information contained in such circular letters.

These lower-level versions of circular letters were intended to aid both sender and recipient in keeping track of instructions.

If circular letters, bulletins, and office orders communicated new information, instructions, and exhortations as they were issued, HOW, the other major mode of downward communication at HEOD, provided a more permanent and comprehensive compilation. This book, the department's "Bible of operations,"[53] compiled all the rules, standard procedures, and job descriptions related to supervisory and clerical employees. It recorded such information as how to distribute costs on the accounting forms used for cost accounting; what daily, weekly, and monthly reports had to be filled out; and what to do in case of injuries.

Although at Scovill circular letters appeared much earlier than manuals, at Du Pont's HEOD, manuals were introduced by at least 1907, almost simultaneously with the numbered series of circular letters.[54] Moreover, instead of using a bound book that had to be reprinted each year, HEOD used a loose-leaf notebook for which new pages could be printed and issued and old ones recalled whenever changes were made.[55] The loose-leaf format allowed the copies of HOW to be personalized (each recipient received only the sections pertinent to that person's job) because "it is believed that economy in printing and mailing revisions will result from having each book contain only those pages of value to the holder for reference in connection with his immediate line of work."[56] References to HOW in other documents indicate that HEOD management clearly expected its employees to study and consult it frequently.[57]

While the types of downward communication discussed so far were sent by the HEOD headquarters to the plant superintendents, or by plant superintendents to their supervisory and clerical subordinates, other methods of downward communication were used to communicate to explosives workers in each plant. They received little booklets of rules and regulations, translated into other languages for immigrant workers, as needed.[58] Moreover, rules, announcements, and other notices were posted in plants. Initially, these posted items primarily concerned safety rules, but gradually they broadened in scope and increased in number. By 1911, as a result of increased corporate welfare programs, the number of such posted notices became so great that headquarters had to regulate the posting:

> The printed notices for posting in our plants have increased in number more than was realized would be the case, and it has now been decided that with the exception of such rules and regulations which are necessary for the house crew to observe in the performance of their duties, and which at the present time consists of the general rules and regulations and special house rules, nothing of this nature should be posted in explosives buildings.
>
> This is following out the obvious rule that nothing superfluous should be allowed in powder houses.
>
> Such notices as pension plan, secret process notice, bonus plan, reward payroll plan, etc. should be posted in offices, change houses and similar places and which should be sufficient information to the men.[59]

Even at this lowest level, downward communication was increasing in quantity and becoming more varied in subject matter.

Within a decade of the HEOD's establishment as a department in 1904, then, it had an extensive and well-developed system of downward communication in place. The increased control offered by this system

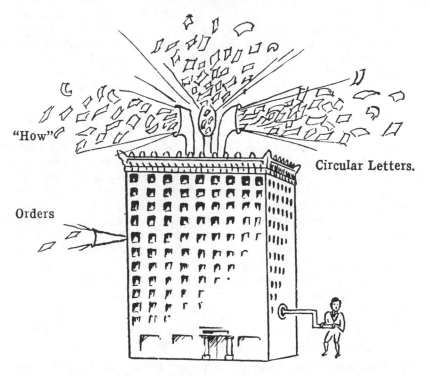

"Bulletins"

"How"

Circular Letters.

Orders

A New Christmas Toy for the Superintendent

A STUDY IN EXPRESSION
The Super at His Morning's Mail

Ha!
An Order

Hump!
A Circular Letter

D——
Another
Circular Letter

H——l
What!
Another

Still Coming!
What's the Use?

8.1. Humorous cartoons about downward communication in the *H.E.O.D. Knocker*, ca. 1909. *(Courtesy of Hagley Museum and Library.)*

did not go unnoticed. Two cartoons in the humorous *H.E.O.D. Knocker* issued in conjunction with the 1909 superintendents' meeting (see fig. 8.1) suggest that plant superintendents sometimes resented the flow of downward communication. "A New Christmas Toy for the Superintendent" pictures the HEOD superintendent in the main office cranking out endless quantities of circular letters, bulletins, and HOW pages. Similarly, "A Study in Expression" shows a plant superintendent's frustration at the predominance of circular letters in his mail.

This reaction was not surprising, given the quantity and effect of such communication. In the first year of the circular letter series, for example, 210 of them—one every day or two—were issued.[60] More important than the quantity of such communication, however, was its effect, for downward communication was a mechanism by which headquarters increased its own powers while simultaneously limiting the plant superintendents' previously extensive powers. Systematizing put more power in the hands of the department as a whole and reduced the power and discretion of plant superintendents to dictate operating methods. With the aid of the superintendents' meetings, however, the downward communication system seems to have been successfully established in spite of this resentment. During these early years, HEOD took immense strides in standardizing procedures and rules in the many acquired plants.

Records and Reports for System and Efficiency

Downward communication alone, of course, was not enough to guarantee efficiency. From the beginning, Barksdale and Haskell recognized that a good system of records and reports was also essential. Numerous routine reports were established to analyze and monitor operating data, and special purpose reports were used to investigate and report on problems or opportunities for improved efficiency. Forms, tables, and graphs were adopted to make the reports themselves more convenient to create and to use. The plant superintendents, however, resented reports even more than they did circular letters, so reports were often at the center of power struggles between the department and its plants.

Records and routine reports on many types of operating information were apparently instituted in the newly acquired plants very soon after they were absorbed.[61] Although the early records and reports themselves have not survived, references elsewhere reveal that by 1907, yearly, monthly, weekly, and even daily reports on everything from materials in stock to labor costs for certain operations were made out from recorded information and sent in to the Wilmington headquarters.[62] As representatives from headquarters explained in a superintendents' meeting, routine reports to HEOD headquarters in Wilmington served two functions, one for the plant superintendent and one for headquarters.[63] First, they gathered information "necessary for the [plant] Superintendent's guidance" in making the day-to-day decisions and adjustments necessary to keeping the plant running efficiently. "It would be impossible," according to the main office, "for the Superintendent to intelligently direct the Plant's activities without this information."

Second, the reports were "for the Main Office records," where they were used by department headquarters for monitoring the department's operations and evaluating the various plants against one another. At headquarters the reports from various plants were combined into comparative analytic reports that were frequently discussed at superintendents' meetings. One of the chemical reports shown at the 1911 meeting, for example, summarized the 1910 nitrate of ammonia reports for each plant.[64] The summary report enabled each plant to see how its yields, strengths, and other measures compared to those of other plants. Discussion of such comparative reports often centered on how the plants with the best statistics for various items achieved their good records, and what the lagging plants could do to improve their records. The comparative reports made up from the plant reports were undoubtedly also used in making personnel decisions about the superintendents and, especially in the early years of consolidation, in deciding which plants to close and which to expand.

The many routine reports were a constant source of friction between the superintendents and departmental headquarters. Headquarters complained that the plants were inaccurate, careless, or late in compiling them. As one circular letter noted, "It is obvious that in order to make these records of real value they should be at all times accurate. It is, in fact, more important that the records on hand be true records than that they be complete records."[65] In another case, headquarters protested that some monthly fire equipment reports were apparently being filled out with no reference to the actual condition of the fire equipment.[66] Finally, tardiness in submitting reports was a chronic problem. At one superintendents' meeting, for example, someone from the main office complained that he was often late in compiling his reports because he had problems in getting daily reports of materials from the various plants.[67] In that same meeting, Barksdale made a plea that superintendents at least inform headquarters when a report was going to be late: "I can readily understand that these reports cannot always be gotten off as the main office wants them to, but I don't understand why, when a man finds he cannot do that, he doesn't come forward and say so and give the reasons."[68]

The superintendents, on the other hand, constantly complained about the number and frequency of reports they were required to submit. In spite of Barksdale's frequent pep talks encouraging plants to cooperate with headquarters in reporting, the superintendents saw the many reports and statements as red tape.[69] Their resentment was expressed satirically in the same humorous newsletter that contained the cartoons about downward communication:

> The Chemical Division [of HEOD] is just about to distribute new
> forms known as the Hourly Operating Reports. Every plant opera-
> tion is covered from the Nitration Process to tool sharpening and
> belt lacing. Each form contains about 600 spaces to be filled in and
> the reports are to be forwarded to the Wilmington Office hourly by
> special messenger, where they will be carefully filed for the benefit
> of Posterity.[70]

This item captures the superintendents' feelings that more and more of
their time was being sacrificed to what they perceived as useless reports
for the main office files.

More serious opposition repeatedly arose at meetings. The following
discussion was typical:

> Mr. Chambers thought as the superintendents are responsible for
> the keeping up of stocks, he did not see why a weekly report would
> not give the Wilmington Office all the information they desired. If
> they have to send in daily reports, they will have to increase the
> clerical force at the mill. Mr. Beers said they had so much trouble
> in getting up reports at Barksdale [Plant] that finally they took on a
> low salaried clerk especially for this work.[71]

This opposition, like that encountered by Stuyvesant Fish when he instituted new records and reports at the Illinois Central Railroad, had
both a practical and a philosophical side. The practical side of the argument was that clerical staff had to be increased to keep up with all of the
reports. Barksdale's response was simply that "the advantage of daily
reports quite offset the expense of increased clerical force necessary at
the mills." Perhaps the more significant underlying issue, however, was
that of control. Mr. Chambers, according to the passage quoted above,
suggested that daily reports infringed on the superintendents' responsibilities. The upward reports, like downward communication, shifted
power from the superintendents to the main office. It gave the main
office the information necessary to evaluate the efficiency of one plant
against another and to make decisions about where to locate production.
And indeed, the efficiency of the plants was improving, so the systematic management that headquarters was imposing in part through those
reports was paying off.[72]

Nevertheless, the complaints about excessive reports were vociferous
enough that Barksdale periodically established joint committees of representatives from headquarters and the plants to investigate whether all
of the required reports were necessary.[73] These committees generally
recommended the elimination or reduction in frequency of a few reports.[74] Here, as at Scovill, the desire for the control afforded by systematic reports occasionally exceeded the point of efficiency, and modest
retrenchment resulted. The majority of reports, however, were left in-
tact.

Given the large number of records and reports, it was particularly important to develop formats that promoted completeness and consistency of information and efficiency in compiling, reading, and analyzing them. Forms, tables, and eventually graphs were used to improve the gathering and compiling of information necessary for systematizing operations.

In many cases, printed forms were created both to assure a standard procedure that would produce complete information and to ease the burden on both compilers and readers of reports. The monthly fire equipment reports, for example, consisted of a form listing various items to be checked off or commented on.[75] As headquarters pointed out to the superintendents, "The present form of report sheet was gotten up especially to call to the inspector's attention all faults that might generally occur, and these general faults are listed thereon." In addition to specifying the procedure and standardizing the content, the form also required less effort from the inspector and less effort from headquarters. Similarly, to undertake a construction project, a plant had to fill out a special form for headquarters entitled "Reasons for Making This Expenditure."[76] This form included fill-in-the-blank items such as "Known loss per annum (if any) under present conditions" and "Estimated saving per annum by making this expenditure," both to guarantee that the plants would systematically analyze such expenditures before proposing them and to make it easy for headquarters to find the critical financial data necessary to decide whether the expenditure was worthwhile.

Finally, some forms were created simply to relieve headquarters of the burden of dealing with information in nonstandard formats: "With a view of relieving this Department of a considerable amount of routine work now necessary when handling correspondence in connection with payroll record cards, we have decided to change our present practice by adopting the use of printed forms."[77]

Many of the routine statistical reports used at the plants and sent to headquarters were made by taking records of individual items (often from forms) and compiling them into tables for ease of analysis. For example: "It is usual, also, to have other detailed information, such as operating labor, unloading costs on raw materials, daily production, orders and shipments, dry-house temperatures, etc., tabulated for the Superintendent's guidance, by the office force. This information involves such a mass of detail that the Superintendent cannot hope to give it careful attention except in summarized form."[78] These tabular reports were used by the plant superintendent as well as sent on to headquarters where they were compressed and recompiled with the statistics of other plants into further comparative tables. These tables were then used at headquarters and distributed to all of the plant superintendents for discussion at the superintendents' meetings.[79] The consolidated tables made it possible to see how the various plants compared with each other at a given time on certain dimensions.

Initially by 1907 and increasingly over the 1909–1914 period, tabular reports and tables included in longer reports presented at superintendents' meetings were supplemented or replaced by graphic reports.[80] While the meetings prior to 1911 averaged between one and two graphs each, the 1912 meeting had ten, and the 1913 and 1914 meetings had twenty and nineteen graphs, respectively. The department even turned to a new copying technology, the Photostat machine, to make copies of those graphs and of some extremely complex tables.[81]

This dramatic increase in the use of graphs suggests a growing awareness of their power as a communication tool. In addition, HEOD reports were increasingly focusing on trends over time, which were well suited to graphic presentation. Initially, summary reports tended to present simple comparative data across plants. Once the reporting system had been in place for a few years, however, the consistent and complete data needed for tracing quantities over time were available. Thus, for example, HEOD headquarters could compile quarterly reports of plant labor costs from 1904 to 1907 into a graphic statement, which it distributed to superintendents.[82] Similarly, a graph of quarterly labor costs for dynamite box packing at the various HEOD plants, presented at a 1910 superintendents' meeting (see fig. 8.2), facilitated comparisons of both absolute costs and cost trends among plants.[83] These data presented in tabular form would only yield their information to careful study, while the graph made the information instantly accessible.

Such graphs were also used for analytical purposes. One of the HEOD Chemical Department operating reports presented in 1913 (see fig. 8.3) plotted average monthly nitroglycerine yield figures for the top five plants, the bottom five plants, and all plants.[84] Graphed this way, the data revealed a consistent dip in July yields, which became the focus of discussion and suggested remedies during the meeting. Similarly, a report on "Prevention of Accidents" at the 1912 meeting contained a graphic analysis of relationships between accidents and various other quantities.[85] That graph demonstrated, without complex mathematical techniques, that the number of nonfatal injuries was *not* related to the quantity of powder packed. Thus, the usefulness of graphs for reporting and for analysis was increasingly being exploited in the HEOD.

Records and reports in a wide variety of forms established a constant flow of information up the HEOD hierarchy. Such complete and consistent information was essential to running the department according to the principles of systematic management. The control afforded by this information was an important part of what allowed the post-1902 company to create an operating company that was more efficient than a holding company. Even more than downward communication, upward reporting was a managerial tool used to achieve system and efficiency.

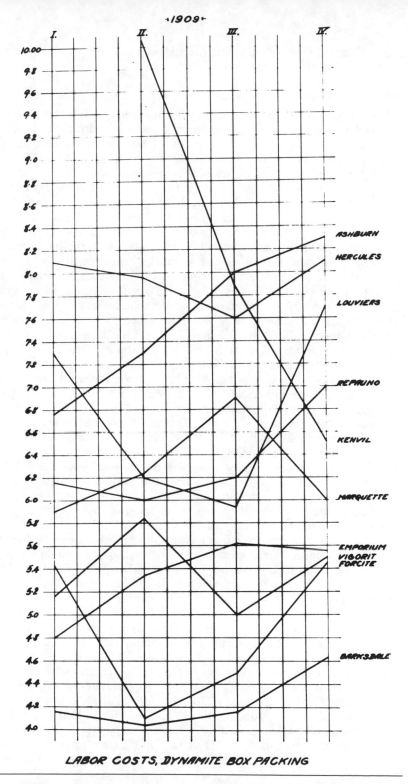

ASHBURN
HERCULES
LOUVIERS
REPAUNO
KENVIL
MARQUETTE
EMPORIUM
VIGORIT
FORCITE
BARKSDALE

LABOR COSTS, DYNAMITE BOX PACKING

8.2.   Comparative graph on labor costs at various HEOD plants. *(HEOD superintendents' meeting 32, 1910, courtesy of Hagley Museum and Library.)*

Efficiency in Internal Correspondence

In addition to vertical flows of communication, much internal correspondence between individuals evidently occurred within the HEOD. Very little of it has survived, but its evolution can be traced through references in circular letters and other documents.

The form and handling of internal correspondence became a major focus of reform late in 1913, but even before then it was being differentiated from external correspondence in the interests of efficiency and cost cutting. As early as 1907, for example, HEOD headquarters issued large envelopes to the plants for sending batches of internal letters to headquarters to save the time of unfolding individual letters sent in small envelopes.[86] In 1908, Haskell issued a circular letter reminding his employees that "it is customary in interdepartmental communications to use letter paper of cheap quality. . . . It is of course advisable to use good stationery when writing to outsiders, but an unnecessary expense to use too good stationery for letters between ourselves."[87] Internal correspondence was further differentiated from external in 1911, when Du Pont's Purchasing Department set standards for stationery.[88] Although high-quality white stationery was chosen for external correspondence, cheaper green stationery was mandated for internal usage. Carbon copies for the files were to be made on white sheets designated "File Copy," and tickler copies on light yellow sheets.

As the systematization of the HEOD created more and more written correspondence of all kinds, it became the target of further standardization. "In order to secure uniformity in the preparation of typewritten letters," a 1913 circular letter announced, "the following has been adopted as a standard to which such letters should conform."[89] The letter went on to specify margins, spacing, paragraphing, and use of pins and clips. Some of the specifications were designed to make letters easier and more efficient to read: "A short line being easier to read than a long one," side margins were made wider; and lines were to be double spaced "on account of the greater ease with which a letter so written can be read, or corrections or notations made between the lines." Furthermore, "Where possible a paragraph should be devoted to each subject or division of a subject," because "this automatically brings these individual points to the reader's attention." Although some of the rules were ludicrously detailed (e.g., "If a pin is used it should be inserted diagonally at an angle of 45 degrees with the top edge of the sheet, and not more than 1-1/4 inch away from the corner"), the purpose of these changes was to apply the principles of efficiency to correspondence.

The ultimate application of efficiency to internal correspondence came when HEOD's Efficiency Division (established in 1911 and disbanded around 1915) took on the subject of efficiency in letter writing.[90] The division originally reported the results of its 1913 study in a letter to Harry Haskell (director of HEOD at the time), then revised it into a report that was ultimately circulated throughout the entire Du Pont

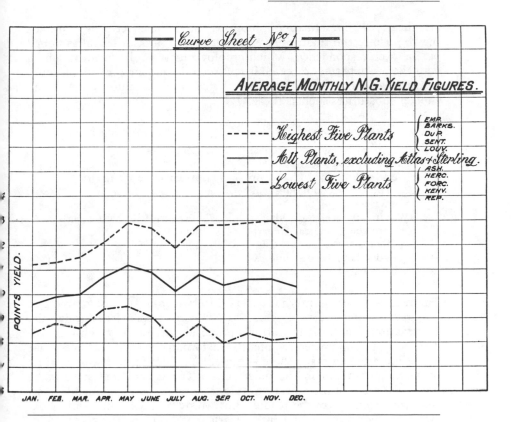

8.3. Comparative graph of nitroglycerine yields. *(HEOD superintendents' meeting 35, 1913, courtesy of Hagley Museum and Library.)*

Company with a note from T. C. du Pont advocating the principles it set out.[91] For internal correspondence, the study recommended eliminating the salutation and complimentary close, simplifying the title and address of the recipient, standardizing the format, and reducing unnecessary words throughout the letter. The first two changes, the report asserted, "would cut out an average of at least 10 words each." The third change, standardizing the format, was designed to aid handling and filing by putting several important pieces of information into the heading: the name of the person or department addressed, the file numbers, the subject, references to previous letters, and listed enclosures.

This format also helped in accomplishing the fourth goal, reducing the number of unnecessary words in the body. The revised report pointed out that the heading information enabled the dictator (indicating that most letters were dictated at this point) to omit such initial phrases as "We have letters dated October 4th and October 17th from ——— as per attached . . ." Even further reductions could "be effected by eliminating the use of the traditional expressions and phrases of oldfashioned business correspondence, which employ many words for ideas which could be much better expressed in fewer words, and many of which could be omitted altogether." The report listed some common phrases and shorter substitutes such as these:

As usually written	*Substitute offered*
We are in receipt of a request from St. Louis office for	St. Louis asks for
We would be glad to have you make an examination and advise	Please examine and report
We are in receipt of advices from the Chemical Department that the Eastern Laboratory have carried out experiments	The Eastern Lab. have experimented

The report classified internal letters into two groups: those replying to a request, which were addressed to someone prepared to receive the information; and those initiating a proposal or inquiry, which were addressed to an unprepared recipient. The former "may be just as concise as the dictator can make them—in memorandum form, if you please," using memorandum in the older sense. The latter "may have to be presented more in the narrative style, as circumstances demand, to attract the attention of the reader," but wordy phrases could still be excised. Using several actual interdepartmental letters, the report demonstrated how radically the number of words could be cut.

The original study sent to the director of the HEOD depended heavily on the Taft Commission report (discussed in Chap. 3) in justifying its attack on standard letter-writing practices. The revised version circulated by T. C. du Pont still cited the Taft Commission report but relied more heavily on arguments from efficiency. Cutting words would, it argued, make "the letters easier to dictate, to write [i.e., type], and, most important of all, to read and understand," thus improving efficiency at every stage. The change in old customs was not going to be easy, the report admitted, but the gains in efficiency were worth it:

> Efficiency has come to stay, and in its application the first great essential is "changing our minds" about the fundamentals of business.
>
> To alter one's point of view regarding the ponderous phrases and expressions of commercial correspondence is necessary. It may be difficult, in some cases more than in others, to overcome the inertia of our long-standing habit of clinging to the traditional forms

and usage, but after the right attitude is attained, there should be
little difficulty, and a constantly diminishing tendency on the part
of the recipient of a letter to "get sore" at the terseness or bluntness of the communication.

T. C. du Pont, in his letter transmitting the report to other departments,
quantified the potential gain by totaling the words in the before and after
versions of real internal letters included at the back of the report: "This
study developed that a number of that department's letters originally
totaling 1,444 words could have carried their message just as well with
755 words, a showing of about 52% efficiency in handling correspondence."

The report was clearly naive in its concept of efficiency. Irénée du
Pont, who was Pierre's brother and a vice-president, pointed out the
problem of defining efficiency in letter writing in terms of word count:
"Laborious consideration to make the letter as logical and short as possible is warranted, if a letter is to be read by many or referred to many
times in the future, but most letters will only be read once or twice so
that it is probably better to dictate quickly rather than take twice the
time to express the thought in fifty percent of the words."[92]

How seriously these stylistic recommendations were taken is impossible to determine, since the correspondence has not survived. Format
was more easily mandated. After circulating the recommended format
for comment, Haskell made a slightly altered version standard for the
department, and T. C. du Pont attempted to do likewise for the company
as a whole.[93]

This report reveals that internal correspondence had become a major
activity that both helped systematize other activities and deserved systematization itself. Moreover, the report and subsequent correspondence indicate how pervasive the principles of system and efficiency
were in the office as well as on the factory floor of the department.[94]

Building on the base developed at Repauno before 1902, Barksdale and
Haskell rapidly developed an elaborate internal communication system
through which to introduce and administer the principles and techniques of systematic management. They had pursued these ideals since
before the HEOD was formed, and the principles had gained increasing
force. They may have been reinforced or strengthened in 1911, when
Barksdale read Frederick Taylor's "Principles of Scientific Management" and wrote him requesting references to other articles.[95] Perhaps
it was under this further impetus that HEOD's Efficiency Division was
established later in 1911. The pattern of systematization, however, had
long been established in the department. The linking of the new methods of management and the formal communication system at Du Pont
was demonstrated most clearly in this department.

The Experimental Station: Special Communication Needs for R&D

The Experimental Station was established on the Brandywine in 1903, shortly after the younger generation took over the firm, to institutionalize research in black and smokeless powder at Du Pont.[96] (The HEOD's Eastern Laboratory, which had been established at Repauno only the year before, carried out research related to high explosives.) Initially, the station was part of the newly created Development Department, then from 1908 to 1911 it reported directly to Pierre S. du Pont, treasurer.[97] In 1911, it was placed under the management of the Chemical Department, run by Charles L. Reese.[98] Since research was by nature less routinized than production, the Experimental Station's routine vertical communication to convey standards and to monitor performance was less developed than that of the HEOD. But the station had some special communication needs of its own, stemming from its identity as a research facility and its relations with the rest of the company.

Centralized Communication for Secrecy and Control

From early in the twentieth century, Du Pont was very concerned to assure the secrecy of its research activities. Early in 1909, Pierre S. du Pont wrote the director of the Experimental Station, asking him about "the safeguards adopted by you for the guarding of the records of the Experimental Station."[99] The director's reply revealed that concern with security had already led to centralized control of producing, copying, and filing the station's communications.[100] He could enumerate, for example, exactly how many carbon copies of station reports were created and where they all went. Moreover, the station's numerically organized filing system had one series of files for regular (internal and external) correspondence, one for private or sensitive correspondence (kept under lock and key at all times), and one for reports, all centrally maintained and locked away at night. The antitrust decree increased the preoccupation with secrecy, since it forced the company to spin off two competing companies.[101]

If the desire for secrecy accounted in part for centralized storage and routing of communications at the Experimental Station, a desire for control—both by the Chemical Department over the station and by the station over its more independent subunits—also played a role. In the Chemical Department's desire to consolidate power over the station in 1911, Reese looked first to the communication system. He could not have the station merge its files with those of the Chemical Department, since the department's office was in the Wilmington headquarters and the station was several miles away. Nevertheless, Reese understood the importance of controlling communication in managing a staff unit of this sort, for which the main products were, in a sense, written documents:

In order to facilitate this department in the management of the
Eastern Laboratory and the Experimental Station, all letters requesting new work or involving the policy of the work being carried
out at the laboratories should be sent to this office. In the future,
all reports from either of the laboratories will be issued from this
office, and duplicate copies will be kept on file in this office for
reference, obtainable by those not regularly receiving these reports.[102]

In the wake of the antitrust consent decree two years later, Reese reinforced this rule in an exchange of correspondence with the station director, eliciting the station director's agreement that "we will observe the
single rule of addressing all such communications [i.e., correspondence
dealing with the Station's research] and transmitting all special reports,
analytical reports, ballistic proof tests reports, etc., directly to you and
transmit all the copies to you without any copies being transmitted
directly by us to any other department."[103] By this centralized routing,
Reese was able to monitor the station's operations and control its interactions with other departments.

Similarly, a certain amount of centralization in the handling and routing of communication within the station allowed its director to maintain control over the various divisions. Until 1917, correspondence and
reports being sent out of the station were first routinely read and approved by the division head and the station director or assistant director.[104] When the Experimental Station took in the Smokeless Powder
Inspection Station (which became known as the Ballistic Division), that
unit had its own set of files. The station director decided that they
should be gradually phased out, for, as he pointed out, "since the consolidation with the Experimental Station, it is of course the wrong principle to run a separate line of files."[105] He wanted a single set of files "in
order to avoid confusion, as well as save time and considerable trouble
in general." These practical, efficiency-related reasons were of course
reinforced (if not outweighed) by his desire to gain control over the unit.

Centralization within the station was not absolute. The Ballistic Division, for example, apparently resisted the director's efforts and continued to maintain its own files internally for at least another five years,
though anything being sent outside the division was filed in the central
station files under the station's designations, as well.[106] Moreover, at
least one of the division heads decided that his approving all outgoing
mail from his division was inefficient and stopped requiring it.[107] In 1917,
a new station director rescinded the approval system on routine matters.[108] Whether there was less resistance from the divisions by this time
or whether the volume had increased to a point that rendered the previous plan completely unwieldy and inefficient, the director was willing
to sacrifice some control for increased efficiency. Although exceptions
remained, motives of both secrecy and control drove the Experimental

Station to maintain predominantly centralized files and routing of reports and correspondence throughout the first two decades of the century.

The control afforded by hierarchical routing, along with the small size of the station and the nature of experimental research, meant that downward communication to standardize operations and equipment was much less important here than in the HEOD. The station was not large—even at the height of its wartime expansion in 1918, it had only 241 employees in nine divisions.[109] Moreover, research work was by nature less standardizable than production work. In 1911, Reese sent the director of the station a copy of Frederick Taylor's "Principles of Scientific Management," querying him about whether the principles presented could be applied to any Experimental Station work.[110] His reply was that the principles applied only to routine work, not to experimental work. Although certain aspects of the station's work could be and were standardized, many others could not. Thus, there were relatively few directives conveying procedures or rules from the station director to the division heads. These few documents, which frequently concerned methods for reporting research, were initially filed with internal correspondence and designated "Memoranda," but beginning in 1913 they were kept in a separate file marked "Circular Letters."[111] Because of the small number of division heads, carbon copies were generally used for dissemination.

The first comprehensive compilation of rules, the "Experimental Station Rules and Regulations," was established in 1917, a full decade after the HEOD began issuing HOW.[112] Even then, the director of the station expressed a certain reluctance to codify rules too extensively: "We do not desire to lay down a multitude of rules for the conduct of our men at the Station but there are certain rules which every new man should be made acquainted with as soon as he arrives." In the small and nonroutinized world of the Experimental Station, downward communication of rules was less critical as a method of control than centralized handling and routing of communication with other departments.

Reporting Research

A system of reports was established early in the existence of the station to document its activities and communicate the results of its research. By at least 1908, it was issuing various types of periodic and investigative reports to allow routine monitoring and control of its activities and to communicate and document the methods and results of its research.[113]

Weekly and monthly reports allowed the directors of the station and of the Chemistry Department to monitor ongoing progress on various projects. Divisional weekly reports were submitted for each major line of research starting by at least 1909.[114] They were not as standardized as the reports in the HEOD or as those in Scovill's Research Department.

Although the printed forms on which they were submitted required that certain basic pieces of identifying information be filled in (date, file number, subject, etc.), no other structure was established. What followed the heading could be simply "Status unchanged," or it could be one or more paragraphs of prose plus occasional tables describing work done on a particular project that week. Monthly reports were slightly longer versions of the same thing, sometimes on the same forms (with "weekly" crossed out and replaced with "monthly").

Monthly, annual, and eventually semiannual reports for the entire station traveled all the way up the hierarchy to Du Pont's Executive Committee.[115] These reports were compiled by the station director from reports on individual lines of research. They documented the station's accomplishments as well as enabling the committee to monitor costs and benefits of the research function and to determine overall funding for it. Over time, the reports came to include increasing amounts of financial data that broke down expenditures into various categories.[116]

The format of these reports to the Executive Committee might have been expected to evolve in ways that would make them more efficient to write and to read, but because of the nature of the research endeavor and the station directors' lack of interest in systematization, the amount of actual evolution was relatively small. The burden of writing was handled by the simple expedient of farming out the sections to various individuals in the station. These sections were relatively long and loose in form, since the output of a month or year of research could not as readily be conveyed in compact quantitative terms as could the output of a production unit. As early as 1909, the Annual Report was 77 pages long, and the 1911 report was a staggering 147 pages long. To make matters worse, the director who took over in that year, Fin Sparre, eliminated the initial summary that had been provided in earlier reports.[117] In 1914 and again in 1918, contributors were exhorted to reduce the length by "the economical use of words," but these efforts did not achieve much.[118] Stronger measures might have been taken to reduce the length of station reports, except that this was the period during which the Executive Committee had a special Reports Committee to summarize all reports for them. Thus, the efficiency of readers was never really taken into account in designing the station's reports.

In addition to the periodic reports, the Experimental Station created various records and reports of the research projects themselves. Its most basic and routine research records included, for example, the results of proof tests of various powders performed by the Ballistic Division.[119] While some of these tests assessed new, experimental formulations, others were performed as a service to the production departments to monitor the quality of the powders being produced and sold. These were perhaps the most standardized of the Experimental Station's records, with quantitative measurements provided in a standard tabular form accompanied by a brief prose interpretation. A 1918 directive to the Ballistic Division staff, noting that sometimes special tests were only reported verbally to the requester, pointed out that "this written report is essential in order that every member of the Ballistic Division organization may have access to the records."[120] Such records provided an organizational memory for the division and for the production departments it served.

In addition to such basic experimental records, investigative reports played a major role in documenting and conveying the results of the station's research to the relevant audiences inside and outside the station. Although the station's reports did not follow a specified progression like that followed at Scovill, each major investigation was the subject of several reports that served different functions. First, in most cases, was an unnumbered document serving as a proposal for a program of research.[121] Then a series of numbered and titled reports (where the numbers were linked to the filing system and the titles showed the relationship of the report to the research program as a whole) discussed specific experiments.[122] At the end of a program of experimentation, a final report summed up the results.[123] All the reports, as a by-product of centralized handling, carried the signatures of at least three individuals: the experimenter, the division head, and the director of the Experimental Station. And, of course, all copies of the report went to the Chemical Department director for distribution through his office.

The reports ranged in length and in coverage, depending on where they stood in the progression of the research, but most were loosely organized according to the chronology of the experiment.[124] Early reports and summary reports typically had several demarcated sections: an introduction that explained the goals of the research, a detailed description of the experiment, a presentation of its findings, and a closing section of summary and conclusions. Reports of experiments in the middle of the research program were sometimes quite short, often including just a brief description of the specific experiment and minimal conclusions. The findings sections of longer reports generally consisted of prose and tables, though in 1909 Irénée du Pont of the Development Department suggested increased use of graphs for this purpose, since "the results are shown much clearer than by a simple tabulation."[125]

As the number and length of the investigative reports grew (some were over thirty pages long), they began to create a burden on readers. In 1913, Reese suggested a way of reducing this burden:

> It has occurred to me in reading over your reports that it might be advisable instead of giving a summary at the end of the report to give it at the beginning, so that the person reading the report can readily see just what is covered therein and conclusions drawn, which will indicate whether or not they will wish to go over the details of the work. If this scheme is adopted, it would seem advisable to open the report with a general statement as to the object of the work; follow this with a summary of the results of the work

and conclusions drawn; and follow this with the regular detailed report.[126]

Station Director Sparre, who had eliminated the summaries in annual reports, resisted this suggestion, stating that such a summary would simply lengthen shorter reports and create repetition in longer ones. "The natural place for conclusions," he argued, "is at the end of the report and we have always supposed that you first read the introduction and then after glancing through the report, read the summary and conclusions. The time lost in turning to the end of the report cannot be very great."[127] Efficiency was clearly not high priority to Sparre.

This issue of efficient organization of reports came up again in 1918, when Hamilton Bradshaw was station director, this time to greater effect. During the summer of that year, Bradshaw circulated, with a general endorsement, a memorandum entitled "Suggestions Relative to the Preparation of Formal Reports," prepared by Dr. Tanberg from the Chemistry Department.[128] The proposed report form, which put summary and conclusions before experimental data and discussion of the literature, was clearly intended to make reports more efficient to read and use. Moreover, Tanberg noted that "the most important requirements of the first and second sections are clearness and brevity." In contrast, the third section, intended to "cover all details of the experimental work in such a way that any well-trained chemist will find it possible by following the report to repeat the work and duplicate the results," was to be quite detailed, with blueprints of the apparatus and tables to present results.

The suggestions seem to have been adopted by many station report writers.[129] Thus, the research reports were finally made somewhat more efficient for readers, but the format was never as precisely defined as the formats used in Scovill's Research Department. At Du Pont's Experimental Station, research was a more independent, less standardized activity than research at Scovill or than operations in other departments of Du Pont.

Internal Correspondence: Documenting Friction

The station's internal correspondence reflected and documented the frictions inherent in the station's status as a staff unit made up of relatively independent groups and performing research related to the work of the production departments. Both the format and the storage and handling system evolved to suit the predominantly documentary purpose.

Most of the surviving correspondence within the station documented various disputes and points of friction between the relatively large and independent Ballistic Division and the station's central management. While many of these subjects were undoubtedly discussed orally, as well, the correspondence recorded the various viewpoints. Sparre, for example, sent a memorandum conveying his desire to have the Ballistic

Division's files gradually phased out.[130] Similarly, on one day in 1919 the head of the Ballistic Division sent a series of memos to the next station director (Dr. Bradshaw) to document complaints about the way in which overhead costs were assigned to the division.[131] In his final item of the series, he concluded:

> The system by which we have a separate charge known as E-2-Bal- listic, and no corresponding separate charge for the other Divisions, is unjust and unsatisfactory.
>
> We have just sent you a number of memorandums, showing, in detail, how this has worked out for specific charges. One remedy for this situation is to abolish the account E-2-Ballistic.
>
> Please advise the reasons, if any, why this state of affairs should continue.

In another case, he complained in writing about two incidents of alleged trespassing on Ballistic Division turf and requested that policies against such actions be established.[132] Bradshaw's written reply explained ex- actly where he agreed and where he disagreed with the division head's complaints and proposals.

Such correspondence clearly existed primarily to document the ten- sions within the station. Many of these memos survived only in the Ballistic Division's files, the one set of local files in the station. Whether the existence of this set of files encouraged or simply preserved such exchanges is not clear. It is clear, however, that the head of the Ballistic Division used such correspondence as an instrument with which to attempt to regain some of the control that he lost when the division was placed within the Experimental Station.

The station also engaged in considerable correspondence with other departments. As a staff department, it had no direct products. Thus, documenting the communication by which the station's ideas were dis- seminated was critical to showing that it was a productive unit. Even oral conferences and conversations with other departments had to be "recorded in the files," the director of the station reminded the division heads in one circular letter.[133] He went on to explain: "Whenever a member of your division makes a suggestion at one of the plants which may lead to improvement in the operations or the product, he should give you a memorandum covering the suggestion made, in order that you may use your judgment as to whether the matter is of sufficient importance to justify writing a letter to the main office of the Chemical Department." The motive behind much of the correspondence, then, was to claim credit for the station's accomplishments. Much of the station's general internal correspondence, as well as many of the re- search reports, performed this function.

This correspondence was also shaped by the centralized control that the Chemical Department maintained over the station's communica- tion. All reports and memoranda concerning research were addressed to

Chemical Director Reese and sent to him with specified numbers of carbon copies so that the Chemical Department could file one copy and send out enough to serve the receiving department's needs.[134] Reese forwarded them to the director of the appropriate department, who then sent them on to the appropriate recipient within the department.[135] Thus, interdepartmental correspondence was routed into a hierarchical path.

This attenuated path had consequences for the style of the correspondence. Because the real recipient was so many steps removed from the writer and because so many other people would read a given memorandum, the style tended to be dry and impersonal, as the following example from the head of the Ballistic Division illustrates:

> 1. Advise that the copies of the reports of ballistic tests furnished us by Carney's Point are not satisfactory, in that very frequently we are furnished the carbon, barely legible.
> 2. These sheets are used for record file, and the attempt is made to make use of them in connection with our work. With such copies as we receive from time to time, this is practically impossible.
> 3. Request that Carney's Point take measures to provide us with clearly legible copies of all tests made.[136]

The impersonality of this complaint is obvious in the structure, the absence of subjects in several sentences, and other features. The style clearly contrasts with that used by the same individual when he complained to the station director about the E-2-Ballistic account.

Before 1914, the format of the station's internal correspondence resembled that of its external correspondence except for the presence of file numbers.[137] This form changed significantly in 1914 after T. C. du Pont disseminated the study on efficiency in internal correspondence written by the HEOD Efficiency Division.[138] Sparre endorsed most of the recommendations in that study, including the new conventions for headings.[139] In the matter of omitting polite but wordy phrases, however, he made a very clear distinction between "all routine letters which are handled almost entirely by clerks in an entirely impersonal way," in which "all polite phrases should be omitted," and those that "will be read by men of high positions, as well as letters in general discussing technical subjects," in which "the brief style cannot very well be used," but in which "it is at the same time requested that shorter words should be used as far as possible, and unnecessary phrases omitted." From 1914 on, with occasional modifications, the new format was followed for internal correspondence, differentiating it from external correspondence.[140] As a reminder note pointed out to station members, following the standard format "is desirable not only for the sake of appearance but for the benefit it gives our filing system to maintain as great a degree of uniformity as possible in this connection."[141]

The station's internal communication reflected both its special needs based on its research function and, to a lesser extent, the principles of efficiency. During the war years, with Bradshaw rather than Sparre in charge, the station's commitment to systematization increased. More activities were routinized and many new forms created as the mechanisms for such routinization.[142] Nevertheless, the station's activities were never systematized as much as those of the production departments because research was inherently less susceptible to routinization.

Further Evidence of System in Du Pont, 1902–1920

The relatively rich surviving records of the HEOD and the Experimental Station provide clear evidence of the development of their communication systems. Although surviving records for the rest of Du Pont are not as extensive, they illustrate some other ways in which the principles of systematic management were being implemented through developments in the communication system.

During the first two decades of the twentieth century, circular letters and manuals were the major media through which equipment and processes were standardized in Du Pont. While many standard-setting communications traveled directly down a departmental hierarchy, as in the HEOD and the Experimental Station, some were issued by staff departments to employees throughout the company. The comptroller, for example, issued directives to all employees establishing procedures for ordering supplies.[143] The Engineering Department's "Standard Practice," first issued in 1920, codified procedures for its employees at the Wilmington offices and in every Du Pont plant throughout the country.[144] Controlling members of a staff department operating within line units was a challenge to standard setting. The preface to the 1920 edition of the manual addressed this issue as follows:

> It is our desire that these instructions be as complete and cover as many conditions as possible. For this reason you will study the sheets [of the loose-leaf standard practice manual] as soon as received and advise us of any discrepancies that exist, bearing in mind that there are at present several different systems in effect at the various Plants covering the same details. It is the intention to standardize our work as much as possible, but if it should be found that the adoption of the methods outlined is impracticable due to local conditions at the particular Plant on which you are stationed, you will advise us immediately and revision of Standard Practice will be made if necessary; otherwise you will be instructed how to proceed.

Such a manual was one of the few control mechanisms available to a dispersed staff department.

Another type of communication used in standardizing procedures was also the first level of the reporting system. Job tickets and similar forms were generated at various points in work processes both to standardize the procedure and to provide the basic data needed for monitoring operations. Sometime before 1906, a shop order system was adopted in the Brandywine facility to track production costs, and in that year a system of forms and cards was established "for keeping a record at each Plant for Construction and Repair work."[145] While the primary point of this new system, from the comptroller's point of view, was to provide accurate records for the Accounting Department and for the superintendent or chief engineer of the plant, the forms themselves also built certain steps into the construction and repair work.

Similar primary records were generated for financial transactions. One of the first challenges that faced Pierre S. du Pont and the Treasurer's Department in 1904, when the company began to consolidate the many powder companies it had taken over into a single operating company, was to create the basic procedures and records needed to manage its finances systematically. As Chandler and Salsbury have noted, "From the start Pierre looked upon his office as the one to provide the information necessary to evaluate, appraise, and plan the company's business. Accurate data on all aspects of costs and income had to be developed."[146] The records and procedures that they developed both standardized the processes for handling such transactions and created records necessary for evaluation and control.

The steps involved in processing a single financial transaction through the system were necessarily quite elaborate. At one point, one of Du Pont's vice-presidents complained to the comptroller that a voucher sent through the system for one $0.20 pint of alcohol had acquired sixteen different approvals, registered by initials, stamps, or signatures, as well as an attached form in duplicate.[147] He raised the question of "the slavery to system that compels, I suppose, a dollar's worth of work, and maybe more, in order to procure an article costing $0.20." In reply, the comptroller explained the process that created these notations, ending with the observation that "if any were omitted the record would be incomplete, and there would be left a loophole through which loss might be incurred."[148] His response made clear the extent to which systematic procedures were intertwined with primary records, as well as the importance of those records to higher-level control of the company's accounts.

Based on such primary records as these, departments throughout Du Pont had built systems of routine reports to pull data up the hierarchy. These reports allowed each level to monitor and control lower levels. A major portion of the Engineering Department's manual on standard practice, for example, was taken up by instructions and forms for various types of reports.[149] For the Mechanical Experimental Division of the Engineering Department, as for the Experimental Station, "The proper recording and reporting of the activities are as important as the actual prosecution of the work in that they form the only permanent record of the accomplishments of the Division." After describing how reports were progressively consolidated and summarized as they proceeded up the hierarchy, the manual concluded with the following statement of their purpose: "From these reports and summaries, it is possible for the Assistant Chief Engineer of the Staff Division to prepare his report to the Chief Engineer and Executive Committee. It is equally possible for him to keep directly informed as to the work being done and results obtained by the individual members of the Division."

Similarly, Du Pont's Sales Department had adopted and elaborated the system of sales reports developed at Repauno in the late nineteenth century.[150] The data provided by the reports aided sales offices in evaluating the salesmen and the main sales department in evaluating sales offices. Thus, at each stage managers used the results to monitor and control the work done at lower levels, as well as to summarize the material further for the next level of the hierarchy.

Much of the quantitative and qualitative information gathered in routine reports from departments traveled all the way up the hierarchy to the Executive Committee. The committee received monthly and annual reports from all departments beginning in 1903/1904.[151] Although the reports from the Experimental Station were composed primarily of prose, reports from sales and production departments, for example, included standardized statistical information used to compare past to present performance. Chandler and Salsbury point out that in the early years of the consolidated Du Pont, "The managers had to become sophisticated in their analyses and understanding of such data. The rise of the modern corporation thus demanded the development of the basic business techniques of control through statistics."[152]

Meanwhile, the Treasurer's Department and other executive bodies were establishing systematic methods for monitoring financial data and controlling appropriations.[153] These, too, depended on the upward flow of reports bringing key data to the top levels of the company. Among other measures instituted, in 1911 a companywide Appropriations Committee required that each department submit semiannual forecasts of their probable expenditures computed from progressive aggregations of forecasts from the bottom up.[154] At the top, the Executive Committee or other high-level committees could monitor the actual performance against the forecast to pinpoint areas that needed attention.

During the war years, as rapid expansion and contemplated future diversification created a need for better financial tools for assessing the value of new projects, Treasurer F. Donaldson Brown developed a new technique for measuring return on investment for proposed projects.[155] This technique used his department's forecasts of financial conditions and the proposing department's forecasts of the project's outcome to calculate a probable return on investment (ROI) for any project, whether a minor modification of a process, a new facility, or a new product line. As with all other executive and financial control at the top, this analytic technique depended on the accuracy and consistency of the information flowing up from lower levels.

By the end of the war, the array of statistical information reaching the Executive Committee was so great and so complex that new methods of assuring its consistency and facilitating its comprehension were developed. The General Statistical Committee was formed "for the purpose of coordinating the methods and routine of our Statistical and Forecast work, as it effects [sic] all departments."[156] The company's general statistician chaired the committee, which also included those in charge of statistics and forecasting in the Production, Sales, Accounting, Purchasing, and Treasurer's departments. This committee, in a role similar to that filled by E. H. Davis at Scovill, worked on standardizing both the forms and the reports that pulled data up to the top level, and the routines that produced them. Treasurer Brown explained the role of the committee as follows: "Just as the Executive Committee serves to coordinate the elements of executive control, so, also, does the General Statistical Committee serve to coordinate the data, which makes executive control more efficient."

But while this committee could improve the quality of the data flowing up the hierarchy, it could not help the Executive Committee members with the problem of information overload, which was growing ever worse. During and after the war, the committee had to monitor incredible growth, shortages and price fluctuations in raw materials, and the beginnings of product diversification, in addition to normal operations. Even under prewar conditions, the number and length of reports and the amount of information the committee had to read and absorb frequently got out of control. Under these new and volatile conditions, they needed to have a grasp of even more information. Sometime before August 1919, an "executive series of diagrams" was inaugurated to present great quantities of information in a form easy to understand and monitor.[157] These large line graphs portrayed forecast, actual, and revised forecast values for various financial measures (such as total construction expenditures or net working capital) and for inventory levels of key raw materials (including soda, sulfur, and glycerin). They made these figures much easier to understand and interpret and thus were an initial step toward combating the problem of executive overload.

A further step occurred in response to the firm's postwar evolution from a functional hierarchy to a multidivisional structure, completed in September 1921.[158] In this new structure, each division's general manager, rather than the Executive Committee, was responsible for the division's financial performance; the committee was in charge of allocating funds to divisions and evaluating the general managers on the basis of the division's financial performance. This restructuring changed the committee's information needs. Now it was specifically enjoined from becoming involved in the operations of the division. Brown's ROI formula (see fig. 8.4), because it could be used to analyze the return produced by any division without regard to its product or its operations, provided the necessary financial tool for evaluating each division's fi-

RELATIONSHIP OF FACTORS AFFECTING RETURN ON INVESTMENT

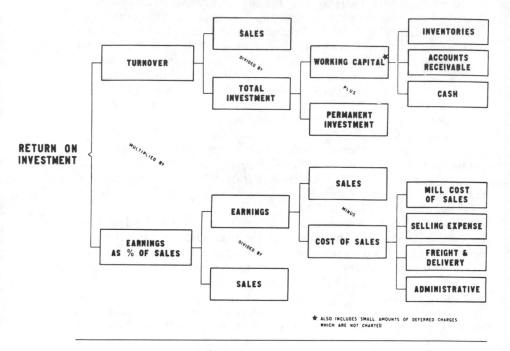

8.4.   Diagram of F. Donaldson Brown's return on investment formula. *(T. C. Davis, "How the Du Pont Organization Appraised Its Performance," presentation at the 1949 conference of the American Management Association.)*

nancial performance in comparison with both its own past performance and the performance of other divisions. The flow of reports up the hierarchy essentially ended at the divisional level. Information was then totally reorganized into the components of the ROI formula in order to allow the committee to monitor financial performance without getting involved in operating details.

The only remaining issue was how to provide this information to Executive Committee members without overburdening them with reports, a chronic problem for the committee since its formation. According to later accounts by Angus Echols, then assistant treasurer, the Du Pont chart room was a direct answer to this problem of information overload.[159] The chart room (see fig. 8.5) was a room on the executive floor of Du Pont's Wilmington office complex specially equipped for displaying graphs.[160] From its ceiling hung large, metal-framed charts displaying the overall return on investment and each of the constituent parts of the ROI formula for each of the divisions. A system of tracks and switches allowed any one of the 350 charts to be moved to the center position. There the committee, seated in a semicircle, could view and discuss trends for a given division or for two or more divisions.

The graphs helped committee members monitor and absorb the information needed to perform their oversight function. In the years since 1904, when a previous Executive Committee had turned down a suggestion that Sales Department information be presented to them in graphic form, Du Pont executives had learned the power of graphics to present information economically and vividly. Moreover, the committee's mandate had changed from overseeing operating details of all departments to overseeing only the financial performance of each division. Although the 1904 committee had feared that charts would obscure details, the 1921 committee wanted to obscure operating details and to focus on the general movement of the financial data. The ROI formula made such monitoring of financial performance possible, but the chart room increased the efficiency and effectiveness with which the committee could perform this function.

Vast quantities of information also had to be dealt with at lower levels of the organization. In the short term, information was often needed in a more compact form than offered in the various reports and other documents. In the long term, even with entire rooms devoted to filing in most departments, space was not infinite and something had to be done to maintain a long-term organizational memory without swamping the company in records. Du Pont developed ways of dealing with both these problems to create accessible organizational memory.

The Sales Department, for example, established the Sales Record Division to maintain an active and accessible base of up-to-date information. As the head of the division said in justifying the unit, "Our sales

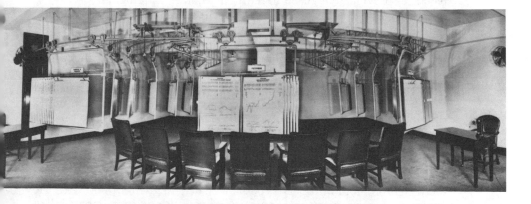

8.5.   Photograph of the Du Pont chart room. *(Courtesy of Hagley Museum and Library.)*

records are entirely a sales proposition comprising as they do *all* information concerning the smallest unit (customer) in *one* place for *Quick reference*—and *Follow-up of trade and salesmen.*"[161] The data on each customer, kept on a set of cards, were extensive, including everything from local credit standing and kind of business to approximate annual consumption of Du Pont products and date when the customer would next be in the market. Before, much of this information would have been carried only in the salesman's head, but now it was systematically recorded in a compact form accessible to all potential users. The cards were organized alphabetically by customer, but tabs were attached to indicate the kind of business and the date the customer would next be in the market, in order to allow easy access to the information along those dimensions.

This vast array of data served many useful functions for the Sales Department. In 1913, the division filled approximately twenty-three thousand routine requests for information from other headquarters divisions of the Sales Department, as well as many special requests from the field and from headquarters to compile certain sets of data from their records. Their information was used both to help monitor and control the Sales Department (*"To check the work of salesmen"*) and to help monitor and aggressively pursue the trade. In the latter role, it allowed the sales force to follow up all trade by type of business and date of expected reentry into the market. It was used to compile block books listing all actual or potential customers in a given territory and to create mailing lists for advertising. Moreover, this corporate data base of sales information fulfilled some special needs that were probably not foreseen. After the San Francisco earthquake destroyed the records of that sales office, the Sales Record Division was able to re-create them from its records. Also, in fighting the various government antitrust suits, the card files provided data used to refute inaccurate statements.

Although the Statistics Office at Scovill served as a central clearinghouse and storage place for data for the whole company, in the much larger postwar Du Pont, the Sales Record Division served that function for Du Pont's Sales Department alone. Similar units, the primary purpose of which was to supply accessible data, were established in other departments of Du Pont. In fact, the executive-level General Statistical Committee was formed in part because Du Pont had several such nodes of information in different areas of the enormous company, the work of which needed to be coordinated.

In addition to assuring short-term access to active and immediately useful data, the company also had to face the problem of maintaining long-term corporate memory. The new, post-1902 company apparently faced that problem quite early, perhaps in deciding what to do with the extensive but unsystematic pre-1902 records. By at least 1907, it had established a functioning records center called the Hall of Records to serve as a repository for old records.[162] This repository was intended to be "a storage place for books, records and valuable papers which have ceased to be of active use but which, for good reasons, it is desirable to keep."[163] To assure the disposal of unimportant records and the survival of important ones, the Executive Committee created the policy that "branch offices, mills and factories be not permitted to destroy any records, but to forward their old records from time to time to the proper departments in Wilmington, such records on receipt to be gone over by the department in Wilmington and the necessary records sent to the Hall of Records and unnecessary ones destroyed."[164] Thus, the company, under the influence of both the du Pont family propensity to save documentation and the systematic management movement's desire to maintain a corporate memory independent of individuals, arrived at this long-term solution to the problem of ever-growing records.

Finally, managerial meetings were widely used throughout Du Pont to reinforce and ease the implementation of other aspects of systematic management. In addition to the Executive Committee and the HEOD superintendents' meetings, regular meetings were also used at the plant level. The smokeless powder plant at Carney's Point, for example, had weekly staff meetings.[165] At these meetings, routine reports monitoring safety and quality were always presented, frequently in graphic form that highlighted deviations from the normal. The plant manager also used the meetings as a forum for reinforcing plant policies and procedures and for addressing other issues of concern to members of the staff. After a year or so of meetings, the plant superintendent formalized the nonroutine items on the agenda by requiring that anyone wanting to raise an issue for discussion submit a memorandum on it in advance.[166] Minutes of these meetings were recorded and saved, providing another form of written documentation. Carney's Point staff meetings, like HEOD superintendents' meetings, allowed multidirectional communication and functioned to humanize some aspects of management, at the same time that they promoted and reinforced systematic management values.

Conclusion

Post-1902 Du Pont used formal internal communication to systematize procedures, to monitor and control operations, to embody organizational memory independent of individuals, and to reinforce new principles and practices in a more personal way. All of this communication was either in writing, or, in the case of meetings, based on and recorded in written documents. Du Pont's internal mail system became increasingly extensive and ambitious in order to handle these flows of paper. By 1919, the Mail Department delivered approximately fifty-five hundred pieces of interdepartmental mail within the Wilmington headquarters daily.[167] The Mail Department delivered items only to the division or department office, from which each unit had to arrange for its own internal distribution. In fact, the subject of the very first efficiency meeting in one division was handling the distribution of mail once it reached the division.[168] An efficient internal mail system was essential to the formal communication system.

In addition, the Planning Division of the Service Department was established for "the promotion of office improvements and standards."[169] Its task of studying and improving clerical work, including procedures for producing, duplicating, storing, and transporting written documents, began informally in 1910 and grew from then. In 1917, an outside efficiency expert was brought in to organize the division and its work. Thus, the communication necessary to systematize the company was itself systematized.

Du Pont's systematization after the 1902 change of leadership had been rapid and extensive. The company was particularly innovative and successful in its use of meetings as a means of easing the introduction of system. By 1920, the firm was coordinated and controlled by a web of communication.

Conclusion

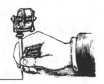

Although their paths differed, the Illinois Central, Scovill, and Du Pont all made the transition from traditional, ad hoc managerial methods to systematic managerial methods during the late nineteenth and early twentieth centuries. In each case, the formal communication system simultaneously emerged as an important control mechanism. Procedures, rules, and financial and operational information were documented at all levels, making organizational rather than individual memory the repository of knowledge. Impersonal managerial systems—embodied in forms, circular letters, and manuals—replaced the idiosyncratic, word-of-mouth management of the foremen and owners of earlier periods. Information and analyses, increasingly in statistical form, were drawn up the lengthening hierarchies to enable upper management to monitor and evaluate processes and individuals at lower levels. Yet, while all three case studies illustrate the general outlines of this development, the chronology and immediate motivations of specific changes varied from firm to firm.

Both the similarities and the differences are revealing. First, the cases all illustrate that growth by itself did not guarantee systematization. Growth might multiply the problems caused by ad hoc management methods, as all three companies discovered at various points, but it did not immediately force the establishment of systematic methods. Although virtually from its incorporation the Illinois Central Railroad was one of the largest railroads in the country, railroads such as the Erie far outstripped it in managerial methods. After it established a rudimentary communication system to assure safety, honesty, and consistent service, it failed to make substantial progress in management methods for another quarter century, in spite of continued growth. Similarly, Du Pont expanded rapidly during the late nineteenth century, but its management methods remained those of midcentury. Both Du Pont and the Illinois Central were operating quite inefficiently during these periods, but whether for lack of strong competition or for other reasons, they survived. Scovill's gradual evolution of management methods more nearly kept pace with its growth during the second half of the nineteenth century. Yet, around the turn of the century even Scovill reached a point when stagnation in managerial technique caused chaos.

Advances in communication technology by themselves were equally incapable of forcing systematization. The Illinois Central had virtually limitless access to the telegraph before the Civil War but used it only to support existing unsystematic patterns of communication. It took wartime urgency and schedule irregularities to induce its management to convert to telegraphic dispatching, and they still did not use the telegraph for other systematic types of reporting. At the conservative Du Pont Company, inventor and scientist Francis G. du Pont was far ahead of the company as a whole in adopting the technologies of written communication. Yet, in spite of his evident fascination with typewriters, filing systems, dictating machines, and other such devices, he was far from progressive in his management methods. Repauno's management had a much clearer sense of the possible uses to which such technological aids could be put. And in the case of Scovill, adoption of technology sometimes lagged behind progress in management methods. For example, the company adopted circular letters as a method of communicating rules and standards before it had adopted the duplicating and filing technologies that best suited such communication. Thus technology alone neither caused nor was essential to the early stages of systematization, though it was an enabling and promoting factor in further development.

Geographical dispersion promoted the early development of flows of internal communication, but again did not necessarily forward real systematization. Both Scovill and Du Pont scattered commissioned and then salaried sales agents around the country, and the earliest regular flows of internal communication in the firms were to and from them, though this correspondence tended to be more voluminous than systematic. Scovill, whose main facilities were the most self-contained of the three firms, was the last to show evidence of internal written communication within the plant. Du Pont's facilities were slightly farther flung, since safety dictated that they be located at intervals along the Brandywine, with buffer zones between. This small-scale dispersal probably accounted for some of the many internal notes and letters that have survived from its early days. Moreover, the establishment of secondary manufacturing centers such as the Iowa Powder Mills and the Wapwallopen plant in Pennsylvania led to the first signs (albeit primitive ones) of regular reporting as a mechanism of control. Because the Illinois Central was, by the nature of its business, spread out over a considerable distance, it had to establish at least rudimentary systems of written communication from its very beginning. Word-of-mouth management was not possible on a railroad. Yet, neither in the Illinois Central nor in the two manufacturing companies did geographical spread cause real systematization.

Perhaps more significant than mere distance were concerns about safety. Written communication coordinating the various railroad employees along the line might have been as haphazard as that between Scovill or Du Pont and their agents, but the dangers of coordinating trains, especially on the early single-track lines, demanded that tighter control be established. After the young industry experienced a few crashes such as those on the Western Railroad, managers learned that they could not afford to be casual about certain types of communication. The Illinois Central thus established circular letters and rules printed on the backs of schedules as regular channels of downward communication at the very beginning of its operations. Du Pont, as a manufacturer of explosives, also faced important safety issues. Here the dangers did not involve coordinating the activities of widely dispersed employees, so safety did not mandate the establishment of an extensive system of downward communication. Yet, it is not by chance that virtually the only piece of mass communication surviving from the firm's first century is a notice concerning the handling of matches. The need to assure safety was a compelling force in the establishment of initial stages of formal communication.

Where it existed, regulation also played an important role. The ICC reporting requirements forced the Illinois Central to collect some kinds of data that it had not previously collected. Once these statistics were gathered, they were available for further analysis. In addition, ICC publication of comparable data from all the country's railroads made it much easier to compare performance across different lines. Such comparisons could highlight inefficiencies where they existed, making companies such as the Illinois Central more open to new managerial techniques and communication mechanisms.

In all three cases, the single factor most immediately related to the emergence of communication as a managerial tool was the intervention of a strong manager championing the new theories. While Stuyvesant Fish was aided in his efforts by ICC-mandated changes in internal reporting, his efforts to improve the reporting system even before becoming president of the Illinois Central in 1887 showed his commitment to the principles and communication mechanisms of systematic management, and the changes he instituted after 1887 exceeded those required by the ICC. At Scovill, John H. Goss, a member of a new generation of managers, came into the firm with a strong belief in the new managerial methods and initiated the period of most intense systematization. Over a decade later, E. H. Davis picked up where Goss had left off to complete the process. At Du Pont, the effects of a generational change in management were even more dramatic. The three cousins and the team that they brought with them were experienced managers who had seen the principles of systematic management at work elsewhere. They came into Du Pont determined to reap the benefits of operational as well as financial consolidation. This team transformed the company, building a new communication system through which to achieve efficiencies that had eluded previous generations of du Ponts.

In each of the case studies, then, one or more committed managers played a significant role in introducing new management methods and communication mechanisms. Although three cases cannot prove that such champions were necessary for thorough and successful systematization, they played an important role for at least two reasons. First, introducing the new methods exerted increased control over subordinates at every level from the Illinois Central general superintendent to Du Pont's HEOD plant superintendents to the Scovill foremen and finally to the workers at all of the companies. Opposition was virtually certain at some of these levels. Furthermore, standardizing procedures and instituting systematic reporting had clear costs. Both Hamilton Barksdale in Du Pont's HEOD and Stuyvesant Fish in the Illinois Central faced opposition based on these added costs. Overcoming these two types of resistance required the courage of conviction; half-hearted attempts might have stalled.

Finally, both Scovill and Du Pont, along with many other companies, discovered the importance of countering the depersonalizing effects of system with some communication mechanisms aimed at humanizing the workplace. With its slower growth and changes, Scovill found the in-house magazine an effective medium for personalizing relations. To achieve its more rapid transformation, Du Pont turned to managerial meetings. These meetings allowed upper management to monitor its introduction of the new methods, responding to problems as they arose rather than after they had reached a crisis stage. By humanizing the otherwise impersonal management system and thus defusing hostility, these modes of communication indirectly reinforced efforts to systematize the firms.

Implications

It is always dangerous to extrapolate from a small number of cases to a general rule, and it is even more hazardous to draw analogies between the past and the present. Nevertheless, the distance and perspective provided by history may enable us to arrive at an improved understanding of contemporary events and relationships.

Perhaps the most obvious implications concern communication and information technology. James R. Beniger has recently argued that the "Control Revolution" that began in the late nineteenth century contained the seeds of today's information society.[1] Certainly, there are some parallels between the revolution in office technology of the 1880–1920 period and the revolution of the last twenty-five years. Recent innovations in computers and telecommunications have been so spectacular that contemporary commentators tend to focus solely on the technology, seeing it as the driving force causing changes in other parts of the organization. The case studies in this book, however, illustrate some of the problems with simple technological determinism. Technologies were adopted, not necessarily when they were invented, but often when a shift or advance in managerial theory led managers to see an application for them. Moreover, technologies were often adopted simply to facilitate existing managerial methods; potentially more powerful applications, such as the use of the telegraph for railroad dispatching, were ignored for long periods. The technology alone was not enough—the vision to use it in new ways was needed as well.

A related implication for contemporary issues concerns both communication technology and geographical dispersion. Just as the telegraph once opened up possibilities for wider domestic markets and more scattered production facilities to companies such as Scovill and Du Pont, worldwide telecommunications systems are now doing the same for international markets. The historical cases suggest, however, that the real potential of these networks cannot be realized through a simple extension of existing patterns of communication. Real gains await innovative thinking about the underlying managerial issues.

One more potentially valuable historical lesson is the benefit Du Pont realized from instituting superintendents' meetings at the same time that it introduced systematic management methods into the plant. This forethought in anticipating potential human problems helped the company absorb and systematize a large number of individual plants into a unified and efficient department in a very short time. As many firms have discovered, introducing a new managerial or technological system without adequate concern for the reaction of the people involved can backfire.

The historical record is, in a sense, the largest data base of all. Documentation of the past for its own sake, as the du Pont family ultimately discovered, is not necessarily useful in the present. Systematic recording and analysis of data, however, can serve as the basis for informed decision making. If my main purpose has been to understand the relationships among the communication system, managerial theory, and communication technology in the late nineteenth and early twentieth centuries, a secondary purpose has been to study and interpret historical events that may illuminate current problems and issues.

Notes

Introduction

1. See, for example, Gary John Previts and Barbara Merino, *A History of Accounting in America: An Historical Interpretation of the Cultural Significance of Accounting* (New York: John Wiley and Sons, 1979); and recently, H. Thomas Johnson and Robert S. Kaplan, *Relevance Lost: The Rise and Fall of Management Accounting* (Boston: Harvard Business School Press, 1987).

2. Alfred D. Chandler, Jr., *Strategy and Structure: Chapters in the History of the American Industrial Enterprise* (Cambridge: MIT Press, 1962); and *The Visible Hand: The Managerial Revolution in American Business* (Cambridge: Harvard University Press, Belknap Press, 1977).

3. Francis X. Blouin, Jr., points out this subtheme of Chandler's work in his review of *The Visible Hand*, "A New Perspective on the Appraisal of Business Records: A Review," *American Archivist* 42 (July 1979): 312–20.

4. See, for example, Ithiel de Sola Pool, ed., *The Social Impact of the Telephone* (Cambridge: MIT Press, 1977).

5. James R. Beniger, *The Control Revolution: Technological and Economic Origins of the Information Society* (Cambridge: Harvard University Press, 1986).

6. For example, the articles in George H. Douglas and Herbert W. Hildebrandt, eds., *Studies in the History of Business Writing* (Urbana, Ill.: Association for Business Communication, 1985); and Kitty O. Locker, "The Development of the Faceless Bureaucrat: The Emergence of Bureaucratic Writing in the Correspondence of the British East India Company, 1600–1800," unpublished manuscript.

7. Richard Edwards, *Contested Terrain: The Transformation of the Workplace in the Twentieth Century* (New York: Basic Books, 1979), p. 17.

8. Beniger, *Control Revolution*, p. 7.

9. Joseph A. Litterer, "Systematic Management: The Search for Order and Integration," *Business History Review* 35 (Winter 1961): 461–76.

10. See the Note on Archival Sources preceding the Index.

1: Managerial Methods and the Functions of Internal Communication

1. Alfred D. Chandler, Jr., *Strategy and Structure: Chapters in the History of the American Industrial Enterprise* (Cambridge: MIT Press, 1962).

2. I use the term *formal communication* to denote not stylistically formal communication but communication that is an established and documented (rather than purely incidental and ephemeral) part of the functioning of the firm. A firm's formal internal communication, taken together, makes up its *internal communication system*.

3. For descriptions of these early firms and their management, see Alfred D. Chandler, Jr., *The Visible Hand: The Managerial Revolution in American Business* (Cambridge: Harvard University Press, Belknap Press, 1977), pp. 50–64; and Daniel Nelson, *Managers and Workers: Origins of the New Factory System in the United States, 1880–1920* (Madison: University of Wisconsin Press, 1975), pp. 3–4.

4. Christopher Densmore, "Understanding and Using Early Nineteenth-Century Account Books," *Midwestern Archivist* 5 (1980): 5–19. A clear contemporary description is provided in Thomas Dilworth's *The Young Book-keeper's Assistant: Shewing Him in the Most Plain and Easy Manner, the Italian Way of Stating Debtor and Creditor* (New York: Thomas Wilson and Sons, 1839).

5. See, for example, Theodore D. Marburg, "Commission Agents in the Button and Brass Trade a Century Ago," *Bulletin of the Business Historical Society* 16 (February 1952): 13.

6. I use masculine pronouns in referring to owners and managers in this book because, during the period studied, they were almost exclusively male.

7. For descriptions of such factories, see Barbara M. Tucker, "The Merchant, the Manufacturer, and the Factory Manager: The Case of Samuel Slater," *Business History Review* 55 (Autumn 1981): 297–313; Nelson, *Managers and Workers*, pp. 3–4; and Chandler, *Visible Hand*, pp. 67–72.

8. Nelson, *Managers and Workers*, pp. 3–4.

9. Daniel Wren, *The Evolution of Management Thought*, 2d ed. (New York: John Wiley and Sons, 1979), p. 49.

10. Tucker, "Merchant, Manufacturer," pp. 308–9. See also Gary John Previts and Barbara Merino, *A History of Accounting in America: An Historical Interpretation of the Cultural Significance of Accounting* (New York: John Wiley and Sons, 1979), p. 62. H. Thomas Johnson and Robert S. Kaplan, *Relevance Lost: The Rise and Fall of Management Accounting* (Boston: Harvard Business School Press, 1987), came out as I was in the final stages of preparing this book. It discusses cost accounting in early textile mills, especially Lyman Mills, on pp. 21–31.

11. In *The Visible Hand* (pp. 72–75), Chandler shows that more advanced theory and techniques of factory management—and a more formal communication system—were developed at the U.S. Army's Springfield Armory in the early nineteenth century. Roswell Lee's principles of control and accountability were similar to those later emerging as part of the systematic management philosophy, and they also resulted in the development of a more formal internal communication system with a regular flow of information up the hierarchy. However, Chandler also shows that this development was an isolated instance not carrying over to private factories of this period.

12. Ibid., pp. 79–80.

13. "Report on the Collision of trains, near Chester," October 16, 1841, Western Railroad Clerk's File #74; in Western Railroad Collection, Case #1, Baker Library, Harvard Business School. For discussion of this incident, see Stephen Salsbury, *The State, the Investor, and the Railroad: The Boston & Albany, 1825–1867* (Cambridge: Harvard University Press, 1967), pp. 185–86.

14. "Report on Avoiding Collisions, etc.," November 30, 1841, Clerk's File #104; in Western Railroad Collection, Case #1, Baker Library, Harvard Business School.

15. Ibid.

16. "Accidents on the Western Railroad," February 1842, Clerk's File, #55; in Western Railroad Collection, Case #1, Baker Library, Harvard Business School.

17. As quoted in Alfred D. Chandler, Jr., comp. and ed., *The Railroads: The Nation's First Big Business, Sources and Readings* (New York: Harcourt, Brace and World, 1965), p. 118.

18. McCallum as quoted in Chandler, *Visible Hand*, p. 101. The summary of McCallum's principles that follows is based on the six points listed on p. 102.

McCallum himself restates his principles in terms of these two issues in a passage quoted on p. 104.

19. Ibid., p. 102.

20. Chandler, Ibid., p. 104.

21. Alfred D. Chandler, Jr., *Henry Varnum Poor: Business Editor, Analyst, and Reformer* (Cambridge: Harvard University Press, 1956), pp. 146–47. The quotations that follow are also from those pages.

22. "A New Perspective on the Appraisal of Business Records: A Review," review of Chandler's *Visible Hand*, in *American Archivist* 42 (July 1979): 319.

23. Chandler, *Visible Hand*, p. 109.

24. Ibid. The following discussion of changes in accounting practices draws on pp. 109–27 of Chandler, as well as on Previts and Merino, *History of Accounting*, pp. 55–62.

25. H. M. Norris, "Shop System," *Iron Age* 54 (November 1, 1894): 746, as quoted in Joseph A. Litterer, "Systematic Management: The Search for Order and Integration," *Business History Review* 35 (Winter 1961): 473.

26. Nelson, *Managers and Workers*, chap. 3, "The Foreman's Empire." See also Robert H. Wiebe, *The Search for Order, 1877–1920* (New York: Hill and Wang, 1967), pp. 20–21.

27. Joseph Litterer, "Systematic Management: Design for Organizational Recoupling in American Manufacturing Firms," *Business History Review* 27 (Winter 1963): 372–73.

28. Mariann Jelinek, "Toward Systematic Management: Alexander Hamilton Church," *Business History Review* 54 (Spring 1980): 69.

29. Joseph A. Litterer has concluded that previous to 1870 there is "little evidence of any literature relevant to management in the United States." ("The Emergence of Systematic Management as Indicated by the Literature of Management from 1870 to 1900," Ph.D. diss., University of Illinois, 1959, as quoted in Nelson, *Managers and Workers*, p. 49.) This statement ignores the specialized literature about railroad management discussed previously but is probably otherwise fairly accurate.

30. Litterer, "Search for Order and Integration," pp. 473–74. For other discussions of systematic management, see also Litterer, "Design for Organizational Recoupling"; Litterer, "Alexander Hamilton Church and the Development of Modern Management," *Business History Review* 35 (Summer 1961): 211–25; Jelinek, "Toward Systematic Management," pp. 63–79; Chandler, *Visible Hand*, pp. 272–81; Nelson, *Managers and Workers*, chap. 4; and Nelson, "Scientific Management, Systematic Management, and Labor, 1880–1915," *Business History Review* 48 (Winter 1974): 479–500.

31. See, for example, Wren's *Evolution of Management Thought*, pp. 111–77, which skips over the earlier proponents of systematic management to focus almost exclusively on Frederick Taylor and his followers.

32. Nelson, "Scientific Management, Systematic Management, and Labor," p. 480.

33. Jelinek, "Toward Systematic Management," pp. 64–65; Litterer, "Design for Organizational Recoupling," p. 389.

34. Taylor, *Scientific Management*, as quoted in David F. Noble, *America by Design: Science, Technology, and the Rise of Corporate Capitalism* (New York: Oxford University Press, 1977), p. 265.

35. Noble, *America by Design*, p. 260.

36. Litterer, "Search for Order and Integration," p. 474.

37. Litterer, "Alexander Hamilton Church," p. 214.

38. Alexander Hamilton Church, "The Meaning of Commercial Organization," *Engineering Magazine* 20 (1900): 395, as quoted by Litterer, "Search for Order and Integration," p. 471.

39. Jelinek, "Toward Systematic Management," pp. 64–65. These forms of communication seem a manifestation of what Richard Edwards calls *bureaucratic control*, though they appeared many decades before the post-World War II period that Edwards claims saw the origin of bureaucratic control. Edwards, *Contested Terrain: The Transformation of the Workplace in the Twentieth Century* (New York: Basic Books, 1979), chap. 8.

40. Litterer, "Design for Organizational Recoupling," pp. 378–79.

41. Horace Lucian Arnold, *The Complete Cost-Keeper* (New York, 1901), p. 9, as quoted in Litterer, "Search for Order and Integration," p. 471.

42. Henry Metcalfe, "The Shop-Order System of Accounts," *Transactions of the American Society of Mechanical Engineers* 7 (May 1886 meeting): 440.

43. Alexander Hamilton Church, "Practical Principles of Rational Management," *Engineering Magazine* 45 (1913): 675, as quoted in Litterer, "Alexander Hamilton Church," p. 223.

44. Jelinek, "Toward Systematic Management," pp. 64–65; Litterer, "Design for Organizational Recoupling," p. 389.

45. Alexander Hamilton Church, "The Meaning of Commercial Organization," *Engineering Magazine* 20 (1900): 391, as quoted in Litterer, "Alexander Hamilton Church," p. 213.

46. Litterer, "Design for Organizational Recoupling," p. 385.

47. The importance of cost accounting to the whole shift in managerial methods is highlighted by the fact that Church, Metcalfe, Slater, and others made major contributions to both cost accounting in particular and systematic management in general. See Litterer, "Alexander Hamilton Church," p. 212; Previts and Merino, *History of Accounting*, p. 116; Chandler, *Visible Hand*, pp. 272–74; Litterer, "Design for Organizational Recoupling," pp. 370, 380–82.

48. Slater Lewis, "Organization as a Factor of Output," *Engineering Magazine* 18 (1899): 67, as quoted in Leland Jenks, "Early Phases of the Management Movement," *Administrative Science Quarterly* 5 (1960–61): 435.

49. Metcalfe, "Shop-Order System," p. 441.

50. Litterer uses the phrase "administrative systems" to denote these systems of upward reports in "Design for Organizational Recoupling," pp. 384–85.

51. W. H. Leffingwell, for example, wrote a book entitled *Scientific Office Management* (New York: A. W. Shaw Co., 1917). Harry Braverman examines the systematization of clerical workers in *Labor and Monopoly Capital: The Degradation of Work in the Twentieth Century* (New York: Monthly Review Press, 1974), pp. 293–348.

52. James B. Griffith, *Correspondence and Filing*, Instruction Paper, American School of Correspondence (Chicago, 1909), p. 7.

53. Oscar Charles Gallagher and Leonard Bowdoin Moulton, *Practical Business English* (Boston: Houghton Mifflin Co., 1918), p. 184.

54. Samuel Haber, *Efficiency and Uplift: Scientific Management in the Progressive Era, 1890–1920* (Chicago: University of Chicago Press, 1964), pp. 51–74.

55. See, for example, "System for Factory Purchases," *System* 3 (January 1903): n.p.

56. For discussions of this movement, see Stuart D. Brandes, *American Wel-*

fare Capitalism, 1880–1940 (Chicago: University of Chicago Press, 1976); Henry Eilbirt, "The Development of Personnel Management in the United States," *Business History Review* 33 (1959): 348–52; Jenks, "Early Phases of the Management Movement," pp. 436–40; Wren, *Evolution of Management Thought*, pp. 202–3; and Noble, *America by Design*, p. 265.

57. Brandes, *American Welfare Capitalism*, pp. 10–11.

58. Eilbirt, "Development of Personnel Management," p. 349.

59. Henry Bruere in a letter to Stanley Fowler McCormick, July 27, 1903, as quoted in Brandes, *American Welfare Capitalism*, p. 30.

60. Frederick W. Taylor, "A Piece Rate System Being a Step toward Partial Solution of the Labor Problem," *Transactions of the American Society of Mechanical Engineers* 16 (1895): 856–903, as quoted in Nelson, "Scientific Management, Systematic Management, and Labor," p. 479.

61. Hugh G. J. Aitken, *Taylorism at Watertown Arsenal* (Cambridge: Harvard University Press, 1960).

62. Jenks, "Early Phases of the Management Movement," p. 437. Brandes sees the welfare movement as competing with the philosophy of scientific management and notes their coexistence in some companies as an anomaly (*American Welfare Capitalism*, p. 32). For the broader systematic management movement, however, the welfare movement functioned as a reinforcing rather than competing ideology. In fact, as Robert Wiebe has pointed out in *The Search for Order*, the progressives were systematizers themselves, with a "bureaucratic vision" of how to help workers (pp. 167–76).

63. Noble, *America by Design*, p. 265.

64. "The Shop Paper as an Aid to Management," *Factory* 20 (January 1918): 68.

65. Harry W. Kimball, "Fostering Plant Spirit through a Plant Paper," *Industrial Management* 57 (March 1919): 245.

66. For further discussion of such plans, see Brandes, *American Welfare Capitalism*, pp. 119–34; and Daniel Nelson, "The Company Union Movement, 1900–1937: A Reexamination," *Business History Review* 56 (Autumn 1982): 335–57. For contemporary treatments of the issue, see, for example, H.F.J. Porter, "The Higher Law in the Industrial World," *Engineering Magazine* 29 (August 1905): 641–55; Dale Wolf, "Successful Industrial Democracy: Participation Board Plan of the Miller Lock Company," *Industrial Management* 58 (July 1919): 67–71; William Leavitt Stoddard, "How Far Should Shop Committees Go? One Answer Is: So Far As Can Be Agreed To," *Industrial Management* 58 (August 1919): 121–28; and E. H. Fish, "Some Dangers in the Shop Committees," *Industrial Management* 58 (September 1919): 205.

67. Nelson, "Company Union Movement," p. 341. Quoted phrases from Ben M. Selekman, *Employee's Representation in Steel Works* (New York, 1924), p. 26.

68. William Leavitt Stoddard, "Committee System in American Shops," *Industrial Management* 57 (June 1919): 473.

69. Henry Roland (H. L. Arnold), "An Effective System of Finding and Keeping Shop Costs. Part 1. Simplicity and Sufficiency of the Job Ticket Method," *Engineering Magazine* 15 (1898): 77–78, as quoted in Litterer, "Search for Order and Integration," p. 472.

70. Lee Galloway, *Organization and Management*, Modern Business Vol. 2 (New York: Alexander Hamilton Institute, 1914), p. 170.

71. William J. Keeley, "Getting More Out of Shop Conferences," *Factory* 17 (July 1916): 68.

72. For example, one advocate said that in his company's foremen's meetings, department expense analyses were compared each month. "Getting More Out of Shop Conferences" (no author), *Factory* 18 (January 1917): 123.

73. Galloway, *Organization and Management*, pp. 170–71. The quotes in the next paragraph are also from p. 171.

74. Fredric A. Parkhurst, *Applied Methods of Scientific Management* (New York: John Wiley and Sons, 1912), p. 56.

2: Communication Technology and the Growth of Internal Communication

1. Information on the early history of the telephone may be found in Frederick L. Rhodes, *Beginnings of Telephony* (New York: Harper and Brothers, 1929); J. E. Kingsbury, *The Telephone and Telephone Exchanges: Their Invention and Development* (London: Longmans, Green, and Co., 1915); and Rosario Joseph Tosiello, *The Birth and Early Years of the Bell Telephone System: 1876–1880* (New York: Arno Press, 1979). History accompanied by analysis of its impact may be found in Herbert N. Casson, "The Social Value of the Telephone," *Independent* 71, October 16, 1911, pp. 899–906; and in essays included in Ithiel de Sola Pool, ed., *The Social Impact of the Telephone* (Cambridge: MIT Press, 1977).

2. For general historical information on the telegraph, see Robert Luther Thompson, *Wiring a Continent: The History of the Telegraph Industry in the United States, 1832–1866* (Princeton: Princeton University Press, 1947); and Alvin F. Harlow, *Old Wires and New Waves: The History of the Telegraph, Telephone, and Wireless* (New York: D. Appleton-Century Co., 1936). For analyses of its impact on business, see Richard B. DuBoff, "Business Demand and the Development of the Telegraph in the United States, 1844–1860," *Business History Review* 54 (Winter 1980): 459–79; and "The Telegraph and the Structure of Markets in the United States, 1845–1890," *Research in Economic History* 8 (1983): 253–77.

3. DuBoff, "The Telegraph and the Structure of Markets," pp. 255–65.

4. Bureau of the Census, *Historical Statistics of the United States: Colonial Times to 1970* (Washington, D.C., 1975), pt. 2, p. 807.

5. Richard John, "Private Mail Delivery in the United States during the Nineteenth Century: A Sketch," *Business and Economic History*, 2d ser., 15 (1986): 142.

6. *Historical Statistics of the U.S.*, pt. 2, p. 790.

7. Thompson, *Wiring a Continent*, pp. 204–7. See also E. A. Marland, *Early Electrical Communication* (London: Abelard-Schuman, 1964), pp. 90, 138.

8. HR doc. 24 (28-2), December 23, 1844, as cited in "The Railroads and the Telegraph, Part I," *P.S.—A Quarterly Journal of Postal History* 22 (June 1984): 7.

9. Edward H. Mott, *Between the Ocean and the Lakes: The Story of the Erie* (New York: John S. Collins, Publisher, 1899), p. 416.

10. Daniel C. McCallum, Superintendent's Report (1856), as cited in Alfred D. Chandler, Jr., *The Visible Hand: The Managerial Revolution in American Business* (Cambridge: Harvard University Press, Belknap Press, 1977), p. 103.

11. Harlow, *Old Wires*, pp. 205–12; Thompson, *Wiring a Continent*, pp. 207–9; George Rogers Taylor, *The Transportation Revolution, 1815–1860* (New York: Holt, Rinehart and Winston, 1951, 1964), p. 152.

12. McCallum, in Chandler, *Visible Hand*, pp. 105–6.

13. Chandler, *Visible Hand*, pp. 391–402.

14. Ibid., p. 396.

15. Ibid.

16. Maygene Daniels, "The Ingenious Pen: American Writing Implements from the Eighteenth Century to the Twentieth," *American Archivist* 43 (Summer 1980): 312–13.

17. W. B. Proudfoot, *The Origin of Stencil Duplicating* (London: Hutchinson and Co., 1972), p. 21. For further discussion of press copying see, for example, William D. Wigent, Burton D. Housel, and E. Harry Gilman, *Modern Filing: A Textbook on Office Systems* (Rochester, N.Y.: Yawman and Erbe Mfg. Co., 1916), pp. 53–55.

18. Proudfoot, *Origin of Stencil Duplicating*, p. 29.

19. Elyce J. Rotella, "The Transformation of the American Office: Changes in Employment and Technology," *Journal of Economic History* 41 (March 1981): 54.

20. For example, the outgoing correspondence of Joseph Bancroft and Son, retained at Hagley Museum and Library, is recorded in press books into the 1930s. Proudfoot (*Origin of Stencil Duplicating*, p. 32) tells us that in London the press book was still used by the Law Society of Chancery Lane in the late 1950s.

21. David W. Duffield, comp., *Progressive Indexing and Filing for Schools* (Tonawanda, N.Y.: Rand Kardex Bureau for Library Bureau, 1926), p. 9.

22. J. Camille Showalter and Janet Driesbach, eds., *Wooton Patent Desks: A Place for Everything and Everything in Its Place* (Indianapolis, Ind. and Oakland, Calif.: Indiana State Museum and the Oakland Museum, 1983), a book produced to accompany an exhibit of Wooton Desks, provides a fascinating chronicle of these desks.

23. Advertising circular, the Wooton Desk Manufacturing Company, ca. 1880, Trade Catalogues, Hagley Museum and Library.

24. Betty Lawson Walters also discusses this issue in "Makers of the King of Desks," in Showalter and Driesbach, *A Place for Everything*, p. 42.

25. Advertisement of Haynes, Spencer & Co., then manufacturers of Wooton Desks, ca. 1885, Trade Catalogues, Hagley Museum and Library.

26. Deborah Cooper, "Evolution of Wooton Patent Desks," in Showalter and Driesbach, eds., *A Place for Everything*, p. 49.

27. Walters, "King of Desks," p. 42.

28. For a full discussion and illustration of these various types of flat files, see Duffield, *Progressive Indexing*, pp. 10–11; E. R. Hudders, *Indexing and Filing: A Manual of Standard Practice* (New York: Ronald Press Co., 1916), pp. iii–iv; and Margaret A. Lennig, *Filing Methods: A Textbook on the Filing of Commercial and Governmental Records* (n.p., 1920), p. 19.

29. See, for example, the Illinois Central Railroad materials described in Chapter 4. This method was probably copied from government methods, since some government agencies appear to have bound their incoming correspondence in this way by the 1820s. Significantly, businesses did not generally adopt the government registry system of keeping track of incoming correspondence stored in this way. The registry system required separate index or register volumes in which a complete entry—including date of letter, date of receipt, number it received on receipt, sender's identity and location, subject, and clerk holding the item—was created and updated for each incoming item. (See, for example, Kenneth F. Bartlett, "Early Correspondence Filing Systems of the Office of the Secretary of the Navy," *National Archives Accessions* 58 [1964]: 4–5. For a more complete discussion of the whole registry system in England and, in its modified form, in America, see T. R. Schellenberg, *Modern Archives: Principles and Tech-*

niques [Chicago: University of Chicago Press, 1956, Midway Reprint, 1975], pp. 67–93.) Although this system kept tight control over every document, it was so time consuming and costly as to be rejected by businesses.

30. Hudders, *Indexing and Filing*, p. iii.

31. William Henry Williams, *Railroad Correspondence File*, revised and supplemented by John L. Hanna (New York: Devinne-Hallenbeck Co., 1902, rev. 1910), p. 8. For discussion of the numerical system's origin in the British registry system, see Schellenberg, *Modern Archives*, p. 84.

32. Allen Chaffee, *How to File Business Papers and Records* (New York: McGraw-Hill Book Co., 1938), p. 3. The attribution to Amberg and the approximate date are supported by Chauncy M. DePew, LL.D., ed., *One Hundred Years of American Commerce, 1795–1895*, 2 vols. (New York: D. O. Haynes and Co., 1895), 2:645.

33. An 1881 catalogue of Cameron, Amberg and Co., located in the Baker Library at Harvard Business School, describes and illustrates the complete line of filing apparatus.

34. Bruce Bliven, Jr., *The Wonderful Writing Machine* (New York: Random House, 1954), p. 34.

35. Isaac Pitman, *A History of Shorthand*, 3d ed. (London: Isaac Pitman and Sons, 1891), traces its development from Roman times.

36. Janice Harriet Weiss, "Education for Clerical Work: A History of Commercial Education in the United States since 1850" (Ed.D. diss., Harvard School of Education, 1978), p. 21.

37. These drawbacks of flat filing are referred to in many early twentieth-century texts on filing as well as in pamphlets for filing company salespeople, such as the Shaw-Walker Co.'s "Sectional Filing Devices and Vertical File Systems" (Muskegon, Mich., 1909), in the Baker Library, Harvard Business School.

38. Bliven, *Wonderful Writing Machine*, pp. 24, 42; Richard N. Current, *The Typewriter and the Men Who Made It* (Champaign: University of Illinois Press, 1954), pp. 22–28.

39. Bliven, *Wonderful Writing Machine*, p. 42.

40. Daniels, "Ingenious Pen," p. 320, shows the drawing for this patent.

41. Current, *Typewriter*, pp. 29–59.

42. Daniels, "Ingenious Pen," pp. 321–22.

43. Current, *Typewriter*, pp. 59–73.

44. Daniel J. Boorstin, in *The Americans: The Democratic Experience* (New York: Random House, 1973), asserts, "The first market for typewriters was among authors, editors, and ministers, and it was assumed that the machine would be mainly a tool for the world of letters" (p. 399). As the following discussion shows, evidence suggests that the developers actually first targeted the more specialized market of court reporters.

45. Current, *Typewriter*, p. 14.

46. Ibid., p. 59.

47. Herkimer County Historical Society, *The Story of the Typewriter, 1873–1923* (Herkimer, N.Y., 1923), pp. 68–71.

48. Current, *Typewriter*, p. 87.

49. Figured from ibid., p. 105. The following figures, including the *Scientific American* estimate, come from p. 110.

50. From the *Pensman's Art Journal*, as quoted in Current, *Typewriter*, p. 110.

51. Current (*Typewriter*, p. 113) cites 1893 as the date of its introduction. G.

Tilghman Richards, in *Handbook of the Collection Illustrating Typewriters* at the Science Museum of South Kensington (London: His Majesty's Stationery Office, 1938), p. 43, dates it as 1897.

52. Bureau of the Census, 12th Census, Bulletin no. 239 (July 28, 1902), Harry E. Barbour, "Typewriters," p. 4.

53. Cited in Bliven, *Wonderful Writing Machine*, p. 66.

54. Current, *Typewriter*, p. 114; Bliven, *Wonderful Writing Machine*, p. 127.

55. For discussion of the change in clerical work ushered in by the typewriter, see, for example, Margery Davies, *Women's Place Is at the Typewriter* (Philadelphia: Temple University Press, 1982), chap. 2.

56. Figured from Weiss, "Education for Clerical Work," pp. 55–56.

57. *Historical Statistics of the U.S.*, pt. 1, pp. 139–40. By this time, the job title "secretary" had a very different connotation than it had had earlier when confidential secretaries were being trained for managerial positions. Most secretaries probably used typewriters by 1900.

58. Current, *Typewriter*, p. 114; Weiss, "Education for Clerical Work," pp. 53–59.

59. Weiss, "Education for Clerical Work," p. 53.

60. Bliven, *Wonderful Writing Machine*, p. 71.

61. Weiss, "Education for Clerical Work," p. 56.

62. Ibid., pp. 74–75.

63. As early as 1857 one commercial school's circular announced that "it is now too late to question the capacity of females for high mental accomplishments.... [W]hy are not these capacities more frequently trained to practical uses?" ("Circular and Catalogue of Burnham's Commercial Institute, Rockford, Illinois," 1857, quoted in Weiss, "Education for Clerical Work," p. 32.) By 1873, just before the first mass-produced typewriters became available, a business school could state, "The idea that a woman shall be forever an alien to business has passed away, and ladies now in all parts of our country are filling the honorable positions which, but a few years ago, it was thought man only could fill" (Bryant and Stratton Business College [Boston], *Index* 10, no. 1 [1873]: 6; quoted in Weiss, "Education for Clerical Work," p. 33). The enrollment and employment figures come from Weiss, pp. 55–56.

64. Many scholars and observers have tried to explain why the typewriter seemingly played such an important role in bringing women into the office. Weiss outlines several of these theories ("Education for Clerical Work," pp. 86–88, n. 114). See also Davies, *Women's Place*, pp. 55–78. In perhaps the most convincing explanation to date, Elyce Rotella argues that the typewriter reduced the firm-specific skills required for clerical jobs, thus encouraging firms to hire lower-salaried women, in spite of their shorter work lives ("Transformation of the American Office," pp. 51–57). See also Rotella's *From Home to Office: U.S. Women at Work, 1870–1930* (Ann Arbor: UMI Research Press, 1981).

65. William Henry Leffingwell, ed., *The Office Appliance Manual* (published for the National Association of Office Appliance Manufacturers, 1926), pp. 331–34.

66. "Report on the Observations of Methods Employed in Handling and Filing Correspondence in Railroad, Manufacturing, and Other Commercial Concerns in the Cities of Boston, New York, Philadelphia, Harrisburg, and Baltimore," in National Archives, Record Group (RG) 56, President's Commission on Economy and Efficiency, box 374, p. 9.

67. Wigent, Housel, and Gilman, *Modern Filing*, pp. 55–57.

68. Proudfoot, *Origin of Stencil Duplicating*, p. 21.

69. The information in this paragraph comes from ibid., pp. 25, 32–33.

70. Rupert T. Gould, *The Story of the Typewriter from the Eighteenth to the Twentieth Centuries*, reprinted from a series of articles in January-September 1948 issues of *Office Control and Management* (London: Office Control and Management, 1949), p. 23.

71. Current, *Typewriter*, p. 12.

72. Proudfoot, *Origin of Stencil Duplicating*, p. 33.

73. President's Commission on Economy and Efficiency, Circular no. 21, "Memorandum of Conclusions Reached by the Commission Concerning the Principles That Should Govern in the Matter of Handling and Filing Correspondence and Preparing and Mailing Communication . . ." (Washington, D.C., 1912), p. 20. The commission's work on record-keeping methods is described in Bess Glenn, "The Taft Commission and the Government's Record Practices" *American Archivist* 21 (July 1958): 277–303. Its papers are in the National Archives.

74. Duffield, *Progressive Indexing*, p. 14.

75. The Keep Commission studied similar issues around 1906. The following references to it are based on its "No. 1. Report of the Subcommittee on Distribution, Record, and Handling of Correspondence to the Committee on Department Methods Outlining a System by Which a Method of Preserving Copies of Letters Sent May Be Selected" (August 11, 1906), found in the files of the Taft Commission, National Archives, Record Group 51, President's Commission on Economy and Efficiency, 1910–1914 (PCEE), file 410.

76. The issue is treated, for example, in the Shaw-Walker Company's 1909 manual for salesmen, "Sectional Filing Devices and Vertical File Systems."

77. National Archives, Record Group 56, PCEE 374, June 6, 1912.

78. Proudfoot, *Origin of Stencil Duplicating*, pp. 34–35.

79. Leffingwell, *Office Appliance Manual*, p. 378; Proudfoot, *Origin of Stencil Duplicating*, p. 36.

80. Proudfoot (*Origin of Stencil Duplicating*, p. 36) dates the development of spirit duplicating to 1923 in Germany. The fact that the new process does not appear in Leffingwell's comprehensive 1926 manual of office appliances suggests that it did not become popular in America until later.

81. In *The Origin of Stencil Duplicating* Proudfoot gives a complete history of the stencil process up to the turn of the century.

82. Ibid., pp. 40–46.

83. Pamphlet for "Edison's Electric Pen and Press," George H. Bless, General Manager, reproduced in Proudfoot, *Origin of Stencil Duplicating*, p. 43.

84. Advertising circular, "Edison's Electrical Pen and Duplicating Press, Chas. Batchelor, General Manager, New York," 1876, from the Edison National Historic Site in Menlo Park, New Jersey.

85. "Catalogue of Telegraph Instruments and Supplies," Western Electric Company, 1883, Trade Catalogues, Hagley Museum and Library.

86. Proudfoot, *Origin of Stencil Duplicating*, pp. 50, 75–78. According to Proudfoot (p. 75), Albert Blake Dick started out in the lumber business, but he became interested in the duplicating business when he began introducing labor-saving devices to make his lumber company office more efficient. His partner in the duplicator and office machine business came from the railroad industry, the earliest major industry to encounter the need for duplicators and other technologies that supported new methods of management. Thus, they were undoubtedly

more aware than Edison himself was of the potential internal as well as external business uses of the mimeograph.

87. Ibid., p. 78.

88. Ibid., pp. 82–104.

89. William Z. Nasri, "Reprography," *Encyclopedia of Library and Information Science*, vol. 25 (New York: Marcel Dekker, 1978), p. 232.

90. Information on the photostat comes from two different commission documents: "Report to the President on the Use of a Photographic Process for Copying Printed and Written Documents, Maps, Drawings, etc.," submitted by the Commission on Economy and Efficiency, December 1911; and an undated memorandum and attached "Photostat" advertising brochure in National Archives, Record Group 51, PCEE box 12, file 045.2.

91. Figured from data in Taft Commission, "Report to the President."

92. Leffingwell, *Office Appliance Manual*, p. 404.

93. Some archivists have considered this issue, but few others. In particular, see Schellenberg, *Modern Archives*, chap. 9. See also JoAnne Yates, "From Press Book and Pigeonhole to Vertical Filing," *Journal of Business Communication* 19 (Summer 1982): 5–26.

94. Schellenberg (*Modern Archives*, pp. 81–82) also notes these two essential conditions.

95. Schellenberg states that the first vertical files were invented by the Amberg File and Index Company in 1868 (ibid., p. 83). A careful reading of his source for that statement (Chaffee, *Business Papers*, p. 3) reveals that 1868 was the year Amberg introduced the first alphabetically indexed file drawers. Although this system was, according to Chaffee, "the grandfather of all modern vertical files" by virtue of being the first alphabetically indexed file cabinet system, it was a *horizontal*, not a vertical, file.

96. Information on the role of the Library Bureau in the development of vertical filing can be found in several places: Duffield, *Progressive Indexing*, p. 11; *Vertical Filing*, a pamphlet put out by the Library Bureau, 1917–1918; George de Leneer, *An Outline of Business Systems Invented by Library Bureau, Boston, U.S.A. and Selected for High School of Commerce, Brussels University* (Boston: Library Bureau, 1904), p. 1; and Mabel E. Deutrich, "Decimal Filing: Its General Background and an Account of Its Rise and Fall in the U.S. War Department," *American Archivist* 28 (April 1965): 200–202. The Library Bureau ultimately became a division of Remington Rand, according to Chaffee, *Business Papers*, p. 4.

97. Chaffee, *Business Papers*, p. 4.

98. Hudders, *Indexing and Filing*, p. iv.

99. Taft Commission, "Report on the Observations of Methods Employed in Handling and Filing Correspondence . . . ," p. 9.

100. Duffield (*Progressive Indexing*, p. 12) discusses these advantages.

101. Shaw-Walker, "Vertical File System."

102. Duffield, *Progressive Indexing*, p. 12.

103. Williams, *Railroad Correspondence File*, pp. 7–8, 18.

104. Hudders, *Indexing and Filing*, p. iv.

105. Blanche Baird Shelp, *Office Methods*, Practical Bibliographies (New York: H. W. Wilson Co., 1918).

106. James B. Griffith, *Correspondence and Filing*, Instruction Paper, American School of Correspondence (Chicago, 1909), pp. 32–38. Except where indicated, the following description of the four systems is from this source.

107. Schellenberg, *Modern Archives*, p. 84.

108. Williams, *Railroad Correspondence File*, p. 8.

109. See Loree's introduction to the first edition of Williams's *Railroad Correspondence File*, p. 5.

110. Robert Cross, "Correspondence Filing," *Railway Age Gazette* 56 (May 15, 1914): 1075.

111. Taft Commission, "Report on the Observations of Methods Employed in Handling and Filing Correspondence . . . ," pp. 1–4.

112. Ibid., p. 9.

113. Chaffee, *Business Papers*, p. 16.

114. Shaw-Walker Co., "Sectional Filing Devices and Vertical File Systems," p. 6.

115. Edward A. Cope, *Filing Systems: Their Principles and Their Application to Modern Office Requirements* (London: Sir Isaac Pitman and Sons, 1913), p. 14.

116. Yates, "From Press Book and Pigeonhole."

117. See, for example, Griffith, *Correspondence and Filing*, and Duffield, *Progressive Indexing*, both of whom assume that filing is always centralized in a filing department.

118. John L. Hanna, "Efficient Methods of Handling Correspondence," *Railway Age Gazette* 54 (January 10, 1913): 61.

119. National Archives, RG 51, PCEE file 500.001A.

120. Leffingwell, *Office Appliance Manual*.

121. David Lockwood, *The Blackcoated Worker: A Study in Class Consciousness* (London: George Allen and Unwin, 1958), p. 89.

3: Genres of Internal Communication

1. *Genre* is a literary term for a generic form or type of discourse such as novels, plays, or poetry. The term can be extended to nonliterary discourse to refer to types of communication that have similar formats and purposes.

2. See, for example, Donald Dickson, "Humanistic Influences on the Art of the Familiar Epistle in the Renaissance," *Studies in the History of Business Writing*, ed. George H. Douglas and Herbert W. Hildebrandt (Urbana, Ill.: Association for Business Communication, 1985), pp. 11–21. Later developments in business communication styles and forms have contrasted with this early dependence on theory. In "The Divergence of 'Bureaucratic' Style from Standard English Prose in the Correspondence of the British East India Company, 1600–1800," a paper presented at the annual convention of the Modern Language Association in Los Angeles, December 28, 1982, Kitty O. Locker has demonstrated that the eighteenth-century development of bureaucratic style in letters written between the central office and distant employees in the British East India Company is "not due to changes in the theory, practice, or teaching of rhetoric from 1600 to 1800. Instead, the causes are to be found in changes in the interpersonal dynamics of the Company" (p. 10).

3. C. A. Burt, "Advantage of Written Orders," in P. W. Lennen et al., *Business Correspondence*, vol. 2 of The Business Man's Library (Chicago: System Co., 1910), p. 166.

4. "General Orders and Circulars of the War Department and Headquarters of the Army, 1809–1860," pamphlet describing Microfilm Publication M1094, National Archives and Records Service (Washington, D.C.: 1982). See also Kenneth Bartlett, "Early Correspondence Filing Systems of the Office of the Secretary of the Navy," *National Archives Accessions* 58 (1964): 5.

5. Daniel Nelson, *Managers and Workers: Origins of the New Factory System in the United States, 1880–1920* (Madison: University of Wisconsin Press, 1975), p. 44.

6. See, for example, "Rules of the Boston & Lowell and Nashua & Lowell Railroad Companies, for the Government of the Executive Service, adopted by the Boards of Directors, February 1857," Nashua and Lowell Railroad Corporation Papers, vol. 181, located in the Baker Library, Harvard Business School. This booklet gives the duties of each employee, from the manager and superintendent to the brakemen and station agents. Specific sections cover such topics as "Rules in Regard to Freight," "Running Directors," and "Signals."

7. See, for example, Stephen Salsbury, *The State, the Investor, and the Railroad: The Boston & Albany, 1825–1867* (Cambridge: Harvard University Press, 1967), p. 161.

8. "Report on the Collision of trains, near Chester," October 16, 1841, Western Railroad Clerk's File no. 74; and "Report on Avoiding Collisions, etc.," November 30, 1841, Clerk's File no. 104, p. 4. Both in Western Railroad Collection, Case no. 1, Baker Library, Harvard Business School.

9. Daniel McCallum's 1856 "Superintendent's Report for the New York and Erie Railroad Company," which created a system of management intended to assure economies of scale in the railroad business, concentrated on establishing the upward reporting system but briefly noted the need for each department head to prescribe rules for his subordinates. In Alfred D. Chandler, Jr., comp. and ed., *The Railroads: The Nation's First Big Business, Sources and Readings* (New York: Harcourt, Brace, and World, 1965), p. 107.

10. Rosario Joseph Tosiello, *The Birth and Early Years of the Bell Telephone System: 1876–1880* (New York: Arno Press, 1979), p. 107.

11. James B. Griffith, *Correspondence and Filing*, Instruction Paper, American School of Correspondence (Chicago, 1909), pp. 7, 47.

12. Burt, "Advantage of Written Orders," p. 167.

13. See, for example, the New York City Board of Education, *English and Business* (Report of the Commercial Research Committee, June 1923), pp. 59–61.

14. See, for example, the "standard practice instructions form" described in Willard S. Worcester, "How to Write and Use Standard Practice Instructions in Your Shop," *Industrial Management* 55 (1918): 402–4.

15. Ibid.

16. *Railway Age* 14 (March 1, 1889): 137. See also *Railway Age* 6 (January 13, 1881): 18; and *Railway Age* 9 (August 28, 1884): 542.

17. William Henry Leffingwell, *Office Management: Principles and Practice* (New York: A. W. Shaw Co., 1927), p. 241. See also Worcester, "Standard Practice Instructions," p. 404.

18. Lee Galloway, *Organization and Management*, Modern Business, vol. 2 (New York: Alexander Hamilton Institute, 1914), p. 325.

19. *Office Manual; Including Policy Book and Standard Practice Instructions* (A. W. Shaw Co., 1917).

20. Ibid., p. 1.

21. H. A. Russell, "Keeping Track of Work in Process," *Iron Age* 98 (July 6, 1916): 8.

22. William H. Barton, "Office Organization Manuals," *Office Manager* 1 (May 1925): 182.

23. Galloway, *Organization and Management*, pp. 325–26.

24. Barton, "Office Organization Manuals," p. 182.

25. James B. Griffith, ed., *Systematizing*, vol. 1, published by the International Accountants Society, Inc. (Detroit: Book-Keeper Press, 1905), p. 19.

26. J. William Schulze, *The American Office: Its Organization, Management, and Records* (New York: Key Publishing Co., 1913), p. 173.

27. Wallace Clarke, *Shop and Office Forms: Their Design and Use* (New York: McGraw-Hill Book Co., 1925), p. 2.

28. Frederick W. Taylor, "Shop Management," *ASME Transactions* 24 (1903): 1393, 1445.

29. See, for example, Ford W. Harris, "Fitting Special Orders into Routine," *Factory* 18 (January 1917): 42-45; and E. A. Clark, "How We Keep Costs in Our 75-Man Plant," *Factory* 17 (October 1916): 405-8.

30. Burt, "Advantage of Written Orders," p. 171.

31. Harry Kimball, "Fostering Plant Spirit through a Plant Paper," *Industrial Management* 57 (March 1919): 245-46; "The Shop Paper as an Aid to Management" was the title of a long series of articles in *Factory* beginning in January 1918 and continuing well into 1919.

32. Peter F. O'Shea, "The Shop Paper as an Aid to Morale," *Factory* 23 (September 1919): 520-21.

33. "The Shop Paper as an Aid to Management," *Factory* 23 (October 1919): 968-69; O'Shea, "Morale," p. 520.

34. Robert Voorhees, "The Factory Newspaper," *Industrial Management* 55 (February 1918): 145.

35. See, for example, Peter F. O'Shea, "The Shop Paper as an Aid to Self-Training," *Factory* 23 (October 1919): 792-93.

36. "The Shop Paper as an Aid to Management," *Factory* 23 (October 1919): 972.

37. Kimball, "Fostering Plant Spirit," p. 246.

38. Hi Sibley, "The Shop Paper as an Aid to Management," *Factory* 20 (April 1918): 776.

39. O'Shea, "Morale," p. 520.

40. See, for example, ibid.; and Kimball, "Fostering Plant Spirit," pp. 245-46.

41. Kimball, "Fostering Plant Spirit," p. 246.

42. Experts regularly identified the two types of reports constituting the genre. Lee Galloway, in *Organization and Management*, stated, "There are two kinds of reports, special or emergency reports and regular reports" (p. 180). The New York City Board of Education identified the two types as the "periodic report" and the "investigation report" (*English and Business*, p. 15). Ralph U. Fitting, author of the first text on reports, spoke of the "period report" and the "examination report," as well as adding a third type, the "research report," which may be considered a type of special report (*Report Writing*, Ronograph Series No. 9 [New York: Ronald Press Co., 1924], p. 5).

43. Fitting, *Report Writing*, p. 11.

44. Many annual reports were published in the *American Railroad Journal* during this time.

45. Western Railroad Clerk's File, 1842, no. 52, "President's 1st Monthly Communication," March 14, 1842; see also nos. 59, 66, 70, and 91.

46. McCallum, in Chandler, *Railroads*, p. 105. The quotes in the following paragraph are from the same source, pp. 106-7.

47. W. J. Johnston, *Telegraphic Tales and Telegraph History* (New York: W. J. Johnston, Publisher, 1880), p. 188.

48. George Burton Hotchkiss, *Business English* (New York: Business Training Corp., 1916–17), vol. 12, p. 57.

49. Ibid., pp. 57–58.

50. Galloway, *Organization and Management*, p. 182.

51. Hotchkiss, *Business English*, pp. 60–61.

52. William Henry Leffingwell, ed., *The Office Appliance Manual* (published for the National Association of Office Appliance Manufacturers, 1926), pp. 352–53. For examples of handwritten and typed tables, see H. A. Russell, "Factory Purchasing System—Methods and Records," *Industrial Management* 53 (August 1917): 694; and W. H. Dennis, "Handling Employment in a 500-Man Plant," *Factory* 18 (June 1917): 825.

53. Galloway, *Organization and Management*, p. 184.

54. See, for example, "System for Factory Purchases," *System* 3 (January 1903), n.p. Here, the "system" is a set of forms and cards for ordering purchases and reporting on the orders.

55. Clarke, *Shop and Office Forms.*

56. Leffingwell, *Office Management*, p. 470.

57. See, for example, H. S. Gooch, "Dovetailing Costs into the Books," *Factory* 16 (June 1916): 577–80.

58. Leffingwell, *Office Management*, pp. 483, 485–86.

59. C. L. Barnum, "The Layout and Arrangement of Printed Forms," *Office Manager* 1 (April 1925): 74.

60. Ibid., p. 75.

61. Galloway, *Organization and Management*, pp. 186–87.

62. Leffingwell, *Office Management*, p. 478; Barnum, "Printed Forms," p. 75; Ford W. Harris, "Fewer Factory Forms for Getting the Same Work Done," *Factory* 18 (April 1917): 481.

63. Barnum, "Printed Forms," pp. 75–76.

64. Harris, "Fewer Forms," p. 482.

65. Leffingwell, *Office Management*, pp. 470–77; L. F. Boffey, "Common Sense in the Purchase of Printing," *Office Manager* 1 (February 1925): 16.

66. Boffey, "Common Sense," pp. 15–16; Harris, "Fewer Forms," pp. 481–82.

67. For discussions of some of the problems created by careless design, see Leffingwell, *Office Management*, p. 483; and Boffey, "Common Sense," pp. 15–16.

68. Clarke, *Shop and Office Forms*, pp. 2–3.

69. I have discussed the evolution of business graphics in "Graphs as a Managerial Tool: A Case Study of Du Pont's Use of Graphs in the Early Twentieth Century," *Journal of Business Communication* 22 (Winter 1985): 5–33.

70. Carl Parsons, *Business Administration* (New York: System Co., 1909), pp. 214–15.

71. Howard Gray Funkhouser, "Historical Development of the Graphical Representation of Statistical Data," *Osiris* 3 (November 1937): 281, 337–42.

72. Ibid., pp. 338–39. For engineering use of graphs for experimental data, see the *Transactions of the American Association of Mechanical Engineers*. From its first year of publication in 1880, it showed graphs of experimental data.

73. Henry R. Towne, "The Engineer as an Economist," *ASME Transactions* 7 (1885–86): 428–32.

74. Willard C. Brinton, *Graphic Methods for Presenting Facts* (New York: Engineering Magazine Co., 1914), p. vii; Funkhouser, "Historical Development," p. 323.

75. See, for example, Allan C. Haskell, *How to Make and Use Graphic Charts* (New York: Codex, 1919), and *Graphic Charts in Business: How to Make and Use Them* (New York: Codex, 1922); and Karl G. Karsten, *Charts and Graphs* (New York: Prentice-Hall, 1923).

76. Leffingwell, *Office Management*, p. 205.

77. Haskell, *Graphic Charts in Business*, p. 5.

78. Brinton, *Graphic Methods*, p. 2.

79. Parsons, *Business Administration*, p. 215.

80. Brinton, *Graphic Methods*, p. 107.

81. Haskell, *Graphic Charts in Business*, p. 1.

82. For details of Gantt's life and work, see L. P. Alford, *Henry Lawrence Gantt: Leader in Industry* (New York: Harper and Bros., 1934). See also Daniel Wren, *The Evolution of Management Thought*, 2d ed. (New York: John Wiley and Sons, 1979), pp. 161–66. One of his followers wrote an entire book on the Gantt chart: Wallace Clarke, *The Gantt Chart: A Working Tool of Management* (New York: Ronald Press Co., 1922).

83. Alford, *Gantt*, pp. 207–23.

84. Wallace Clarke, *The Gantt Chart*, as quoted in ibid., p. 213.

85. Graphs used for forecasting apparently came into common use between 1914 and 1919, since the generally comprehensive Brinton did not deal with this use of charts, while Haskell did include it in 1919. And indeed, one graphics expert commented that "during the World War graphics were the main and fundamental mechanism through which control was secured," both in the armed forces and in the factories (Winfield Savage, *Graphic Analysis for Executives* [New York: Codex, 1924], p. 5). However, one of the most interesting uses of graphs in forecasting occurred long before graphs were commonly used for any business applications, much less for the more advanced forecasting. In 1841 Henry Varnum Poor's *American Railroad Journal* used a graph to project the Erie Canal's future revenues ("Examination of the Report of the Canal Board of New York, Respecting the Enlargement of the Erie Canal, etc., Made April 11, 1840," *American Railroad Journal, and Mechanics' Magazine*, no. 373 [January 1841]: 41–42.) As far as I can tell, this use of a graph for forecasting business outcomes preceded any similar attempts by decades. It is also the earliest use of a graph for managerial purposes that I located, though it appears to have been an isolated instance.

86. Haskell, *Graphic Charts in Business*, p. 6; see example, p. 183.

87. Ibid., p. 7.

88. Brinton, *Graphic Methods*, pp. 1–2.

89. Funkhouser, "Historical Development," p. 323.

90. The rules are quoted from Haskell, *Graphic Charts in Business*, p. 9.

91. Fitting, *Report Writing*, p. 23.

92. George B. Hotchkiss, head of the Department of Business English at New York University School of Commerce, published *Business English*, a 12-volume set of lessons used there, of which vol. 12 centered on report writing. MIT's course in report writing is described in "Report Writing Course," (no author) *Electric Railway Journal* 46 (December 18, 1915): 1204.

93. Fitting, *Report Writing*, p. 23.

94. Hotchkiss, *Business English*, 12: 61–62. The following quote is from p. 65.

95. Lester Bernstein, "Putting It Up to the President," *Industrial Management* 55 (January 1918): 46–47. All the quotes in this paragraph are from p. 47.

96. Fitting, *Report Writing*, pp. 11–12.

97. Ibid., p. 94.

98. Leffingwell, *Office Management*, p. 220.

99. Hotchkiss, *Business English*, pp. 61, 76.

100. N.Y.C. Board of Education, *English and Business*, p. 17. The final phrase probably implies extensive use of passive voice.

101. Leffingwell, *Office Management*, p. 249.

102. Harry B. Weiss, *American Letter-Writers: 1698–1943* (New York: New York Public Library, 1945. Reprinted from the *Bulletin* of the New York Public Library, December 1944 and January 1945.) While many of the listed books dealt more with social than with business letters, others focused on commercial correspondence. In early years, the model letters in American books were often copied directly from English sources, but gradually American models and instruction began to replace them. *The New Century Standard Letter-Writer* (Chicago, 1900), p. 23.

103. Carter Daniel, "Sherwin Cody: Business Communication Pioneer," *Journal of Business Communication* 19 (Spring 1982): 10.

104. Sherwin Cody, *Success in Letter Writing* (Chicago: McClurg, 1906), p. 27, as quoted in ibid., p. 10.

105. See, for example, a 1923 model letter ending "Regretting that you should feel dissatisfied, but assuring you we are only billing you for the actual amount of current consumed, we are, /Very truly yours,/Auditor." (N.Y.C. Board of Education, *English and Business*, p. 59.)

106. See, for example, Herbert Watson, *Applied Business Correspondence* (Chicago: A. W. Shaw Co., 1922).

107. William Henry Williams, *Railroad Correspondence File*, rev. and suppl. John L. Hanna (New York: DeVinne-Hallenbeck Co., 1902, rev. 1910), p. 19.

108. Henry W. Belfield, "Interhouse Correspondence," *System* 5 (February 1904): 113–14.

109. President's Commission on Economy and Efficiency, Circular no. 21, "Memorandum of Conclusions Reached by the Commission Concerning the Principles That Should Govern in the Matter of Handling and Filing Correspondence and Preparing and Mailing Communications in Connection with the Work of the Several Departments of the Government . . ." (Washington, D.C.: 1912), pp. 27–28.

110. Schulze, *American Office*, p. 172.

111. Oscar Charles Gallagher and Leonard Bowdoin Moulton, *Practical Business English* (Boston: Houghton Mifflin Co., 1918), p. 184.

112. N.Y.C. Board of Education, *English and Business*, p. 16.

113. Ibid., p. 17.

114. Gallagher and Moulton, *Practical Business English*, p. 187.

115. Ibid., p. 184.

116. N.Y.C. Board of Education, *English and Business*, pp. 17, 53.

117. Griffith, *Correspondence and Filing*, p. 7.

118. Gallagher and Moulton, *Practical Business English*, p. 187.

119. Locker, "Divergence of Style," p. 10 and elsewhere.

120. "Getting More Out of Shop Conferences" began with *Factory* 16 (February 1916): 144–46. It continued through 1919.

121. The questionnaire findings are reported in "Getting More Out of Shop Conferences," *Factory* 17 (October 1916): 490–500. Quoted passages are from p. 494.

122. "Getting More Out of Shop Conferences," *Factory* 22 (April 1919): 859.

123. M. D. Strong, "Shop Conferences That Bring Results," *Factory* 18 (March 1917): 430.

124. "Getting More Out of Shop Conferences," *Factory* 17 (October 1916): 492, 494.

125. "Getting More Out of Shop Conferences," *Factory* 17 (August 1916): 166.

126. Griffiths, *Administrative and Industrial Organization*, p. 44.

127. Galloway, *Organization and Management*, p. 174.

128. Ibid., p. 173.

129. "Getting More Out of Shop Conferences," *Factory* 18 (January 1917): 122.

130. See, for example, "Getting More Out of Shop Conferences," *Factory* 18 (January 1917): 118.

131. Brinton, *Graphic Methods*, p. 304.

132. M. D. Strong, "Shop Conferences That Bring Results," *Factory* 18 (March 1917): 438.

4: The Illinois Central before 1887

1. W[illiam] K. Ackerman, "History of the Illinois Central Railroad Company," in Ackerman, ed., *History of the Illinois Central Railroad Company and Representative Employes [sic]* (Chicago: Railroad History Co., 1900), p. 22. See also Paul W. Gates, *The Illinois Central Railroad and Its Colonization Work* (Cambridge: Harvard University Press, 1934; rpt. New York: Johnson Reprint Corp., 1968), especially chap. 8.

2. Gates, *Illinois Central*, p. 149.

3. For this reason and because the surviving records provide less complete information on construction and sales, I will not consider the communication system of the departments involved in these activities. Instead, I will concentrate on that of the Transportation and Freight departments and the executive offices. For more information on the operations of the Land Department, see ibid., chap. 8.

4. John F. Stover, *History of the Illinois Central Railroad* (New York: Macmillan Co., 1975), pp. 42–43.

5. John F. Stover, *American Railroads* (Chicago: University of Chicago Press, 1961), p. 45; Stover, *History of the ICR*, p. 81.

6. 1855 Annual Report, IC/ +2.1, Newberry Library (see the Note on Archival Sources). Further references to annual reports are by year only.

7. William K. Ackerman, *Historical Sketch of the Illinois Central Railroad: Together with a Brief Biographical Record of Its Incorporators and Some of Its Early Officers* (Chicago: Fergus Printing Co., 1890), p. 58.

8. Information on Neal, Mason, and Calhoun is from ibid., pp. 48–51, 82–83.

9. See Carlton J. Corliss, *Main Line of Mid-America: The Story of the Illinois Central Railroad* (New York: Creative Age Press, 1950), p. 68.

10. April 4, 1857, Silas Bent to freight train conductors, for example, made an exception to the general rule against riders on freight trains for a particular individual (IC1/C5.1/vol. 1).

11. 1855 annual report, p. 7.

12. Corliss, *Main Line*, p. 64.

13. See, for example, March 29, 1857, Silas Bent to Osborn, IC1/C5.1/vol. 1.

14. 1864 annual report.

15. February 20, 1863, Osborn to Caton, IC1/06.1/vol. 5.

16. In the early 1860s, telegraphic dispatching was certainly used under special

circumstances to speed up the movement of military troops and supplies to critical areas (Stover, *History of the ICR*, p. 94). The Illinois Central was not afraid to trust the wires in this capacity, as the Erie had been before Charles Minot's daring experiment in 1851 (see Chap. 2). Nevertheless, they apparently did not use it routinely.

17. December 22, 1863, Osborn to Caton, IC1/06.1/vol. 5.

18. October 1, 1854, from R. B. Mason, general superintendent of the Illinois Central, found in the Newberry Library's Chicago, Burlington and Quincy Railroad records, CB&Q1.2.

19. April 2, 1859, J. W. Foster to station agents, IC2.8/vol. 1. The first circular of instructions has apparently not survived, but the April 2 circular refers to it.

20. March 3, 1857, J. C. Clarke (master of transportation) per Silas Bent, IC1/C5.1/vol. 1.

21. June 9, 1857, Silas Bent to passenger train conductors, IC1/C5.1/vol. 1.

22. See, for example, September 10, 1860, from W. P. Arthur (general superintendent), announcing the appointment of W. K. Ackerman (later to become president of the company) as local treasurer and William Vernon as auditor (IC2.8/vol. 1).

23. September 12, 1856, from J. C. Clarke, IC1/C5.4.

24. See IC2.8, Circulars, 1858–1906.

25. 1856 annual report.

26. See, for example, November 4, 1858, Clarke to Jacobs and Clarke to Arthur, each transmitting twenty-five circulars with instructions that "you will please put them in the hands of such of your agents as you deem necessary" (IC1/C5.1/vol. 4).

27. In fact, beginning perhaps as early as 1852 (according to Corliss, *Main Line*, p. 76), when the first short segment of track was opened, and certainly by 1856 (according to Stover, *History of the ICR*, p. 69), when the main line was completed, the railroad carried U.S. mail, as well.

28. See Thomas C. Cochran, *Railroad Leaders, 1845–1890: The Business Mind in Action* (New York: Russell & Russell, 1965), chap. 4, for a discussion of communication between executives and directors in several railroads, including the Illinois Central.

29. March 22, 1856, Osborn to the executive committee, IC1/06.1/vol. 1.

30. See, for example, Newell's monthly reports to the local treasurer, IC1/N6.1/vol. 1.

31. These bound volumes are in IC/+5.6. The volumes now designated 1–4 cover the first decade, starting around 1855. (Vol. 2 has only one numerical series that starts in 1855, while the various other series of data in it start in 1883. Thus, it was probably compiled starting in 1883, with retrospective data added for one series.)

32. Alfred D. Chandler, Jr., *The Visible Hand: The Managerial Revolution in American Business* (Cambridge: Harvard University Press, Belknap Press, 1977), p. 110.

33. The most likely source for his knowledge of the Erie's control systems was Henry Varnum Poor's *American Railroad Journal*, which publicized McCallum's innovations beginning in 1854 (Alfred D. Chandler, Jr., *Henry Varnum Poor: Business Editor, Analyst, and Reformer* [Cambridge: Harvard University Press, 1956], pp. 147–49). Osborn subscribed to this journal by at least 1857 (February 28, 1857, Osborn to Poor, IC1/C5.1/vol. 1) and undoubtedly read issues of it even earlier, especially since Poor wrote attacks on the Illinois Central's bond issues in 1852.

34. March 22, 1856, IC1/06.1/vol. 1.

35. Both Clarke's incoming and his outgoing correspondence were bound in chronological order and continuously numbered (see IC1/C5.3 and C5.4), thus making it almost impossible for any such reports to have been lost. Moreover, the frequent letters to Osborn made no reference to any other reports. For examples of these letters, see March 7, 1857, or April 23, 1857, IC1/C5.1/vol. 1.

36. September 5, 1855, J. B. Calhoun (position at time unknown) to Wartrick, found in CB&Q1.2 with Calhoun's papers from later years when he was employed by that railroad. Since neither Calhoun's nor Wartrick's correspondence has survived in the Illinois Central records, I could find no examples of this report. But starting in 1857, most annual reports contained a statement listing the operating costs and costs per mile for each engine, so some mechanism for reporting this information was indeed established.

37. Stover, *History of the ICR*, p. 88.

38. The first one is dated July 28, 1858, IC1/C5.1/vol. 3.

39. At about this time, Clarke left the Illinois Central to go to another railroad (Stover, *History of the ICR*, p. 83).

40. Transmitted to various individuals by Clarke (per Remmer) in letters dated September 20, 1858, IC1/C5.1/vol. 4.

41. Chandler, *Visible Hand*, p. 110.

42. February 1860, John Newell to L. H. Clarke, IC1/N6.1/vol. 1.

43. Monthly reports from Newell (division engineer for the Second Division) to L. H. Clarke (chief engineer) for 1860 are in IC1/N6.1/vol. 2.

44. 1856–1860 annual reports.

45. See, for example, July 10, 1855, President Griswold to Nortich (agent in Amboy), CB&Q1.2.

46. See, for example, November 25, 1856, Clarke to H. L. Hall (agent at Mattoon), IC1/C5.1/vol. 1.

47. See, for example, April 30, 1857, J. C. Clarke (master of transportation) to A. Mitchell (assistant superintendent), IC1/C5.1/vol. 1; or August 15, 1859, Newell (division superintendent) to Jas. Y. Kennedy, IC1/N6.1/vol. 1.

48. IC1/M3.1.

49. See, for example, June 21, 1857, Clarke per Bent to S. I. Hayes, IC1/C5.1/vol. 1.

50. These are his major correspondents in IC/M2.17/vol. 8, dated March to September 25, 1859.

51. IC1 has several sets of press books for various company officers.

52. IC11/C1.1.

53. See, for example, IC1/C5.3, which contains all of J. C. Clarke's correspondence while he held several different positions.

54. McClellan's travels are discussed in Stover, *History of the ICR*, p. 88. Vol. 4 of IC1/M2.1 overlaps vols. 2 and 3, and internal references make it clear that the letters copied in it were written while he was traveling along the line.

55. See, for example, the letters from J. C. Clarke to W. H. Osborn in IC1/C5.3. While these are now included in Clarke's outletters, they were obviously the originals received by Osborn. They have been folded into the small size used in pigeonholes and annotated with Clarke's name, the date, and their subject.

56. Clarke's in-letters, for example, are pasted into a bound book of gummed, one-inch stubs, "Smith and Butler's Improved Adhesive Letter, Invoice, and Music File" (IC1/C5.4).

57. Edward W. McGrew, "Corporate History of the Illinois Central Railroad

Company and Its Controlled and Affiliated Companies up to June 30th, 1915," written "in substantial compliance with Valuation Order No. 20 of the ICC, dated 13 May 1915."

58. Stover, *History of the ICR*, pp. 134, 170.

59. Ibid., pp. 144–71.

60. Ibid., pp. 123–25.

61. Ibid., chap. 5, pp. 85–107.

62. Robert L. Brandfon, "Political Impact: A Case Study of a Railroad Monopoly in Mississippi," in Bruce Mazlish, ed., *The Railroad and the Space Program: An Exploration in Historical Analogy* (Cambridge: MIT Press, 1965), p. 185.

63. Alfred D. Chandler, Jr., and Stephen Salsbury, "The Railroads: Innovators in Modern Business Administration," in ibid., pp. 156–57.

64. The 1862 annual report lists an appropriation of just over $5,000 for the new telegraph line. According to Ackerman, *History of the ICR*, p. 61, Marvin Hughitt held the titles of train dispatcher and superintendent of telegraph during the period 1862 to 1866.

65. Annual reports, 1860–64. While these annual reports do not explain the origin of the operating expenses listed as "Telegraphing," in the 1875 annual report and occasionally after that the expense is broken into the two categories of "Superintendent and Operations" and "Supplies and Repairs of Telegraph."

66. Many of the Western Union telegrams from the mid-1870s are on forms designated Western Union, and some are marked, "Half Rate Messages" (IC11/N1.4/vol.1). In addition, a May 18, 1897, "Report of the President" on the ten years from 1886 to 1896 (IC/+2.1) referred to a contract with Western Union, in effect in 1886, which allowed free service up to a certain amount in exchange for right of way along the rail line and railroad maintenance of these telegraph lines. That report also provided the mileage statistics cited.

67. See Iowa Division timetable No. 4, August 29, 1869, IC/+3.8. No timetables dated from 1860 to 1868 have survived in the Newberry Library's archival collection of Illinois Central materials, but a section on telegraphic movement of trains probably appeared by at least 1864.

68. The daily reports of January 1866 (ICII/C1.1/vol. 1) frequently omit a section, stating that these items are not available because snow and ice have brought the telegraph lines down.

69. April 2, 1883, John Dunn (assistant to president) to Stuyvesant Fish (second vice-president), IC1/F2.2/vol. 1.

70. July 14, 1884, circular no. 97, E. T. Jeffery (general superintendent) to Officers and Employees, IC2.8/vol. 4.

71. ICII/N1.4/vol. 1, telegrams received by the New York office, beginning in 1876, contains dozens of telegrams concerning these negotiations.

72. March 22, 1884, E. T. Jeffrey to Officers and Employees, IC2.8/vol. 4.

73. See, for example, November 30, 1876, Osborn to Fentress, ICII/N1.4/vol. 1.

74. The first examples of code that I have been able to locate in company records are in a March 22, 1883, letter from W. K. Ackerman (president) to Stuyvesant Fish (then second vice-president), IC1/F2.2/vol. 1. However, a letter dated January 20, 1883, Ackerman to Fish, refers to the existence of such a code (IC1/F2.2/vol. 1). Various commercial and specialized codes are mentioned or used in Fish's in-letters (IC1/F2.2) and his out-letters (IC1/F2.1). In the in-letters, see, for example, the letter cited above and a January 31, 1885, letter from J. C. Clarke (president) to Fish (first vice-president). In the out-letters, see, for example, a

February 20, 1895, telegram from Fish to the lawyers Pirtle and Traburg and its translation in a letter of the same date to Fentress, as well as a December 18, 1895, telegram from Fish to J. C. Welling.

75. April 2, 1883, John Dunn (assistant to president) to Fish (second vice-president), IC1/F2.2/vol. 1; and written onto an undated routine report form bound into Fish's in-letters after a letter dated January 31, 1885.

76. May 15, 1893, Harahan (second vice-president) to several officers, marked "Confidential," IC1/F2.2/vol. 95.

77. January 1, 1877, "Rules Governing the Condition, Inspection, and Repairs of Freight Cars . . . ," IC2.8/vol. 3.

78. IC2.8/vols. 1, 5, and 6 contain circulars predominantly from the general freight agent, while vols. 3 and 4 contain those from the general superintendent. Vol. 1 is more varied and spotty, including circulars to stockholders as well as to employees. It appears to have been assembled much later than its earliest items (1858), probably in the 1870s when vols. 2 through 6 were begun.

79. Circular no. 1 from the Freight Department is dated July 1, 1873, and issued by J. F. Tucker (IC2.8/vol. 2); the earlier numbered circulars from the General Superintendent's Office have not survived, but the notice dated November 18, 1876, is designated circular no. 26 and was issued by J. F. Tucker, who became general superintendent in December 1874 (IC2.8/vol. 3). The obvious assumption is that Tucker himself carried this practice from the Freight Department to the General Superintendent's Office.

80. E. T. Jeffery took over from J. F. Tucker in April 1877. The first circulars he issued have not survived, but his circular dated October 15, 1877, is designated no. 10, indicating that a new series had been started.

81. August 14, 1872, Ackerman to agents, IC2.8/vol. 1.

82. February 28, 1872, circular no. 32, J. F. Tucker to agents, IC2.8/vol. 2.

83. December 15, 1875, circular no. 68, Horace Tucker to agents, IC2.8/vol. 2.

84. September 20, 1872, J. F. Tucker to agents, IC2.8/vol. 1.

85. July 21, 1873, Freight Department circular no. 5, IC2.8/vol. 2.

86. October 15, 1877, circular no. 10, IC2.8/vol. 3.

87. June 27, 1873, circular from J. F. Tucker to agents, IC2.8/vol. 2.

88. Returning the transmittal circular meant that the recipient could no longer have a complete set of circulars. In this case, however, the tariff or rate list itself was the important part of the communication. A variation on this technique, which appears in one of the other case studies, involved returning a tear-off receipt from the bottom of the page, allowing the recipient to keep a copy of the circular itself.

89. IC2.8/vol. 3, from the General Superintendent's Office, has many examples of both types.

90. The characteristic spidery dotted lines of the Electric Pen first appear on a January 10, 1876, notice, IC2.8/vol. 3.

91. See, for example, General Freight Circular no. 71, April 22, 1876, IC2.8/vol. 2.

92. See, for example, January 10, 1876, IC2.8/vol. 3.

93. A rate circular dated January 15, 1885, from Horace Tucker to all agents, IC2.8/vol. 6, refers specifically to the "hektograph" circulars.

94. See IC2.8/vol. 6 for the frequently issued rate schedules.

95. May 18, 1897, "Report of the President," 1886–1896, IC/+2.1 (hereafter, ten-year report), p. 15.

96. April 16, 1883, Clarke to Sheafe and Schlacks, IC1/C5.2/vol. 5.

97. Ten-year report, p. 14.

98. Stover, *History of the ICR*, pp. 184, 190.

99. See, for example, "Business of the Illinois Central Company in March," 1863, IC1/06.1. The weekly reports are in ICii/C1.1/vols. 1–3.

100. Although the weekly statements from this period have not survived, the daily letter accompanying the weekly statement usually contained a brief summary of it. Beginning by at least July 11, 1870 (Douglas to Osborn, ICii/C1.1/vol. 4), the weekly statement included the comparative weekly estimates of earnings.

101. See, for example, "Estimated Earnings for the week ending 30 April 1875," ICii/C1.1/vol. 5.

102. November 25, 1875, Douglas to Osborn, ICii/C1.1/vol. 8.

103. The 1875 Annual Report discusses these innovations.

104. IC/+5.6/vol. 6.

105. Chandler, *Visible Hand*, pp. 116–17.

106. IC1/C5.1/vol. 3.

107. ICii/C1.1.

108. November 25, 1875, Douglas to Osborn, ICii/C1.1/vol. 8.

109. ICii/C1.1/vol. 8. There was no indexing at all up through vol. 6. Vol. 7 included a very few index items, but the index was by no means complete or comprehensive. Beginning with vol. 8, the volumes were thoroughly indexed by subject.

110. February 20, 1877, ICii/C1.1/vol. 9.

111. Report for week ending February 28, 1877, ICii/C1.1/vol. 9.

112. September 20, 1877, ICii/C1.1/vol. 10.

113. Stover, *History of the ICR*, p. 540.

114. Carolyn Curtis Mohr, comp., *Guide to the Illinois Central Archives in the Newberry Library* (Chicago: Newberry Library, 1951), p. 173; Corliss, *Main Line*, p. 209.

115. March 12, 1883, Ackerman to Fish, IC1/F2.2/vol. 1. The first such form is March 6, 1883, ICii/C1.1/vol. 18.

116. See IC/+5.6/vols. 13, 14, 17, and 18. Vol. 18 has the new monthly form for 1883 to 1886. Vol. 13 has the same form for 1882, but since it overlaps vol. 10 (which covers 1879–82 using the old form), I suspect that vol. 13 was compiled retrospectively so the figures for the annual comparisons would be available.

117. Ackerman acknowledged and replied to the question on March 26, 1883, Ackerman to Fish, IC1/F2.2/vol. 1.

118. IC/+5.6/vol. 26.

119. May 3, 1883, Ackerman to Fish, IC1/F2.2/vol. 2.

120. Stover, *History of the ICR*, p. 177.

121. Ibid., p. 540.

122. May 14, 1883, Welling to Clarke, IC1/F2.2/vol. 2.

123. June 6, 1883, Clarke to Fish, IC1/F2.2/vol. 2.

124. May 29, 1885, Fish to Clarke, IC1/C5.5/vol. 3.

125. September 8, 1876, Jeffery to Clarke, IC2.14.

126. April 15, 1882, Ackerman to board of directors, IC7.22.

127. January 23, 1878, Ackerman to Clarke, IC1/A2.1/vol. 2.

128. See Stover, *American Railroads*, pp. 126–31.

129. January 30, 1885, Clarke to Fish, IC1/F2.2/vol. 10.

130. See Stover, *American Railroads*, pp. 131–42.

131. January 30, 1885, IC1/F2.2/vol. 10.

132. January 31, 1885, Clarke to Fish, IC1/F2.2/vol. 10.

133. In an 1886 letter to President Clarke, for example, General Manager E. T. Jeffery stated, clearly for the record, his opposition to a decision Clarke had made: "I sincerely hope that this company will not take the step indicated in the papers handed to me a day or two ago by you" (September 8, 1886, Jeffery to Clarke, IC1/C5.5/vol. 7).

134. During the 1880s, Clarke and Fish both began to add occasional subheadings to separate different topics in their letters. See, for example, April 2, 1883, Clarke to Osborn, IC1/C5.2/vol. 4; and February 28, 1885, Fish to Clarke, IC1/C5.5/vol. 1.

135. The first subject entries in the index were in Clarke's in-letters, April 1–June 30, 1885 (IC1/C5.5/vol. 3), but these entries were scattered. By 1887 (vol. 10), the first volume in which Fish is president, indexing by subject has become thorough and elaborate, with author entries in black and subject entries in red.

136. Corliss (Main Line, p. 405) dates the first use as February 27, 1882, in a letter from S. B. McConnico. I have been unable to locate any of McConnico's letters from before January 3, 1883 (IC1/F2.2/vol. 1).

137. See IC1/C5.5/vol. 6, May 1–July 31, 1886, for first typed letters from Fish in New York; see IC1/F2.2/vol. 20, January 14–February 20, 1886, for first typed letters from Chicago.

138. The local treasurer's report from New Orleans for February 1885 was copied on a roller copier, as is evident from the nonstandard length and uneven upper and lower edges of the tissue copy (IC1/C5.5/vol. 3). A carbon copy of a typed letter was included in a letter dated April 10, 1885, Clarke to Frank S. Bond, IC1/C5.5/vol. 3.

139. In 1883, President Ackerman even went so far as to ask Fish not to use such technologies when sending him copies: "If possible I would be glad to have copies of any papers you send me written rather than to have tissue copies, which are inconvenient to file away or to refer to" (March 21, 1883, Ackerman to Fish, IC1/F2.2/vol. 1).

5: The Illinois Central after 1887

1. John F. Stover, History of the Illinois Central Railroad (New York: Macmillan Co., 1975), p. 540.

2. John F. Stover, American Railroads (Chicago: University of Chicago Press, 1969), pp. 131–35.

3. Interstate Commerce Act of 1887, sec. 20, as quoted in Meyer Fishbein, "Special Study of the Periodic Reports Submitted by Carriers to the Interstate Commerce Commission and the Civil Aeronautics Board," unpublished report, General Services Administration, National Archives and Records Service, Office of Federal Records Centers, Records Appraisal Division, 7/22/66.

4. January 8, 1887, Clarke to Fish, IC1/C5.2/vol. 11.

5. September 4, 1888, Ayer to Fish, IC1/F2.2/vol. 49. In reply to August 28, 1888, Fish to Ayer, IC1/F2.1/vol. 4.

6. October 8, 1888, Fish to ICC, IC1/F2.2/vol. 4.

7. December 3, 1888, Fish to Jeffery, IC1/F2.1/vol. 4.

8. 3 December 1888, Fish to Welling, IC1/F2.1/vol. 4.

9. April 27, 1891, Fish to E. A. Moseley, attached to April 22, 1897, Moseley to Fish, IC1/F2.2/vol. 180.

10. January 21, 1890, Fish to Welling, IC1/F2.1/vol. 6.

11. August 31, 1888, Jeffery to Fish, IC1/J4.1/vol. 2.

12. November 8, 1890, Fish to ICC, quoted in April 27, 1891, Fish to ICC, attached to April 22, 1897, E. A. Moseley, secretary of the ICC, to Fish, IC1/F2.2/vol. 180. In May 23, 1896, Fish to Fentress, IC1/F2.1/vol. 26, Fish mentions the "somewhat lengthy correspondence with the Commission on this subject."

13. "The Use and Abuse of Statistics," *Railway Review*, April 23, 1892, found in IC1/F2.2/vol. 88.

14. May 23, 1896, Fish to Fentress, IC1/F2.1/vol. 26.

15. "Proceedings of a National Convention of Railroad Commissioners Held at the Office of the Interstate Commerce Commission," May 19, 1896 (Washington, D.C.: Government Printing Office, 1896), p. 29.

16. January 21, 1890, Fish to Welling, IC1/F2.1/vol. 6.

17. Annual reports up through the end of Osborn's presidency in 1864 were signed by the president in New York, but the 1865 report was signed by Douglas in Chicago. Correspondence also reveals the change in address.

18. The Code of Rules that Fish had issued in 1890 (discussed in a later section) said that the President's Office "shall be at Chicago" (IC2.22.) Probably based on that, Chandler locates the president in Chicago at this time (Alfred D. Chandler, Jr., *The Visible Hand: The Managerial Revolution in American Business* [Cambridge: Harvard University Press, Belknap Press, 1977], p. 185). Nevertheless, the correspondence is clear on this point: The origin and destination of the many letters in IC1/F2.2 and IC1/F2.1 place Fish in New York most of the time, though a "President's Office" with an assistant was maintained in Chicago, and he used it when he went out there.

19. October 14, 1887, Dunn to Fish, IC1/F2.2/vol. 40.

20. September 27, 1888, Jeffery per Dunn to Welling, IC1/J4.1/vol. 3.

21. A comparison of dates that letters were written in Chicago with dates that they were stamped received in New York in IC1/F2.2/vol. 50 (two months before the new policy was established) shows an average transit time of two days for regular mail. An earlier March 3, 1885, letter from Fish (when he was still vice-president) to Dunn in the President's Office (then still in Chicago) mentioned that express mail took one day between Chicago and New York (IC1/C5.5/vol. 1).

22. July 11, 1889, Harriman to Ayer, IC1/F2.1/vol. 6.

23. August 13, 1889, Harriman to Jeffery, IC1/F2.1/vol. 6.

24. October 19, 1888, and November 28, 1888, Fish to Bruen, IC1/F2.1/vol. 4.

25. October 19, 1888, Fish to Bruen, IC1/F2.1/vol. 4.

26. February 5, 1890, Beck per Bruen to Fish, IC1/F2.2/vol. 60.

27. August 23, 1888, Fish to Welling, IC1/F2.1/vol. 4.

28. See Jeffery's reply, August 31, 1888, Jeffery to Fish, IC1/J4.1/vol. 2.

29. August 25, 1888, Welling to Fish, IC1/F2.2/vol. 51.

30. August 27, 1888, Fish to Welling, IC1/F2.1/vol. 4.

31. See August 31, 1888, Fish to Welling, IC1/F2.1/vol. 4.

32. Oral instructions repeated in Welling's reply, March 22, 1889, Welling to Fish, IC1/F2.2/vol. 60.

33. Oral instructions repeated in Welling's reply, March 26, 1889, Welling to Fish IC1/F2.2/vol. 60.

34. December 3, 1888, Fish to Welling, IC1/F2.1/ vol. 4.

35. March 26, 1889, Welling to Fish, IC1/F2.2/vol. 60.

36. Carolyn Curtis Mohr, *Guide to the Illinois Central Archives in the Newberry Library* (Chicago: Newberry Library, 1951), p. 175.

37. "Code of Rules, adopted by the Board of Directors, December 16, 1889. In effect from and after January 1, 1890," IC2.22. (Hereafter, Code of Rules.)

38. May 18, 1897, "Report of the President," 1886–96, IC/+2.1. (Hereafter, ten-year report.) See also Code of Rules, p. 10.

39. The bound instruction pamphlet issued by that office to all employees, "Classification of Operation Expenses" (July 1, 1894, IC1/F2.2/vol. 119), provided detailed instructions for filling out such forms and reports.

40. Ten-year report, p. 28.

41. August 13, 1891, Harahan to Sullivan, IC1/F2.2/vol. 80.

42. Mohr, *Guide to the ICR*, p. 173.

43. September 12, 1892, Welling to Fish, IC1/F2.2/vol. 90.

44. August 31, 1888, Jeffery to Fish, IC1/J4.1/vol. 2.

45. October 11, 1888, Jeffery to Jacobs, IC1/J4.1/vol. 3.

46. January 28, 1889, Fish to Jeffery, IC1/F2.1/vol. 5. Jeffery's reply to Fish, as quoted below, is dated April 7, 1889, in IC1/J4.1/vol. 5.

47. Chandler *(Visible Hand*, p. 186), makes the point that railroads did not develop systematic ways of allocating capital until well into the twentieth century.

48. April 6, 1889, Fish to Jeffery, IC1/F2.1/vol. 5.

49. April 20, 1889, Jeffery to Fish, IC1/J4.1/vol. 6.

50. October 11, 1888, Jeffery to Jacob, IC1/J4.1/vol. 3.

51. February 15, 1888, Jeffery to Beck, IC1/J4.1/vol. 1.

52. Stover, *History of the ICR*, pp. 208–9.

53. September 15, 1889, Fish to Harriman, IC1/F2.1/vol. 6.

54. February 15, 1894, Bruen to Fish, IC1/F2.2/vol. 104.

55. See, for example, August 31, 1891, "Comparative Statement of Cotton Receipts at New Orleans" by L. Herzog (agent), IC1/F2.2/vol. 80.

56. January 29, 1892, Schlacks and Wallace to Fish, IC2.14.

57. Although Du Puy's original letter has not survived, Fish's response to it (December 22, 1894, Fish to Du Puy, IC1/F2.1/vol. 19) and much subsequent correspondence about it (cited below) has survived.

58. December 22, 1894, Fish to Welling, IC1/F2.1/vol. 19.

59. December 21, 1894, Harry P. Robinson to Fish, IC1/F2.2/vol. 115.

60. December 22, 1894, Fish to Du Puy, IC1/F2.1/vol. 19.

61. January 2, 1895, Welling to Fish, IC1/F2.2/vol. 117.

62. February 18, 1895, Fish to Welling, IC1/F2.1/vol. 19.

63. March 2, 1895, Welling to Fish, IC1/F2.2/vol. 118.

64. March 4, 1895, Fish to Welling, IC1/F2.1/vol. 20.

65. November 29, 1895, Du Puy to Fish, IC1/F2.2/vol. 132; reply in December 9, 1895, Fish to Du Puy, IC1/F2.1/vol. 23.

66. December 12, 1895, IC1/F2.2/vol. 133.

67. Ten-year report. A third graph showing growth at New Orleans rather than at Chicago has been omitted here.

68. May 9, 1898, Fish to Wallace, IC1/F2.2/vol. 35.

69. May 9, 1898, Fish to Wallace, IC1/F2.2/vol. 35; and May 16, 1898, Wallace to Fish, IC1/F2.2/vol. 222.

70. September 17, 1898, Fish to Wallace, IC1/F2.1/vol. 37; and September 27, 1898, Wallace to Fish, IC1/F2.2/vol. 232.

71. The line had grown from just over twenty-one hundred miles to almost forty-three hundred miles, according to a chart in Edward W. McGrew, "Corporate History of the Illinois Central Railroad Company and Its Controlled and

Affiliated Companies up to June 30th, 1915" (written "in Substantial Compliance with Valuation Order No. 20 of the ICC, Dated 13 May 1915").

72. Not surprisingly, Fish and Mahl found graphs a particularly effective tool for carrying out their comparisons of several different railroads. Since the basis for the data was not always identical, the numbers themselves could be misleading. In a letter accompanying comparative diagrams, Fish commented to Mahl, "With you I agree that this is the only way in which comparisons between various railroads may be made of value, by carrying the comparison through a series of years and showing the tendencies" (November 2, 1903, Fish to Mahl, IC1/F2.1/vol. 64). Mahl noted in his reply that Fish's diagram of traffic statistics "impressed me as an admirable arrangement for showing the results to which it is desired to call attention strikingly. I believe in diagrams, particularly when it is desired to contrast the relations of certain results strikingly with others" (November 24, 1903, Mahl to Fish, IC1/F2.1/vol. 64). This exchange illustrates that both Fish and Mahl had come to appreciate the value of the graphic presentation technique in facilitating comparisons.

73. November 25, 1903, Fish to Mahl, IC1/F2.1/vol. 64.

74. November 27, 1903, Mahl to Fish, IC1/F2.2/vol. 359.

75. See, for example, December 21, 1889, Fish to T. E. King, announcing the code (IC1/F2.1/vol. 6).

76. Ten-year report, p. 14.

77. Code of Rules, p. 26.

78. Ibid., p. 27.

79. Ibid., p. 26.

80. For example, the comptroller's responsibilities included furnishing the president and board of directors standard statements each month and other statistics as needed (ibid., pp. 9–10).

81. Ibid., p. 15.

82. October 15, 1891, circular from General Superintendent A. W. Sullivan to Chicago Division, IC2.8/vol. 7.

83. February 10, 1893, circular no. 36 from general superintendent, IC2.8/vol. 7.

84. "Classification of Operation Expenses," July 1894, IC1/F2.2/vol. 119.

85. Ten-year report, p. 34.

86. Ibid., p. 15.

87. See October 22, 1897, circulars no. 143 and 144, IC2.8/vol. 14.

88. August 13, 1891, Harahan to Sullivan, second press copy pasted into Fish's incoming letters, IC1/F2.2/vol. 80.

89. On October 17, 1887, Jeffery wrote seven letters to Fish, one of which was a long status report on projects and operations in Chicago (IC1/F2.2/vol. 40).

90. July 13, 1888, Jeffery to Fish, IC1/F2.2/vol. 50.

91. These first appeared in December 1887 (IC1/F2.2/vol. 42).

92. The position of the stamps and the folds on the letters indicate that the letters were kept folded up in a pigeonhole for a period of time (probably until enough accumulated to fill a volume) before they were flattened out and pasted into the bound books of incoming letters.

93. October 2–26, 1887, IC1/F2.2/vol. 40. The previous ten volumes, beginning in January 1887, covered only ten months, thus averaging about a month each.

94. See, for example, index to IC1/F2.2/vol. 60, March 22–April 19, 1889.

95. See index to S. Fish's in-letters, 1895–1905, IC1/F2.2.

96. For example, one press book covered the period from June 21 to December

8, 1888, IC1/F2.1/vol. 4. Ten years later, in May 1898, the press volumes had reached vol. 31, averaging just over three volumes a year.

97. Vol. 4, cited above, shows the combined index. Vol. 15, beginning February 9, 1894, is the first volume with two separate indexes. Both in IC1/F2.1.

98. Code of Rules, p. 2.

99. March 15, 1895, John Dunn (assistant to the president in Chicago) to W. E. Ruttan (assistant to the president in New York), IC1/F2.2/vol. 119.

100. November 19, 1896, Dunn to Ruttan, IC1/F2.2/vol. 165.

101. See circulars in IC2.8/vol. 9.

102. W. K. Ackerman, ed., *History of the Illinois Central Railroad Company and Representative Employes* (Chicago: Railroad History Company, 1900), p. 131.

103. Ten-year report, p. 9.

6: Gradual Systematization at Scovill

1. General historical information on Scovill comes from several formal and informal histories, many of which are found in the Scovill collections at the Baker Library of Harvard Business School. Most important among the historical accounts are the following: Philip W. Bishop, unpublished manuscript history (ca. 1950), case 59, Scovill 2; E. H. Davis, "Recollections of Scovill and Waterbury, 1916–1968," informal memoir, case 59, Scovill 2; "Scovill—The Story of 150 Years of Ingenuity," unpublished historical sketch, no author, case 59, Scovill 2; Theodore F. Marburg, "Management Problems and Procedures of a Manufacturing Enterprise, 1802–1852: A Case Study of the Origin of the Scovill Manufacturing Company," Ph.D. diss., Clark University, 1945; and Theodore Marburg, "Commission Agents in the Button and Brass Trade a Century Ago," *Bulletin of the Business Historical Society* 16 (February 1952): 8–18. The authors of these accounts had access to primary materials, including some that have not survived to the present.

2. Marburg, "Commission Agents," p. 10.

3. Marburg, "Management Problems and Procedures," p. 341.

4. Bishop, pp. 182–84.

5. Ibid., pp. 107, 188, 200.

6. Correspondence from this period is in vol. 254, Scovill 1; and case 13, Scovill 2 (see Note on Archival Sources).

7. Bishop, pp. 461–62.

8. Vol. 254, Scovill 1, is a hand-copied letter book, dated 1824–1830. The next available letter book is the New York store's press book beginning in 1854.

9. Because incoming letters (case 13, Scovill 2) were unfolded and perhaps rearranged in the twentieth century, their original organization is not apparent.

10. Postmarks and internal references suggest that one-day service to New York was possible if the letter was sent off in time for the morning post. The normal time to Philadelphia seems to have been two days.

11. A January 24, 1840, letter from J. M. L. Scovill to W. H. Scovill mentions that a letter was received from the steamboat captain of the first boat to get in to New York during an icy week (case 13, Scovill 2).

12. Figured from Bureau of the Census, *Historical Statistics of the United States: Colonial Times to 1970* (Washington, D.C.), pt. 2, p. 807.

13. Notations on typescripts of handwritten letters (probably made in the mid-twentieth century when the records were first used for serious historical study) indicate that most letters had postage stamps, but some did not (case 13, Scovill 2). A few indicate that the letter had been carried by a specific individual.

14. These letters usually began with a reference to a letter recently received from the other party. See, for example, February 12, 1829, and January 12, 1840, J. M. L. Scovill to W. H. Scovill, both in case 13, Scovill 2.

15. See, for example, February 12, 1829, J. M. L. to W. H. Scovill, case 13, Scovill 2.

16. Marburg's "Commission Agents" traces Scovill's association with various agents during this period. These agents, of course, served other companies as well as Scovill. The New York agent, who served the firm's largest market, mentioned in one letter that he served eight to ten other persons or companies, "and some to as great an extent as you" (January 3, 1829, Gad Taylor to the Scovills, case 13, Scovill 2).

17. Marburg, "Management Problems and Procedures," p. 403.

18. Marburg, "Commission Agents," p. 13; and case 13, Scovill 2.

19. Averages computed using a four-month period from January through April 1829, folder "1829-thru June," case 13, Scovill 2.

20. Agents often took the initiative in doing preliminary research on an idea or purchasing sample material to send to Waterbury. See, for example, February 13, 1829, A. Goodyear and Sons to Scovill, in which the Philadelphia agent sent Scovill a sample English button and urged the firm to improve its products to meet that standard (case 13, Scovill 2).

21. See, for example, March 11, 1829, C. Goodyear to Scovill, case 13, Scovill 2.

22. March 11, 1829, Gad Taylor to Scovill, case 13, Scovill 2.

23. March 16, 1829, Taylor to Scovill, case 13, Scovill 2.

24. December 1, 1840, W. H. to J. M. L. Scovill, case 13, Scovill 2.

25. Marburg, "Management Problems and Procedures," p. 455.

26. *Historical Statistics of the U.S.*, pt. 2, p. 807.

27. Bishop, pp. 461–62.

28. February 27, 1854, and March 10, 1854, W. B. Holmes at New York store to Scovill at Waterbury (hereafter designated simply as "Waterbury"), vol. 456, Scovill 1.

29. April 25, 1854, New York store to Waterbury, vol. 456, Scovill 1.

30. Marburg, "Management Problems and Procedures," pp. 461–62.

31. Over a four-month period from February 24 to June 29, 1854, vol. 456, Scovill 1.

32. Oddly enough, an 1856 letter book shows a considerable drop in the number of letters from New York to Waterbury, to about four per month. I suspect that this decrease is not a real one, however, but reflects either a change in copying that correspondence or increased travel and telegraphing between the two locations. The trend toward more and shorter communications with other Scovill locations continues later in the century, suggesting that the 1856 figures are an abnormality.

33. See, for example, March 31, 1854, New York to Waterbury (vol. 456, Scovill 1), which includes the statement that New York "shall telegraph you tomorrow on arrival of steamer." No copy of the telegram or further reference to it appears in the press book.

34. Bishop, p. 15.

35. Ibid., p. 52.

36. Ibid., p. 200; and "Scovill—The Story of 150 Years of Ingenuity," p. 29.

37. Bishop, p. 28.

38. Ibid., p. 111.

39. Quoted in ibid., p. 70.

40. Ibid., pp. 64–65, 71.

41. Ibid., p. 73.

42. List of Scovill forms from the 1870s and 1880s, compiled in 1945 by E. H. Davis, company historian, now part of the Scovill collection at Baker Library. Most of these forms themselves have not survived.

43. Sperry, 1887, as quoted in Bishop, p. 205.

44. New York to Waterbury, vol. 47, Scovill 2; Waterbury to New York, vol. 257, Scovill 1.

45. See, for example, Sperry's four letters of June 28, 1879, vol. 257, Scovill 1.

46. Only one of the four letters just noted covered more than one topic.

47. See, for example, June 26, 1879, Sperry to New York store, vol. 257, Scovill 1, with order number only; or January 5 and 7, 1885, Sperry to New York store, vol. 285, Scovill 1, for the two phrases cited.

48. June 28, 1880, vol. 261, Scovill 1.

49. Vols. 456–85, Scovill 1, New York press books, contain at least four different series, one of which is to the factory. Another series contains correspondence with S. Peck and Co., a company that seems to have been a major supplier for Scovill. Still another series, labeled "Upstairs," contains the same general mix of external correspondence that the main, unlabeled series contains. The "Upstairs" label may simply indicate that the New York store always kept one press book downstairs and another upstairs to save trips between the two floors.

50. In his 1945 list of Scovill forms, E. H. Davis lists a form addressed to the New York office, as well as one leaving the location of the office (New York, Boston, or Chicago) to be filled in the blank, as appearing in the 1870s and 1880s. These forms first show up in the press books in the 1880s, but since not all forms are printed in inks that show up in press copies, they could have been used earlier.

51. Such forms first appear in vol. 46 (April 1875–May 1876), Scovill 2.

52. November 26, 1888, Fuller of the New York store to Sperry, vol. 480, Scovill 1.

53. See, for example, April 21, 1876, Sperry to New York store, vol. 47, Scovill 2.

54. February 15 to July 13, 1877, vol. 461, Scovill 1.

55. All three dated April 19, 1877, vol. 461, Scovill 1.

56. "Our Factory Telephone," *Scovill Bulletin* 4, no. 9 (January 1919): 8 (in vol. 314, Scovill 2). The first direct evidence of this internal telephone system comes from an undated entry in Sperry's personal memorandum book (vol. 238, Scovill 1), where he kept his own notes on a wide variety of things. The note, probably from around 1882 or 1883, lists the following locations and rates for telephones in the company:

South Mill	$ 5
Button Shop	5
North Mill	5
Local—here	10
Carrington	10 (taken out May 83)
Central office	60
All paid to Jan. 1, 1884	

. . . .

Apl 15/90 Long Distance Central Office 125.00 an[nually].

This notation suggests that Scovill rented equipment from the telephone company, though the nature of the system remains vague. The first three buildings or areas of the plant probably had one telephone each. "Local—here" may refer to a telephone that connected with the local Waterbury exchange. Carrington, a company that Scovill had purchased, must initially have been directly connected, as well, but in 1883 that private connection was given up, perhaps because the companies could communicate through public exchanges by then. The $60 figure for the Central Office is tantalizingly ambiguous. It suggests either that the office building had several telephones from which Goss, Sperry, and other executives could communicate with each other as well as with those in the other buildings, or that the company had a small private exchange.

57. Vol. 315 (20 April–1 June 1888), Scovill 1.

58. A page of a typed document has more words than a handwritten page, but since most letters, handwritten or typed, were less than a full page, the difference was probably not significant.

59. Bishop, p. 460.

60. See letters in vol. 360, Scovill 1.

61. Bishop, p. 205.

62. Ibid., pp. 205–6.

63. For evidence of this hierarchy, see February 18, 1908, from C. P. Goss, in "Orders and Instructions, 1905–1913 and 1914 from E. G. Main's Letter Files," case 34, Scovill 2 (hereafter referred to as Main's "Orders and Instructions").

64. Undated classification list with amendments dated July 24, 1911, attached to folder "C" of F. J. Gorse files, case 26, Scovill 2.

65. Most of the surviving records concern the manufacturing functions, so less can be known about the other functions.

66. February 22, 1905, "Report Made to the General Manager on Timekeeping in the Departments," from Burner Department [J. H. Goss, superintendent], in "Notices, 1905–1907," case 34, Scovill 2 (hereafter referred to as "Notices, 1905–1907").

67. Signatures and titles on notices in Main's "Orders and Instructions" indicate that J. H. Goss was superintendent of manufacturing departments by at least June 18, 1908, and general superintendent by June 7, 1909.

68. J. H. Goss, "Report . . . on Timekeeping in the Departments," "Notices, 1905–1907."

69. Bishop, pp. 209–10.

70. Main's "Orders and Instructions"; "Notices, 1905–1907"; and Gorse files, case 26, Scovill 2. Although the first such notice that has survived is dated January 1905, it is designated "Superintendent's Order #137," indicating that he had been issuing such orders for some time.

71. "Notices, 1905–1907" contains originals and carbon copies marked "file." Main's "Orders and Instructions" contains notices Goss issued after becoming general superintendent.

72. The holes punched at the top of the notices in "Notices, 1905–1907" indicate that they were once filed on a Shannon file.

73. Main's "Orders and Instructions."

74. Superintendent's Order no. 137, January 28, 1905, Main's "Orders and Instructions."

75. February 25, 1908, case 34, Scovill 2.

76. June 30, 1911, Main's "Orders and Instructions."

77. December 31, 1910, Main's "Orders and Instructions."

78. December 17, 1909, Main's "Orders and Instructions."

79. February 18, 1908, Main's "Orders and Instructions."

80. The original directors' order, dated September 16, 1905, is quoted in full in September 18, 1905, J. H. Goss to Burner Department foremen, Main's "Orders and Instructions."

81. See, for example, May 14, 1907, notice on new hours, printed on heavy cardboard for posting, in "Notices, 1905–1907."

82. For examples, see Gorse files, case 26, Scovill 2.

83. See, for example, January 19, 1914, Instruction Sheet, Gorse files, case 26, Scovill 2.

84. Davis, "Recollections of Scovill and Waterbury," p. 13.

85. March 17, 1908, Main's "Orders and Instructions."

86. January 28, 1905, Main's "Orders and Instructions."

87. See, for example, March 2, 1910, Classification no. 1 "Instruction," Gorse files, case 26, Scovill 2.

88. September 25, 1915, Gorse files, case 26, Scovill 2.

89. February 18, 1908, Main's "Orders and Instructions."

90. July 20, 1909, Main's "Orders and Instructions."

91. December 27, 1912, Main's "Orders and Instructions."

92. Vol. 333, Scovill 2. In 1907 the column for the previous year's figures was left blank, suggesting that 1907 was the first year such records were compiled.

93. These continuations of the Burner Department reports are also in vol. 333, Scovill 2.

94. Vol. 328, Scovill 2. A penciled note by E. H. Davis indicates that these were supervised by John B. Kendrick, company auditor, until 1920.

95. The graph described here was not the first to be used at Scovill, though it was the earliest surviving chart used to analyze the operational data collected by the early systems of records. The earliest graph extant is a primitive line graph of sales at the Chicago store from 1887 to 1904, presumably drawn in 1905 (vol. 318, Scovill 2). To make this graph, the compiler did not plot points on a grid, then connect them. Rather, he drew vertical lines of height proportionate to the sales volume for each year, then connected the tops of the lines.

96. Vol. 328, Scovill 2. Some of their efforts continued to be primitive. For example, in one chart labeled "Mfg. Dept. Graphic Moduli as determined from operations of the year 1911," points measuring a single quantity for a series of rooms or classifications for that year are misleadingly connected into a line graph (vol. 328, Scovill 2).

97. For example, in a note from J. H. Goss to Main dated April 19, 1911 (in Main's "Orders and Instructions"), Goss refers to reports on all motors over one horsepower, apparently inspection reports.

98. For example, a 1911 communication announcing the raising of pressure in the low pressure air system and requesting a corresponding lessening of nozzle size on each air nozzle in order to maintain that pressure ended with the following request: "I should like to have a report from each foreman who uses the low pressure air as to what he will do in this respect" (June 30, 1911, in Main's "Orders and Instructions").

99. See, for example, July 23, 1907, E. G. Main to Mr. Ketchum, in "Notices, 1905–1907."

100. See examples of these stamps and initials in "Notices, 1905–1907."

101. Summing Up Report for Research no. 249, May 14, 1918, case 49, Scovill 2.

102. I have discussed the development of vertical filing at Scovill previously in "From Press Book and Pigeonhole to Vertical Filing: Revolution in Storage and

Access Systems for Correspondence," *Journal of Business Communication* 19 (Summer 1982): 5–26.

103. An instructional note on handling the old records, written by P. W. Bishop and dated February 17, 1948 (case 59, Scovill 2), reveals that most of the incoming correspondence was stored in Amberg box files.

104. Unfortunately, the main series of press books breaks off in 1903. After that date, only the volumes of letters addressed to the New York store remain. The series of letters to the New York store begins with vol. 510, Scovill 1. Vols. 456–509 are letters *from* the New York store, while vols. 510–58 are letters *to* the New York store.

105. Some of these are missing, but the original numbering of the volumes and the gaps in dates reveal how many must have been used.

106. November 1, 1910, Waterbury (Mr. Warner) to New York store, vol. 530, Scovill 1.

107. July 11, 1911, J. H. Goss to E. G. Main, in Main's "Orders and Instructions."

108. December 26, 1913, H. B. Riggs to Mr. Rubino, vol. 558, Scovill 1.

109. See F. J. Gorse files, case 26, Scovill 2. These early notes lack the holes made by Shannon files, so they were probably saved in letter boxes.

110. See, for example, July 3, 1912, Geo. A. Goss, in Main's "Orders and Instructions"; and February 4, 1911, J. H. Goss to E. H. Davis, Main's "Orders and Instructions," case 34, Scovill 2.

111. See copies in vol. 530, Scovill 1. We have seen that J. H. Goss was already using carbon copies for his notices.

112. December 12, 1912, case 26, Scovill 2.

113. December 27, 1913, Waterbury to the New York store, vol. 558, Scovill 1.

114. December 31, 1913, E. O. Goss to Mr. Twining, vol. 558, Scovill 1.

115. An index of notices issued by the general superintendent includes a 1920 notice on "Inter-office correspondence form" (case 60, Scovill 2). While that notice has not survived, internal correspondence after that time had a memo heading with *To, From,* and *Subject* lines at the top. The complimentary close was also omitted.

116. The files are in case 26 of Scovill 1, and include those of William H. Monagan, foreman of the Casting Shop, in the Mills Department; F. J. Gorse, foreman of the Store Room; and E. H. Davis, statistician. While a very few pre-1914 documents appear in Gorse's files, the other two sets of files begin around 1917 or 1918.

117. See table 1 in my paper, "The Development of Internal Correspondence in American Business: A Case Study," in *Cultural Crossroads in the 80s,* Proceedings of the 1982 ABCA International Convention, p. 168.

118. Bishop, p. 136.

119. May 1918 letter to War Department, case 26, Scovill 2.

120. Research no. 259, June 20, 1917, case 49, Scovill 2.

121. See, for example, October 20, 1920, case 26, Scovill 2.

122. See, for example, July 28, 1921, E. O. Goss to C. P. Goss, case 26, Scovill 2.

123. Yates, "Development of Internal Correspondence."

124. March 25, 1918, Monagan to Bahney, case 26, Scovill 2.

125. See, for example, March 9, 1918, D. L. France to Monagan; September 27, 1918, T. F. Dunnigan to Monagan; January 31, 1919, Dunnigan to Monagan; March 4, 1919, Ferguson to Monagan; and November 7, 1919, W. P. Ferguson to Monagan, all in case 26, Scovill 2.

126. See, for example, February 11, 1922, E. H. Davis to R. C. Jetter, case 26, Scovill 2.

127. The impulse toward lateral documentation was not completely consistent: frequently the files revealed the beginning but not the end of a dispute (case 26, Scovill 2).

128. File labeled "Reports to General Superintendent and Superseded Lists" in E. H. Davis's files, case 26, Scovill 2. The examples mentioned in this paragraph come from the Monagan files, case 26, Scovill 2.

129. August 13, 1918, addressed to Mr. R. S. Sperry, in Davis files, case 34, Scovill 2. Underlines are in original. Quotes in the following three paragraphs of text are from the same source.

130. November 8, 1918, special report addressed to R. S. Sperry, from E. H. Davis. The following discussion is based on that second report.

131. William Henry Leffingwell, ed., *The Office Appliance Manual* (published for the National Association of Office Appliance Manufacturers, 1926), p. 167.

132. See references to the filing equipment in October 10, 1918, Davis to Sperry, and August 18, 1921, Davis to Thompson, both in case 34, Scovill 2.

133. "Graphs in the Presentation of Business Statistics and Reports," Modern Business, Report no. 84, in case 34, Scovill 2.

134. Dated October 12, 1917, but its place in the files and other external indications suggest that 1917 should be 1918. In Davis files, case 34, Scovill 2.

135. See September 5, 1918, Davis to Monagan; and October 17, 1918, R. S. Sperry [Davis's boss, head of the Field Research Division of the Research Department] to C. E. Woods, both in case 34, Scovill 2.

136. February 25, 1919, Davis to L. W. Bahney, case 34, Scovill 2.

137. The study itself has not survived, but it was summarized by Davis, July 28, 1939, case 34, Scovill 2.

138. See, for example, December 12, 1918, short report on "Statistics Work to Be Transferred to Research Department from Cost Office"; and February 25, 1919, note from Davis to L. W. Bahney, both in case 34, Scovill 2.

139. May 16, 1919, Davis to Sperry, case 34, Scovill 2.

140. See file labeled "Reports to General Superintendent and superseded lists" in case 34, Scovill 2.

141. October 5, 1925, E. H. Davis to A. J. Wolff, case 34, Scovill 2.

142. March 8, 1920, J. H. Goss to A. J. Wolff, case 34, Scovill 2.

143. April 25, 1921, J. H. Goss to Miss M. Murnane, case 24, Scovill 2.

144. See, for example, December 21, 1921, J. H. Goss to E. H. Davis, case 34, Scovill 2.

145. December 22, 1921, C. L. Dodge to Leggett, case 34, Scovill 2.

146. Computed from the lists cited in note 140.

147. Bishop, pp. 213, 216.

148. It is hard to pinpoint exactly when formal procedures for reporting were established; the earliest surviving research reports were dated 1917, but the project was numbered 200 (case 49, Scovill 2).

149. All numbered research reports are from case 49, Scovill 2.

150. See, for example, Research no. 200, Special Report, "Comparative Tests between Price and Cotton Lacquers," January 31, 1919, case 49, Scovill 2.

151. Case 26, Scovill 2.

152. See, for example, Research no. 249, May 14, 1918. In this summing up report for a study of the internal mail service at Scovill, the author left "Recommendations"; "Observations"; and "Data, Curves, etc." blank. The section en-

titled "Procedure" contained the entire report, in which he followed a chronological order ending with the recommendation.

153. For examples of standard-setting notices from staff departments, see September 14, 1919, Thomas F. McCarthy (head of the Cost Office), on procedures for calculating cost figures (case 26, Scovill 2); and undated Instructions no. 95, A. S. Ketcham, (head of the Department of Supply), on filling out requisitions (case 26, Scovill 2).

154. Bishop, p. 230.

155. The volumes of the *S.F.A. News* and the *Scovill Bulletin* are in vols. 313A-320B, Scovill 2. Further references will be to the volumes of the magazine itself.

156. *S.F.A. News* 1 (May 1915): 1.

157. The percentages were calculated from column space devoted to each type of item or article. Percentages do not add up to exactly 100 because of rounding.

158. "Miss O'Brien's Letter," *S.F.A. News* 1 (December 1915): 1.

159. *S.F.A. News* 1 (March 1916): 4.

160. *S.F.A. News* 1 (July 1915): 3.

161. *S.F.A. News* 1 (May 1915): 1.

162. *S.F.A. News* 1 (June 1915): 4.

163. *Scovill Bulletin* 5 (January 1920): 1.

164. *Scovill Bulletin* 2 (February 1917): 6.

165. See, for example, *Scovill Bulletin* 4 (October 1918): 2.

166. War coverage begins in *Scovill Bulletin* 3 (May 1917).

167. *Scovill Bulletin* 2 (March 1917): 6.

168. *Scovill Bulletin* 5 (January 1920): 10.

169. *Scovill Bulletin* 6 (May 1920): 4-9.

170. *Scovill Bulletin* 6 (June 1920): 3.

171. Bishop, p. 235.

172. The employees' and foremen's manuals are in case 34, Scovill 2.

173. December 2, 1921, E. H. Davis to W. H. Monagan, case 26, Scovill 2.

7: Du Pont's First Century

1. For simplicity, I will refer to the original company and all of its various legal manifestations (such as the E. I. du Pont de Nemours Powder Company) as "Du Pont," and the family as "du Pont."

2. Alfred D. Chandler, Jr., and Stephen Salsbury, *Pierre S. du Pont and the Making of the Modern Corporation* (New York: Harper and Row, 1971), p. 14.

3. Chandler and Salsbury (*Pierre S. du Pont*, pp. 5-6) discuss this relationship in the second half of the century. The company documents scattered through various family collections of records, and vice versa, also give evidence of the merging of family and company.

4. March 9, 1850, Du Pont Company to Daniel Rogers in Newburgh, in Longwood MSS, group 5, series A, outfile, 1844-91, Hagley Museum and Library (hereafter LMSS 5/A/1844-91). See the Note on Archival Sources for a description of Hagley's Du Pont archival materials.

5. For example, the Belins, a family that produced two head bookkeepers as well as the manager of the Wapwallopen mills, were connected with the du Pont family when Mary Belin married Lammot du Pont in 1865 (Chandler and Salsbury, *Pierre S. du Pont*, pp. 6-7).

6. Of course, the trumpet facilitated downward communication more than upward. For any real exchange of information and opinions, the owner would

have had to go down among the foremen and workers themselves, which presumably he did as well.

7. Chandler and Salsbury, *Pierre S. du Pont*, p. 5.

8. Letters between Alfred du Pont and his son Lammot in 1850–51, for example, and between Alfred and his brother Henry in 1853, have no address and bear no postmark (LMSS 5/A/1844–91).

9. These technical papers are in LMSS 5/C/Special Papers.

10. June 3, 1853, Henry to Alfred du Pont, LMSS 5/A/1844–91.

11. Letters from the Du Pont Company to Kemble can be found in the company letter books, acc. 500, series I, part 1, series A (hereafter 500/I/1/A), Hagley Museum and Library; letters from Kemble to the company are in acc. 500, series I, part 1, series B (hereafter 500/I/1/B), Hagley Museum and Library. (Volume or box number preceded by "#" will follow the series letter in specific references.)

12. January 20, 1845, Du Pont to Kemble, 500/I/1/A/#795.

13. In a December 13, 1841, letter to Kemble, the company (the letters were signed with the company name, rather than an individual name) complained that "the arrangement of the mails is now such as will make it impossible to answer our Northern letters [on the day after they were sent] except when we chance to take them out of the post office ourselves" (500/I/1/A/#17). The schedule must have been changed soon after that, since internal references continue to suggest a one-day travel period much of the time.

14. For example, 500/I/1/A, from Kemble to the company, has 15 letters in July 1841. In box 176 of series B, from August through December 1845, Kemble averaged a letter every 2.2 days, and for an overlapping period from April 1845 to January 1846 (vol. 796 of series A) the company averaged the same to Kemble.

15. 500/I/1/A/#802 (November 28, 1849–May 20, 1850) has one letter from the company to Kemble every 1.4 days. Correspondence with the company's agent in Philadelphia, even more frequent because Philadelphia was the closest major city to the Brandywine, showed a similar increase in the late 1840s and early 1850s, from just under one letter a day to significantly over one letter a day (500/I/1/A/#799 and #802).

16. Richard J. Ruth, "F. L. Kneeland of New York: A Study of the Changing Role of the Company Agent, 1849–1884," M.A. thesis, University of Delaware, 1967.

17. Ruth, "Kneeland of New York," p. 22.

18. See, for example, March 1, 1850, Alfred V. du Pont to Kemble, LMSS 5/A/1844–91.

19. April 24, 1850, Du Pont to John J. Twells, LMSS 5/A/1844–91. The annotation also suggests that its author may have written the letter primarily to think out the issues on paper before discussing them orally. Several other items in the records may also have served this purpose. See, for example, a document designated only "Remarks relative to the New Store," with no indication of who wrote it, and for whom, in LMSS 5/C/Special Papers 1846.

20. J. W. Macklem, "Old Black Powder Days," pt. 1, *Du Pont Magazine* 21, nos. 8 & 9 combined (1927): 11.

21. Kneeland's press books, 500/I/1/B/#816, begin August 31, 1854. The company's hand copy books are in 500/I/1/A. The main series from this date on has not survived, but evidence suggests that the company may have switched to press copying shortly after this time. P. C. Welsh ("The Old Stone Office Building, 1837–1891," internal report, Hagley Museum and Library, p. 27), cites an 1860 bill

for a copying press. Oddly enough, a December 31, 1857, "Inventory of Sundries" for the office included a letter press (500/I/2/M/#486). When this press was purchased and why it was not used at that time is not clear. Perhaps it was a new purchase, and they were waiting to finish a hand copy volume before switching. Possibly, the conservatism noted above extended to the point of resisting the use of this "new" technology even after they had bought the equipment.

22. Welsh, "Old Stone Office Building," p. 1.

23. See, for example, December 13, 1847, memorandum of a letter to Grant and Stone, 500/I/1/A/#799.

24. John W. Macklem described this set-up as he remembered it from the 1880s to Pierre S. du Pont in a letter dated September 18, 1944, quoted in Welsh, "Old Stone Office Building," p. 74.

25. Early company correspondence appears, for example, in Pierre S. du Pont's papers (the Longwood Manuscripts) and in Francis G. du Pont's papers (acc. 504, series A, Hagley Museum and Library, hereafter 504/A/box #), as well as in the company's papers, acc. 500.

26. As far as the records indicate, the first telegrams it received were dated May 29 and 31, 1847, from Kemble in New York, 500/I/1/B/#177. It first sent one dated October 15, 1847, to Isaac Comie in Louisville, 500/I/1/A/#799.

27. December 3, 1849, to Charles Smith, 500/I/1/A/#802.

28. February 24, 1848, to T. Burgoyne in Cincinnati, enclosed in letter to F. G. Smith in Philadelphia, 500/I/1/A/#799.

29. December 17, 1855, Du Pont to Kemble and Warner (quote below is from this letter); December 29, 1855, Du Pont to C. I. and A. V. du Pont, 500/I/1/A/#812. See also Welsh report, quoting an article in the December 21 issue of the *Delaware State Journal*.

30. In installing a line during this period before the printing telegraph was popularized in the 1870s, Du Pont took on the challenge of providing telegraph operators who could send and receive Morse code. They initially used a registering Morse telegraph machine, which printed the dots and dashes of Morse code on paper, allowing the receiver to decode messages at his leisure (September 18, 1944, Macklem to Pierre S. du Pont, reproduced in Welsh, "Old Stone Office Building," p. 72). This machine was bulky and the process slow, however, so sometime in the 1880s the registering machine was replaced by a small Morse key, the operators of which had to be skilled enough to receive messages by sound only. To assure that they had the necessary skilled operators, according to Macklem, "The Du Pont Co. made a rule early in the telegraphic days to hire [as clerks] telegraph operators or those who would learn the art for the office."

31. January 2, 1856, Du Pont to Kemble and Warner in New York; January 3, 1856, Du Pont to J. W. Donohue in Cincinnati; January 4, 1856, Du Pont to W. J. Grant in Philadelphia—all in 500/I/1/A/#812.

32. January 2, 1856, Du Pont to F. G. Smith in Philadelphia, 500/I/1/A/#812.

33. J. W. Macklem, "Brandywine Reminiscences," unpublished personal memories of Du Pont, Hagley Museum and Library, p. 13.

34. Macklem, "Brandywine Reminiscences," pp. 12-13.

35. Norman B. Wilkinson, *Lammot du Pont and The American Explosives Industry 1850-1884* (University Press of Virginia, 1984), pp. 67-69; Chandler and Salsbury, *Pierre S. du Pont*, pp. 6, 59. The company had also gotten involved with ventures in coal mining, railroads, and soda lakes, but these ventures went nowhere (Wilkinson, pp. 137-65).

36. Chandler and Salsbury, *Pierre S. du Pont*, p. 60.

37. J. W. Macklem, "Old Black Powder Days," pt. 2, *Du Pont Magazine* 22 (January 1928): 1–2.

38. Wilkinson, *Lammot du Pont*, p. 271.

39. Welsh, "Old Stone Office Building," illustration and caption, p. 58.

40. Ibid.

41. In 1885, the Du Pont Company arranged with the Delaware and Atlantic Telegraph and Telephone Company to have a new phone system installed throughout the Du Pont works (August 19, 1885, Delaware and Atlantic Telegraph and Telephone Co. to F. G. du Pont, 504/A/#17). Only two years later, the Du Pont Company was apparently once again considering changes in its telephone system, as a passage from a letter from E. S. Rice, the company's Chicago agent, indicates: "During my recent visit at your place, in conversation with you, the telephone question was considered and at that time you expressed a wish that Du Pont & Co. might at some time be able to buy and own out-right the instruments in their use" (Rice to F. G. du Pont, January 7, 1887, 504/A/#16). In response to this conversation, Rice did some research on the quality and possible patent controversies of the Cushman telephone and recommended it to Francis G. du Pont. Unfortunately, the correspondence did not indicate whether the company followed up on Rice's lead. Whatever the outcome of Rice's 1887 research, he was involved in considerably more detailed research into new telephone exchanges for the company in 1895, this time looking at systems with two fifty-nine-line switchboards (see January 22, 1895, and February 4, 189[5] [the letter is dated 1896, but internal evidence suggests that the year was typed incorrectly and should be 1895], Rice to F. G. du Pont, both in 504/A/#16).

42. Macklem asserts that "Mr. Henry never used the 'Phone and did not have one in his home" in "Brandywine Reminiscences," p. 14.

43. Macklem, "Old Black Powder Days," pt. 2, p. 1.

44. Macklem, "Brandywine Reminiscences," p. 12.

45. September 18, 1944, Macklem to Pierre S. du Pont, as quoted in Welsh, "Old Stone Office Building," p. 75. Sometime before he left the company in 1880, Lammot du Pont proposed regular monthly meetings with a formal agenda, but there is no evidence that these were ever adopted at the company (Wilkinson, *Lammot du Pont*, p. 64).

46. Rice's first typed letter to the company is dated August 25, 1886, 500/I/1/B/#307.

47. September 3, 1886, Rice to Du Pont Company, 500/I/1/B/#307.

48. January 17, 1888, 504/A/#16.

49. January 17, 1888, 504/A/#16.

50. A series of exchanges with stationery and office equipment companies in 1890 in Francis G.'s own name make it clear that he maintained his own fully equipped office at home. One letter even explicitly pointed out that an equipment order was in his own personal name, not in the company's (February 24, 1890, Francis G. du Pont to Office Specialty Mfg. Co., 504/A/#40).

51. He referred to his typing efforts in many letters, including November 28, 1890, Francis G. du Pont to Alfred I. du Pont, 504/A/#1; May 27, 1890, Francis G. du Pont to F. G. Thomas, acc. 1462, Hagley Museum and Library, hereafter acc. 1462; and January 12, 1889, Francis G. du Pont to Thomas, acc. 1462.

52. January 1, 1889, Francis G. du Pont to F. G. Thomas, acc. 1462.

53. September 17, 1890, and November 20, 1890, Francis G. du Pont to F. G. Thomas, 504/A/#11.

54. Macklem, "Old Black Powder Days," pt. I, p. II.

55. Ibid.

56. General Henry was apparently not willing to take this process further by dictating his letters. The company's first stenographer was hired in 1891, after Eugene du Pont had taken over (Macklem, "Brandywine Reminiscences," p. 16).

57. Macklem has described the old office, including duplicating and filing equipment, in great detail in his various writings. My description, except where otherwise indicated, comes from P. C. Welsh's "The Old Stone Office Building: 1837–1891," drawing principally on Macklem's descriptions. Thus, even though the main series of letterbooks of outgoing correspondence is missing from 1857 on, we know a great deal about how correspondence was handled in the 1880s.

58. June 24, 1936, John W. Macklem to Mr. Kerr, reproduced in "Brandywine Reminiscences," pp. 16–17.

59. The Lower Yard list, dated December 31, 1896 and found in 504/A/#5, is filled in on a duplicated form.

60. See, for example, letters from E. S. Rice to the company in 500/I/1/B/#307–10.

61. Photograph of the second office, S20-11, Hagley Museum and Library.

62. January 1, 1897, Pierre Gentieu to Francis G. du Pont, 504/A/#5.

63. The letters from E. S. Rice cited in note 60, for example, are all together.

64. The line between personal and company paperwork was not always clear. Francis G. du Pont's papers (acc. 504) include some correspondence that is clearly noncompany, but also include his exchanges with E. S. Rice, Pierre du Pont, F. G. Thomas (at the Iowa Powder Mills), and others at Du Pont. All of this correspondence, however, is to and from him personally, rather than to and from the company itself, as personified by Gen. Henry and then Eugene du Pont.

65. An examination of the papers found in 504/A/#1–8 and 10–11 shows that incoming correspondence up to 1885 was always folded and abstracted. Starting with some 1885 correspondence, however, Shannon file holes begin to appear in the tops of incoming letters. Letters in the 1885–88 period had been folded and abstracted before they were placed on the Shannon file. When he bought Shannon files in 1888, he apparently unfolded and placed some back correspondence on them to allow easy reference. Starting with a letter to F. G. Thomas dated December 25, 1888, Francis G.'s correspondence ceased to bear signs of folding and abstracting (504/A/#10).

66. May 15, 1889, Francis G. du Pont to F. G. Thomas, acc. 1462.

67. September 20, 1890, Francis G. du Pont to Office Specialty Mfg. Co., 504/A/#3.

68. February 24, 1890, Francis G. du Pont to Office Specialty Mfg. Co., 504/A/#40.

69. In the files of correspondence from various people or companies in 504/A/#1–4, for example, interfiled press copies appear beginning in 1890. The disappearance of Shannon file holes in 1895 is also evident in these files.

70. Thomas's letters to Francis G. du Pont in 504/A/#10 are occasionally typewritten and sometimes mention his difficulties with the machine. A May 18, 1889, letter mentions his Shannon file (504/A/#11).

71. September 17, 1890, Francis G. du Pont to Thomas, 504/A/#11. In fact, Francis G. du Pont even expressed a desire for Thomas to get a machine identical to his own so that they could send cylinders rather than letters back and forth (November 25, 1890, 504/A/#11). Various compatibility problems, not to mention the inconvenience and expense of shipping cylinders, barred that project.

72. In April 1, 1892, Francis G. du Pont to Rice, Francis G. referred to "yours containing carbon copy of your letter to Mr. Thomas" (504/A/#16).

73. LMSS 10/files 81, 84, and 85, for example.

74. Because the main series of company letter books from 1857 on is lost, we cannot know with certainty whether any additional types of internal communication emanated from the main office that are not represented in the other company and family records. Since internal communication was not kept in the main copy books in the earlier period, however, we are probably safe in drawing tentative conclusions on the basis of the surviving internal documents.

75. See, for example, notes from Charles I. du Pont at the Upper Yard to Francis G. du Pont at Hagley in the late 1880s (504/A/#1).

76. A notation on the 1896 Lower Yard annual report (January 1, 1897, 504/A/#5) indicated that the person filling it out had been making out annual lists since 1853.

77. 504/A/#5.

78. 504/A/#17. The earliest packing list is dated 1884, but a consecutive series starting in 1896 was the first such series to survive. The surviving series of pattern lists begins in 1897.

79. 504/A/#20. In an oral interview, Luther D. Reed (who joined the company in 1903) stated that the match rules were then the only printed rules at the main facility. (Oral interview by Walter J. Hancock, Joseph P. Monigle, Faith K. Pizor, John Scafidi, and Norman B. Wilkinson, on May 28, 29, and June 6, 1968, at Eleutherian Mills Historical Library, now Hagley Museum and Library, where transcript is located.)

80. March 22, 1892, letter from Francis G. to the Steel Works Club announcing the formation of the Brandywine Club (504/A/#2).

81. Chandler and Salsbury, *Pierre S. du Pont*, p. 34.

82. See September 18, 1888, Francis G. du Pont to F. G. Thomas, acc. 1462.

83. January 20, 1897, Francis G. du Pont to Mr. Haley, 504/A/#40.

84. F. G. Thomas's letters to Francis G. du Pont, 504/A/#10.

85. These early records are in acc. 501, series D, #21, Hagley Museum and Library (hereafter 501/D/#21); 504/A/#19; and LMSS 10/#410.

86. Chandler and Salsbury, *Pierre S. du Pont*, p. 35.

87. William S. Dutton, *Du Pont: One Hundred and Forty Years* (New York: Charles Scribner's Sons, 1942), p. 140.

88. The letters from Francis G. du Pont to Thomas up to 1890 are in acc. 1462, then begin to be interfiled with those from Thomas to Francis G. in 504/A/#10–11. The kinship between the two men is revealed in a personal letter from Francis G. to Thomas, July 2, 1890 (504/A/#11), which is signed "Your affectionate cousin."

89. September 17, 1890, Francis G. du Pont to Thomas, 504/A/#11.

90. January 15, 1889, Thomas to Francis G., 504/A/#10.

91. Francis G. wrote to Thomas every 1.9 days (from September 7, 1888, to October 8, 1889, acc. 1462). Thomas wrote to Francis G. every 2.3 days (from September 7, 1888, to October 8, 1889, 504/A/#11).

92. September 17, 1888, Francis G. to Thomas, acc. 1462.

93. By January 1889, Francis G. had requested that Thomas set up a system for classifying expenses so that the amount spent on each building could be ascertained (January 15, 1889, Francis G. to Thomas, acc. 1462). Such a detailed system for classifying expenses was at least a prerequisite of cost accounting and apparently went beyond anything operating in the Brandywine plants. Nevertheless,

it was a special system set up for the construction period, not for regular operations.

94. September 22, 1888, Francis G. to Thomas, acc. 1462; December 30, 1889, Francis G. to Thomas, acc. 1462; December 4, 1890, Francis G. to Thomas, 504/A/#11.

95. September 28, 1888, Thomas to Francis G., 504/A/#10.

96. September 17, 1890, Francis G. to Thomas, 504/A/#11.

97. November 25, 1890, Francis G. to Thomas, 504/A/#11.

98. May 15, 1889, Francis G. to Thomas, acc. 1462. Response May 18, 1889, Thomas to Francis G., 504/A/#11.

99. September 25, 1888, Thomas to Francis G., 504/A/#10.

100. September 17, 1888, Francis G. to Thomas, acc. 1462.

101. September 19, 1888, Thomas to Francis G., 504/A/#10.

102. May 29, 1889, Francis G. to Thomas, acc. 1462; January 7, 1890, Thomas to Francis G., 504/A/#11.

103. See letters from Thomas to Francis G. on October 25, 1888, December 11, 1888, January 2, 1889, all in 504/A/#10; and Francis G. to Thomas on October 30, 1888, and November 13, 1888, both in acc. 1462.

104. May 6, 1890, Francis G. to Thomas, acc. 1462.

105. Although these reports were all to be sent to the company, Francis G. requested that duplicate copies be sent to him, "as I am watching the output very closely, and want to have a record of my own" (May 1, 1890, Francis G. to Thomas, acc. 1462).

106. May 1, 1890, Francis G. to Thomas, acc. 1462.

107. May 24, 1890, Thomas to Francis G., 504/A/#11.

108. Bessie G. du Pont, *E. I. du Pont de Nemours and Company: A History, 1802–1902* (Boston: Houghton Mifflin Co., 1920), p. 140; Chandler and Salsbury, *Pierre S. du Pont*, pp. 71–72.

109. March 1, 1888, contract between Kohl and Warner General Store of Centralia, Illinois, and E. S. Rice, 500/I/1/B/#309.

110. See, for example, the correspondence with E. S. Rice, 500/I/1/B/#307-10.

111. See, for example, "Compendium of Rules in Force May 10th, 1897," 501/D/#17. See Wilkinson, *Lammot du Pont*, chap. 9, and Chandler and Salsbury, *Pierre S. du Pont*, pp. 57–61, for a discussion of this association.

112. Lucius F. Ellsworth, "Strategy and Structure: The Du Pont Company's Sales Organization, 1870-1903," in *Papers Presented at the Annual Business History Conference*, ed. J. van Fenstermaker (Kent, Ohio, 1965), p. 110.

113. March 5, 1888, Rice to Du Pont, 500/I/1/B/#309.

114. See, for example, August 29, 1887, Rice to Du Pont, 500/I/1/B/#308.

115. His correspondence is in 500/I/1/B/#307-10.

116. September 1, 1886, Rice to Du Pont, 500/I/1/B/#307.

117. His reaction to the post-1902 tightening up of the sales organization illustrates how important he considered himself (Chandler and Salsbury, *Pierre S. du Pont*, pp. 72–73).

118. For example, from October 18 to October 24, 1886, he wrote twelve letters to the company, and from July 1 to July 7, 1887, he wrote eleven (500/I/1/B/#307 and #308).

119. In fact, according to Chandler and Salsbury, retrospective analysis performed after the 1902 change in management found that Rice's unit costs were the highest of any major branch (*Pierre S. du Pont*, p. 628, n. 90).

120. In a letter complaining about the Hazard Powder Company's infringe-

ment on his territory (he evidently did not know that Du Pont had secretly bought Hazard in 1876), he closed with this statement of his frustration: "I try honestly to avoid any feeling in these matters, and yet in the face of all that I am brought to contend with and the information submitted by you while at the Brandywine regarding the falling off of my sales, I find it very hard sometimes to write 'Sunday-school' letters when stating these points to you" (September 22, 1888, Rice to Du Pont, 500/I/1/B/#310).

121. These letters are in 504/A/#16.

122. April 28, 1888, Rice to Francis G. du Pont, 504/A/#16.

123. See letters from Rice to Francis G. du Pont, June 23, July 3, July 6, and July 19, 1888, 504/A/#16. The examples come from September 8, 1888, F. G. Thomas to Francis G. du Pont, 504/A/#10.

124. June 23, 1888, 504/A/#16.

125. Thomas to Francis G., September 18, 1888, 504/A/#10, and September 22, 1888, Francis G. to Thomas, acc. 1462. See also September 24, 1888, acc. 1462.

126. July 10, 1889, Francis G. to Thomas, acc. 1462.

127. July 3, 1888, Rice to Francis G. du Pont, 504/A/#16.

128. July 19, 1888, Rice to Francis G. du Pont, 504/A/#16.

129. September 7, 1888, Thomas to Francis G. du Pont, 504/A/#10.

130. July 9, 1888, Rice to Francis G. du Pont, 504/A/#16.

131. July 6, 1888, Rice to Francis G. du Pont, 504/A/#16.

132. Chandler and Salsbury, *Pierre S. du Pont*, pp. 36–38.

133. This story is found in detail in Wilkinson, *Lammot du Pont*, pp. 231–47.

134. Accession 1600, #8–13, Hagley Museum and Library.

135. Wilkinson, *Lammot du Pont*, p. 62.

136. Accession 384, #27, Hagley Museum and Library (hereafter acc. 384/#27), described in Wilkinson, *Lammot du Pont*, p. 174.

137. Wilkinson, *Lammot du Pont*, p. 166.

138. Ibid., p. 172. See also Norman B. Wilkinson, "In Anticipation of Frederick W. Taylor: A Study of Work by Lammot du Pont, 1872," *Technology and Culture* 6 (Spring 1965): 208–21.

139. Accession 384/#32, as described in Wilkinson, *Lammot du Pont*, p. 64.

140. December 1877, "Criticism on the Articles of Co-Partnership of 1858," acc. 384/#29, as cited in Wilkinson, *Lammot du Pont*, p. 237.

141. Dutton, *Du Pont*, p. 144.

142. Wilkinson, *Lammot du Pont*, p. 270.

143. Ellsworth, "Strategy and Structure," p. 111.

144. September 6, 1893, J. Amory Haskell to G. F. Hamlin, Hamilton Barksdale Papers, 500/II/2/#989, book A8.

145. September 7, 1893, Haskell to H. S. Ely, 500/II/2/#989, book A8.

146. See September 6, 1893, Haskell to Hamlin; September 13, 1893, Haskell to Ely, both in 500/II/2/#989, book A8.

147. September 7, 1893, Haskell to J. W. Willard, 500/II/2/#989, book A8.

148. Circular letters are referred to in February 20, 1895, Haskell's assistant (signature not readable) to J. Amory Haskell, 500/II/2/#991.

149. E. S. Rice noted disparagingly that "Repauno's Chicago Agent is unable to make necessary prices, and in fact under his instructions can not meet competition without first communicating with the home office" (August 30, 1888, 580/I/1/B/#309). Although this tight control seemed uncompetitive to Rice, in fact it produced more consistent results than Rice's methods.

150. As noted in May 5, 1896, G. H. Kerr to Lieber Publishing Co., 500/II/2/#992.

151. April 15, 1896, G. H. Kerr to Lieber Publishing Co., 500/II/2/#992.

152. May 5, 1896, G. H. Kerr to Lieber Publishing Co., 500/II/2/#992.

153. Materials on the code are in 500/II/2/#986, file 25–17.

154. Ernest Dale and Charles Meloy, "Hamilton MacFarland Barksdale and the Du Pont Contributions to Systematic Management," *Business History Review* 36 (Summer 1962): 130.

155. June 16, 1896, Haskell to O. R. Jackson, 500/II/2/#992. Unfortunately, none of the reports seem to have survived.

156. In one case, for example, Amory Haskell (then president of Repauno) sent a carbon copy of a letter to another interested party, while press copying the letter in Repauno's own press book (September 8, 1893, Haskell to Edward Green, 500/II/2/#989).

157. In a letter to a Philadelphia stationer about an order of letter paper, Haskell complained that he had "asked you to engrave the word 'subject' and leave a space to be filled in; this you have not done" (December 8, 1893, Haskell to Craig, Finley and Co., 500/II/2/#989).

158. See, for example, the notation between letters dated September 16 and September 17, 1895, 500/II/2/#992.

159. "Letters to various Departments/Buildings," 500/II/2/#1001.

160. See files in 500/II/2/#986.

161. May 18, 1901, Barksdale to Ramsay, 500/II/2/#986, box 25-16, file 57:G:61.

8: Du Pont, 1902–1920

1. Chapters 3 and 4 of Alfred D. Chandler, Jr., and Stephen Salsbury, *Pierre S. du Pont and the Making of the Modern Corporation* (New York: Harper and Row, 1971) describe the events and negotiations leading up to and following this change in leadership and are the main source for the summary in this and the next paragraph. See also William S. Dutton, *Du Pont: One Hundred and Forty Years* (New York: Charles Scribner's Sons, 1942), pp. 169–91; and Ernest Dale, "Du Pont: Pioneer in Systematic Management," *Administrative Science Quarterly* 2 (1957): 43–44.

2. October 13, 1903, T. C. du Pont to A. J. Moxham, 500/II/2/#805.

3. August 3, 1903, Moxham to T. C. du Pont, 500/II/2/#808.

4. The Executive Committee was joined by a Finance Committee, formally a subcommittee of the former.

5. Chandler and Salsbury, *Pierre S. du Pont*, p. 134.

6. Ibid., p. 128.

7. April 3, 1903, Moxham to T. C. du Pont, 500/II/2/#808.

8. Because the Executive Committee minutes are no longer available to researchers, my description of the workings of the committee is based on the account in Chandler and Salsbury (who had access to the minutes) and on surviving correspondence between members of the committee as cited.

9. Chandler and Salsbury, *Pierre S. du Pont*, pp. 135–36. The schedule of meetings must have reverted to one a month sometime during the next five years, because a September 2, 1910, communication to the committee from L. R. Beardslee, secretary to the Executive Committee, announced the splitting of the monthly meeting into two such meetings, one with regular reports and one to

consider appropriations (500/II/2/#205/XES-4). The 1911 and 1914 reorganizations are described in detail in Chandler and Salsbury, chap. 11.

10. Chandler and Salsbury, *Pierre S. du Pont*, p. 132.

11. The issue of adequate filing space near the Executive Committee meeting room and executive offices as the Wilmington office was built, then progressively expanded, is dealt with in the following letters: October 3, 1904, T. C. to Pierre du Pont; October 31, 1905, Raskob to Executive Committee; and March 4, 1907, T. C. du Pont to W. H. Fenn, all in 500/II/2/#808.

12. See, for example, June 13, 1904, T. C. du Pont to Moxham, 500/II/2/#805.

13. Chandler and Salsbury, *Pierre S. du Pont*, p. 133.

14. 1911 Annual Report of the Experimental Station, 500/II/2/#167.

15. Chandler and Salsbury, *Pierre S. du Pont*, p. 133. I have explored Du Pont's use of graphs from this initial rejection to the installation of the Chart Room in 1921 in "Graphs as a Managerial Tool: A Case Study of Du Pont's Use of Graphs in the Early Twentieth Century," *Journal of Business Communication* 22 (Spring 1985): 5–33.

16. Chandler and Salsbury, *Pierre S. du Pont*, p. 133.

17. January 31, 1911, Report of the Committee on Reorganization, 500/II/3/#131. For an example of their work, see the reports attached to a February 14, 1913, letter from Beardslee, secretary to the Executive Committee, to T. C. du Pont, 500/II/3/#131.

18. February 21, 1914, T. C. du Pont to heads of departments, "New Organization—Form of Reports," 500/II/3/#131.

19. February 18, 1913, Executive Committee report to the Board of Directors of E. I. du Pont de Nemours Powder Co. on proceedings from November 20, 1912, to February 18, 1913, 500/II/3/#131. For more on the antitrust suit and the resulting break up of the company, see Chandler and Salsbury, *Pierre S. du Pont*, pp. 259–300.

20. See, for example, May 31, 1904, Moxham to T. C. du Pont; June 8, 1904, Moxham to T. C. du Pont; June 13, 1904, T. C. du Pont to Moxham; September 8, 1904, Moxham to Haskell; and October 13, 1904, T. C. du Pont to Moxham, all in 500/II/2/#805.

21. February 3, 1906, J. A. Haskell to Moxham, 500/II/2/#805.

22. May 31, 1904, Moxham to T. C. du Pont, 500/II/2/#805.

23. June 13, 1904, T. C. du Pont to Moxham, 500/II/2/#805.

24. Of course, documentation for its own sake did not vanish. Alfred du Pont began one letter as follows: "For the reason that my views, as to its merits, are so different from the rest of the Executive Committee, I have decided to express them in letter form; which, owing to the fact that the rest of the members of the Executive Committee approved of the plan and it has been practically adopted, will naturally have no value other than as a record. The future will, undoubtedly, show that my conclusions are correct" (January 27, 1911, Alfred to T. C. du Pont, 500/II/3/#131).

25. Post-1902 materials in the Du Pont collections at Hagley Museum and Library are generally in vertical file folders, with carbon copies interfiled. Some of the sets of files (e.g., Pierre S. du Pont's own files, in LMSS/10) have brads inserted through each manila folder and the enclosed papers, a precaution that would prevent loss of papers and make it possible to store the vertical files on edge like books, but would also make filing and removing papers much more time consuming. The brads may have been added later, however, when the records were not in active use.

26. T. C. du Pont's files are in 500/II/2/#804–16 and in 500/II/3/#123–41. Pierre S. du Pont's files are in LMSS/10. In a 1903 letter, Haskell complained to T. C. du Pont that "communication pertaining to Executive Committee matters [is] being sent to my office in New York," and that consequently "confidential communications are . . . being there opened by the Sales Department force" (September 14, 1903, 500/II/2/#806/18).

27. See, for example, T. C. du Pont's files in 500/II/3/#131. A 1911 circular letter to the High Explosives Operating Department announces that Du Pont's Purchasing Department has set standards for stationery that include green for internal correspondence and onionskin marked "File Copy" in red for copies.

28. February 7, 1914, T. C. to Alfred I. and Pierre S. du Pont, 500/II/3/#131.

29. January 19, 1914, Moxham to T. C. du Pont, 500/II/3/#131.

30. March 14, 1904, Haskell to T. C. du Pont; April 2, 1904, Haskell to T. C. du Pont; and April 4, 1904, T. C. du Pont to Haskell, all in 500/II/2/#806. Also June 8, 1904, Moxham to T. C. du Pont; and June 13, 1904, T. C. du Pont to Moxham, both in 500/II/2/#805.

31. Chandler and Salsbury, *Pierre S. du Pont*, p. 138.

32. Dale, "Pioneer in Systematic Management," pp. 34–40. Details of the early meetings are from Harry Haskell's opening address at HEOD meeting 33, 1911, 500/II/2/#577. Minutes of superintendents' meetings from no. 30 on are in this collection, boxes 566, 570, 573, 577, 581, 585, and 589. Hereafter, references to these minutes will be given only by date and number of meeting.

33. Minutes, HEOD meeting 33, 1911.

34. Program 20, "Superintendents Troubles: Main Office Troubles," meeting 28, November 20–22, 1907, in LMSS/10/A/418/#13. Surviving records of these early meetings up to no. 30 are in this location. Hereafter, references to them will be given by date and number of meeting only. Similarly titled sessions appear in later meetings.

35. Program 13, HEOD meeting 31, October 12, 1909.

36. HEOD meeting 33, April 20, 1911.

37. HEOD meeting 36, 1914.

38. Program 5, HEOD meeting 31, October 1909; and program 11, HEOD meeting 35, 1913.

39. HEOD meeting 33, 1911.

40. Harry Haskell described that committee's formation in the opening address from meeting 33, 1911. A report from the commission is included in each set of meeting minutes.

41. See programs 7a, 7b, and 7c in HEOD meeting 33, 1911.

42. Programs 12 and 19, HEOD meeting 35, 1913, and program 4, HEOD meeting 31, October 1909.

43. See *H.E.O.D. Knocker*, January 28, 1909, Hagley Museum and Library.

44. Numbered bulletins are in 500/II/2/#553–54. A complete set of circular letters beginning in 1907 is in 500/II/2/#550–51. They are reproduced on forms designated with a form number and the date of the form's creation. The date on early ones is 1906, indicating that the forms had existed since at least the previous year, if not earlier. Further references to circulars and bulletins will be to their dates and numbers only.

45. HEOD circular letter 814, October 11, 1911, for example, has a subject line but also still has a letter salutation and closing. Circular 1144, June 15, 1914, which has neither salutation nor closing, is identified by a subject title and the issuer's name at the bottom.

46. See, for example, HEOD circular letter 164, February 6, 1908; and circular letter 352, March 15, 1909.

47. HEOD circular letter 17, May 30, 1907.

48. HEOD circular letter 219, June 8, 1908.

49. HEOD circular letter 354, March 31, 1909.

50. Program 20, HEOD meeting 27, June 19–21, 1907.

51. HEOD circular letter 572, July 8, 1910.

52. HEOD circular letter 730, April 8, 1911.

53. Oral interview with Luther D. Reed by Walter J. Heacock, Joseph P. Monigle, Faith K. Pizor, John Scafidi, and Norman B. Wilkinson on May 28, May 29, and June 6, 1968, transcript in Hagley Museum and Library.

54. HOW is first referred to in the minutes of a 1907 superintendent's meeting. Program 20, HEOD meeting 27, June 19–21, 1907.

55. One version of HOW survives in the imprint collection at the Hagley Museum and Library.

56. HEOD circular letter 1149, July 31, 1914.

57. See, for example, HEOD circular letter 354, March 31, 1909.

58. HEOD meeting 19, August 17, 1906; and meeting 20, September 21, 1906.

59. HEOD circular letter 818, October 28, 1911.

60. Starting from May 16, 1907, when the first numbered circular letter was issued, we find the following number of circular letters: 1907–8, 210; 1908–9, 157; 1909–10, 184; 1910–11, 197; 1911–12, 168; 1912–13, 136; 1913–14, 83; 1914–15, 32; 1915–16, 8; 1916–17, 6.

61. The systematic recording of accidents, for example, began in 1903, even before the consolidation had been completed (HEOD meeting 30, January 1909).

62. For example, one of the first numbered circular letters in the series started in 1907 requested that Physical Progress Reports be submitted in triplicate rather than duplicate (circular letter 7, May 23, 1907).

63. "Operating a High Explosives Plant," HEOD meeting 36, 1914.

64. HEOD meeting 33, 1911.

65. HEOD circular letter 218, June 8, 1908.

66. HEOD circular letter 825, October 26, 1911.

67. Program 15, HEOD meeting 31, October 1909.

68. Program 13, HEOD meeting 31, October 1909.

69. See, for example, HEOD meeting 21 (October 1906), meeting 31 (October 1909), and meeting 32 (1910).

70. *H.E.O.D. Knocker.*

71. Program 15, HEOD meeting 31, October 1909.

72. Haskell pointed out, for example, the reduction in the cost of manufacturing explosives from 1909 to 1910 in his opening address to HEOD meeting 33, 1911.

73. See, for example, HEOD meetings 27 and 28, both in 1907.

74. See, for example, HEOD meeting 29, June 1908; and program 9, HEOD meeting 32, 1910.

75. HEOD circular letter 825, October 26, 1911.

76. Attached to HEOD circular letter 716, February 25, 1911.

77. HEOD circular letter 965, February 21, 1912.

78. "Operating a High Explosives Plant," HEOD meeting 36, 1914.

79. See, for example, the "Summary of Nitrate of Ammonia Reports" discussed in program 6, HEOD meeting 33, 1911.

80. I have explored this issue in more detail in "Graphs as a Managerial Tool."

The number of graphs was computed from HEOD meetings 30–36, 1909–1914 (two meetings in 1909).

81. The minutes of HEOD meeting 31, October 1909, for example, include negative-image photocopies of graphs. The department may have had these copies done externally because the company acquired a Photostat on trial in 1912 and bought one by at least 1914 (January 5, 1912, assistant comptroller to list and July 13, 1914, Ramsay to list, both in 500/II/2/#1005/2).

82. HEOD circular letter 60, August 12, 1907.

83. HEOD meeting 32, 1910.

84. Program 5, HEOD meeting 35, 1913.

85. HEOD meeting 34, 1912.

86. HEOD circular letter 41, July 3, 1907.

87. HEOD circular letter 198, April 10, 1908.

88. HEOD circular letter 834, November 7, 1911.

89. HEOD circular letter 1089, September 16, 1913.

90. Haskell's opening address, HEOD meeting 33, April 20, 1911, mentions the Efficiency Division's establishment. Donald R. Stabile has dated its dismantling to sometime early in 1915 in "The Du Pont Experiments in Scientific Management: Efficiency and Safety, 1911–1919," *Business History Review* 61 (Autumn 1987): 365–86.

91. The earliest form of the study, a letter dated November 17, 1913, is attached to HEOD circular letter 1113, December 8, 1913. The later report, dated December 22, 1913, is in 500/II/2/#1005/2, as is T. C. du Pont's letter to all departments dated Jan 6, 1914. Quotations below are from these documents, as indicated in the text.

92. Irénée du Pont to HEOD Efficiency Division, May 19, 1914, acc. 228, papers of Irénée du Pont, series H, VP files, box 99, file ID-3.

93. Haskell to plants, July 2, 1914; T. C. du Pont to all departments, May 16, 1914, both in 500/II/2/#1005/2.

94. Another indication of the focus on systematizing office activities was a survey of the methods of office workers in the Wilmington office of the HEOD (HEOD circular letter 867, January 18, 1912).

95. March 28, 1911, unsigned copy presumably from Barksdale, 500/II/2/#1005/16. In his recent article ("Du Pont Experiments in Scientific Management"), Donald R. Stabile attributed the company's interest in efficiency to this contact, but the interest, as I have shown in this and the previous chapter, clearly predates this contact.

96. Chandler and Salsbury, *Pierre S. du Pont*, pp. 142–43; Charles F. Rideal, *The History of the E. I. du Pont de Nemours Powder Company* (New York: Business America, 1912), p. 170.

97. February 26, 1909, C. M. Barton to P. S. du Pont, 500/II/2/#205/XES-4.

98. February 10, 1911, Irénée du Pont to C. M. Barton, 500/II/2/#205/XES-4.

99. February 24, 1909, P. S. du Pont to C. M. Barton, 500/II/2/#205/XES-4.

100. February 26, 1909, C. M. Barton to P. S. du Pont, 500/II/2/#205/XES-4.

101. See, for example, April 12, 1913, Reese to Sparre; and April 15, 1913, Sparre to Reese, both in 500/II/2/#118/ES-57B. Under the court decree, when Du Pont split off the Atlas and Hercules Powder Companies certain staff departments such as the Chemical Department were required to continue to serve them in limited capacities for a designated length of time. The results of certain types of routine research, such as ballistics tests, were to be shared with the new com-

panies, but further analysis based on those results was not to be passed on. See Chandler and Salsbury, *Pierre S. du Pont*, pp. 291–300.

102. September 30, 1911, Reese to list of Du Pont executives, 500/II/2/#205/ XES-4. In the same file, see also August 22, 1911, Reese to Barton.

103. April 15, 1913, Sparre to Reese, 500/II/2/#118/ES-57B.

104. April 26, 1917, Bradshaw to Kaighn, Woodbridge, and Calvert, 500/II/2/ #193/BD-41D. Items preceding October 1907 in 500/II/2/#133/ES196-B are stamped, while items after that are punched with holes in the shape of initials to indicate approval.

105. January 31, 1914, Fin Sparre to Ballistic Division, 500/II/2/#193/BD-41-A.

106. The BD files have correspondence dated into 1919. See, for example, March 5, 1919, Henning to Bradshaw (then director of the station), on which is typed "Our File ES-136," a reference to the *Station* file number, but has "BD-41" written on top by hand. Such items abound in the BD-41 files, 500/II/2/#193. In June 1919 a new filing system was established for the entire station, with a research file, a subject file, and an alphabetic file (see June 1, 1919, unaddressed, 500/II/2/#345). At this point the BD files end, suggesting that perhaps the Ballistic Division went over to the new designations.

107. May 24, 1911, Grubb (head of Torpedo and By-Product Division) to list of subordinates, 500/II/3/#142. He announced, "I find that serious delay is occasioned in this office due to my personal inspection of outgoing mail and in order to avoid same and be prompt in closing, have decided to discontinue this practice." Thus, efficiency won out over control. He went on to remind them, however, that the control was still there: "Everyone writing letters will be held strictly accountable for the contents of same and any errors will be dealt with according to the serious nature of same."

108. April 26, 1917, Bradshaw to Kaighn, Woodbridge, and Calvert, 500/II/2/ #193/BD-41D.

109. June 17, 1918, chart, 500/II/2/#118/ES-57.

110. November 17, 1911, Reese to Sparre and Comey, 500/II/2/#205/XES-4; reply dated December 5, 1911, Sparre to Reese, same file.

111. These circular letters are in 500/II/2/#118/ES-57. See, for example, September 13, 1917, Bradshaw to division heads. The station had a stencil duplicator by 1914, but it was not needed for downward communication. Report B79-34-XES-189-1, serial 1905, dated August 25, 1914, 500/II/2/#307/B, is duplicated by a stencil process.

112. June 29, 1917, Bradshaw to division heads, 500/II/2/#118/ES-57.

113. February 26, 1909, [Barton] to P. S. du Pont, 500/II/2/#205/XES-4.

114. 500/II/2/#145/ES-263 contains divisional weekly and monthly reports beginning in April 1909.

115. These reports are all in 500/II/2/#167.

116. See, for example, the 1914 Annual Report, which has much more financial analysis than earlier ones.

117. Two years later, Reese suggested that summaries be added, but the director, Fin Sparre, rejected the idea (April 11, 1913, Reese to Comey and Sparre, and April 30, 1913, Sparre to Reese, both in 500/II/2/#205/XES-4). This exchange referred more to investigative reports than to annual reports, but the principle was the same.

118. January 13, 1914, Henning to Director, 500/II/2/#193/BD-41A; January 2, 1918, Woodbridge to division heads, 500/II/2/#118/ES-57.

119. The filing of these test sheets is discussed in April 28, 1913, and November

26, 1913, Kaighn to Lloyd, 500/II/2/#193/BD-41. See also April 25, 1918, Coxe to list, 500/II/2/#193/BD-41D.

120. April 25, 1918, Coxe to list, 500/II/2/#193/BD-41D.

121. For example, "Stabillite: A Program for Stabillite Tests" laid out a multi-year program of research (November 5, 1907, 500/II/2/#307/B3).

122. See, for example, "Stabillite: Storage Tests with Stabillite: Heat Tests on Samples from Constant Temperature Magazines A," February 14, 1908, 500/II/2/#307/B3-17-ES397/serial 525.

123. See, for example, October 5, 1910, "Investigation of Cost of Yellow Compound," 500/II/2/#307/B18-10-XES-39/serial 1138.

124. The general description that follows is drawn from reports in the "B" files in 500/II/2/#307+.

125. July 21, 1909, Irénée du Pont to C. M. Barton, 500/II/2/#205/XES-4.

126. April 11, 1913, Reese to Comey (director of the Eastern Laboratory) and Sparre, 500/II/2/#205/XES-4.

127. April 30, 1913, Sparre to Reese, 500/II/2/#205/XES-4.

128. July 29, 1918, Bradshaw to list, cover memo to the extended memo by A. P. Tanberg, June 24, 1918, 500/II/2/#118.

129. See, for example, November 25, 1918, "Drying Smokeless Powder, 500/II/2/#308/B-74-24-XES-121, serial 2278.

130. "Memorandum for Ballistic Division," January 31, 1914, 500/II/2/#193/BD-41A. Also in 500/II/2/#118/ES-57.

131. Memoranda all dated March 5, 1919, Henning to Dr. Bradshaw, 500/II/2/#193/BD-41-E.

132. January 26–31, 1918, Henning to Bradshaw and Bradshaw to Henning, 500/II/2/#193/BD-41D.

133. January 15, 1919, circular letter no. 1, Bradshaw to division heads, 500/II/2/#118/ES-57.

134. April 19, 1916, Henning to chemical director, for example, was ultimately intended for Carney's Point (500/II/2/#134). On copying policy, see, for example, September 29, 1915, H. F. Brown (director of Smokeless Powder Operating Department) to Chemical Department asking for more copies (500/II/2/#134/ES-196); or undated (ca. April 1919) document entitled "Miscellaneous Data Regarding Number of Copies of Letters Written to Various Departments" (500/II/2/#193/BD-41E).

135. For example, October 26, 1916, Brown to Porter at Carney's Point, 500/II/2/#134/ES-196.

136. April 19, 1916, Henning to chemical director, 500/II/2/#134.

137. The exceptions, written by one of the ballistics engineers, omitted inside address and salutation and were headed "Memorandum for Mr. S. C. Lloyd," followed by a centered subject heading, a form that foreshadowed some aspects of what was to come. See, for example, November 26, 1913, Kaighn to Lloyd, 500/II/2/#193/BD-41A.

138. January 6, 1914, T. C. du Pont to all departments, included in January 12, 1914, Reese to Comey and Sparre, 500/II/2/118/ES-57.

139. January 13, 1914, Sparre to list, 500/II/2/#118/ES-57.

140. Slight modifications to aid the filer were introduced in September 7, 1917, Bradshaw to division heads, 500/II/2/#118/ES-57; and January 26, 1917, Bradshaw to division heads, 500/II/2/#118/ES-57.

141. June 27, 1919, William Sweetman (no addressee listed), 500/II/2/#118/ES-57.

142. See, for example, July 16, 1917, and October 20, 1918, both Bradshaw to division heads, 500/II/2/#118/ES-57; and September 9, 1918, Coxe to members of Ballistic Division, 500/II/2/#193/BD-41.

143. See, for example, October 28, 1914, Turner to all departments, 500/II/2/#1005/2.

144. "Standard Practice," 1920–1922 edition, Engineering Department, 500/II/2/#III.

145. April 30, 1906, [unreadable signature] to R. H. Dunham, comptroller, LMSS 10/418/2.

146. Chandler and Salsbury, *Pierre S. du Pont*, p. 123.

147. September 17, 1921, Wm. Coyne to J. B. Eliason, acc. 1662/#19.

148. September 27, 1921, Eliason to Coyne, acc. 1662/#19.

149. "Standard Practice," 500/II/2/#III. The quotations in this paragraph are from section IA.

150. In 500/II/3/#127 is an untitled and undated description and justification of the activities of the Sales Record Division which describes the trade reports filed by salesmen.

151. Chandler and Salsbury, *Pierre S. du Pont*, p. 132.

152. Ibid.

153. See ibid., chaps. 6 and 8, for many details of this developing system.

154. May 27, 1911, Reese to Comey and Barton, quoting from the May 23, 1911, meeting of the Appropriations Committee, 500/II/2/#205/XES-4. This letter illustrates the process by which the task of data collection was pushed progressively down the hierarchy.

155. Much has been written about Brown's return on investment formula, a brilliant innovation in financial control. See Alfred D. Chandler, Jr., *Strategy and Structure: Chapters in the History of the American Industrial Enterprise* (Cambridge: MIT Press, 1962), pp. 66–67; Thomas Johnson, "Management Accounting in an Early Multidivisional Organization: General Motors in the 1920s," *Business History Review* 52 (Winter 1978): 508–9; and T. C. Davis, "How the Du Pont Organization Appraises Its Performance," presented at the 1949 annual meeting of the American Management Association, reprinted in *Financial Management Series* no. 94 (New York: AMA, 1950), pp. 3–23.

156. August 23, 1919, Treasurer Brown to heads of departments, acc. 1662/#78. The quotation below is also from this memorandum.

157. Brown's August 23, 1919, memo on the General Statistical Committee refers to this executive series. They are almost surely either the same graphs now labeled "Treasurer's Book Diagrams" in 500/II/3/#181, or very similar ones. The following description is based on these graphs.

158. This discussion is based on Chandler, *Strategy and Structure*, pp. 91–113.

159. I have explored this issue in greater detail in "Graphs as a Managerial Tool." The exact dating of the establishment of the chart room itself is unclear. Probably the executive chart series was developed in 1919, before the change to the multidivisional structure, while the chart room itself was established in 1921 or early 1922.

160. This description is based on several sources: Davis, "How Du Pont Appraises Its Performance"; Lawrence P. Lessing, "The Story of the Greatest Chemical Aggregation in the World: Du Pont," *Fortune* 42 (October 1950): 86–118+; and a March 28, 1983, personal interview with L. T. Alexander, who was a clerk in the Secretary's Office at Du Pont from 1931 to 1934, where he helped prepare materials for the Executive Committee.

161. All information on the Sales Record Division in these two paragraphs is from an untitled and undated document (ca. 1914) describing and justifying its role in the department, 500/II/3/#127.

162. HEOD circular letter 3, May 20, 1907.

163. A statement of the original purpose of the Hall of Records, as stated in an April 11, 1922, circular letter to all departments from Charles Copeland, LMSS 10/418/#5.

164. As quoted in HEOD circular letter no. 475, November 8, 1909.

165. The minutes of these meetings are in 500/II/2/#134/196-1.

166. December 11, 1916, meeting 54.

167. "The Ideal Mail Department," by John C. Slyhoff, ca. July 1919, acc. 1662/#1.

168. March 1, 1912, Torpedo & By-Product Division—Efficiency Meeting no. 1, 500/II/3/#142.

169. A historical sketch of the division was given in a March 23, 1938, address by H. A. Piper, manager of Du Pont's Planning Division, to the National Office Management Association, New York Chapter, acc. 1662/#73.

Conclusion

1. James R. Beniger, *The Control Revolution: Technological and Economic Origins of the Information Society* (Cambridge: Harvard University Press, 1986).

A Note on Archival Sources

I used three archival collections extensively in preparing the three case studies in this book. In addition, I consulted a few other collections at certain points in my research. This note describes these sources.

The Illinois Central Railroad

The Special Collections Department of the Newberry Library in Chicago houses a large collection of records from the Illinois Central Railroad (designated in the Notes by "IC" plus a Newberry Library call number). These records start in the 1850s, at the founding of the railroad. While the collection includes a few items from after the turn of the century, most of the materials date from before 1900. In general, the surviving correspondence and records come from high-ranking individuals or, in a few cases, specific offices. Fortuitously for the researcher, several of the individuals worked their way up from lower positions, and their records include their earlier correspondence and materials. Thus, some records at the divisional level or below have survived.

The most extensive materials are those of Stuyvesant Fish during his vice-presidency and presidency (in IC1/F2.1 and IC1/F2.2). They comprise hundreds of bound volumes of incoming and outgoing correspondence. (Incoming correspondence was pasted into bound volumes of blank pages.) These volumes provide a particularly complete record of the important changes initiated by Fish.

The Newberry Library also houses a collection of records from the Chicago, Burlington, and Quincy Railroad. In a few cases, that collection contained documents related to the Illinois Central Railroad, which I used to supplement the Illinois Central's records (designated by "CB&Q" plus the Newberry Library's designation).

The Scovill Manufacturing Company

The Manuscripts and Archives Department of the Baker Library, Harvard Business School, houses the records of the Scovill Manufacturing Company. This large collection consists of two groups of records (designated Scovill 1 and Scovill 2, plus a volume or case number), starting at the beginning of the nineteenth century and continuing into the first half of the twentieth. Many of the records of the 1840s, 1850s, and 1860s were destroyed by a fire, according to a letter in the collection. The footnotes to this case study also indicate other gaps in company letter books. Most of the surviving internal correspondence from 1914 (when the company abandoned letter books in favor of vertical files) into the early 1920s can be found in the files of three men—William H. Monagan, foreman of the Casting Shop in the Mills Department; F. J. Gorse, foreman of the Store Room; and Statistician E. H. Davis—all preserved in case 26, Scovill 1.

Cases 59 and 60 of Scovill 2 contain some very useful historical materials. In the mid-twentieth century E. H. Davis, the company statistician who played an important role in the later stages of the company's systematization, extended his record-keeping role into historical as well as contemporary records. He compiled lists and collections of old forms, old circular letters, and historical documents (as well as sets of old buttons from Scovill's earliest product lines). The lists may be found in these cases, along with several informal and formal company chronologies and histories, several of which were prepared in conjunction with the company's one hundred fiftieth anniversary celebration in 1952. Most valuable of the unpublished historical materials is a lengthy manuscript history written by Philip W. Bishop shortly before the 1952 anniversary, cited with Bishop's permission. Since Bishop had access to people and documents that no longer exist, this thorough and scholarly study is an exceptionally valuable source on the company.

E. I. du Pont de Nemours and Company

The Hagley Museum and Library in Wilmington, Delaware, houses several extensive collections of Du Pont company and family materials. The largest collection, made up of two major series, is accession 500, the papers of E. I. du Pont de Nemours and Company. Most of the records in series I date from the nineteenth century, while those in series II date from the twentieth. (References to these materials include the accession, series, group, and box or item numbers, as in 500/II/2/#989. Some references also include a second series designation after the group number, as in 500/I/1/A/#799.) Accession 500 includes several subgroups of records originally bearing separate accession numbers, such as the Hamilton Barksdale Papers (originally accession 518 but now part of 500/II/2) and the T. C. du Pont Papers (some of which were originally accession 1075 but are now part of 500/II/3). The consolidation of several sets of records into the two series making up accession 500 has occurred in recent years, so some of my designations will differ from those of earlier scholars studying Du Pont.

Other groups of Du Pont records useful in this study have not been absorbed into accession 500 but retain separate designations. The papers of Pierre S. du Pont are designated as the Longwood Manuscripts. References to these materials take the form of "LMSS" followed by a group, series, and box or item label. These include old family and company papers that he inherited as well as his own personal and business files. Accessions 504 and 501 contain the personal and laboratory papers of Francis Guerney du Pont; accession 1462, Francis G. Thomas's correspondence from the Iowa Powder Mills; accessions 1600 and 384, Lammot du Pont's papers; accession 228, Irénée du Pont's papers; and accession 1662, papers from the Office of the President (principally under Lammot du Pont, Jr.), 1914–48.

In addition to these collections, Hagley holds several useful articles and internal reports otherwise difficult or impossible to obtain: J. W. Macklem, "Old Black Powder Days," parts 1 and 2, *Du Pont Magazine* 21 (1927): 11, 46; and 22 (1928): 1–3; J. W. Macklem, "Brandywine Reminiscences," unpublished personal memories of Du Pont; the *H.E.O.D. Knocker*, a humorous newsletter printed for one meeting of the High Explosives Operating Department, January 28, 1909; transcription of an oral interview with Luther D. Reed by Walter J. Heacock, Joseph P. Monigle, Faith K. Pizor, John Scafidi, and Norman B. Wilkinson on May 28, May 29, and June 6, 1968; HOW, a Du Pont Company employee manual; and P. C. Welsh, "The Old Stone Office Building: 1837–1891," internal Hagley report.

Other Archival Materials Consulted

Although the collections discussed above were my major archival sources, I also consulted a few other collections for subjects covered in the first three chapters of the book. Stephen Salsbury's discussion of certain Western Railroad reports in *The State, the Investor, and the Railroads* (Cambridge: Harvard University Press, 1967) prompted me to consult the Western Railroad Collection in the Baker Library, Harvard Business School, for the full reports. In my research on communication technology for Chapter 2, I found several archival collections useful. The Hagley Library has an extensive collection of trade catalogues from the nineteenth and early twentieth centuries. These provided useful information about what was available at different times, as well as serving as the source for the majority of my illustrations in that chapter. I found a few other such catalogues in the Baker Library of Harvard Business School. The records of the President's Commission on Economy and Efficiency (called the Taft Commission) in the National Archives, Record Group 51, provided invaluable descriptions of office methods and technologies in the second decade of the twentieth century, as well as more trade catalogues. Advertising circulars of the Edison Electric Pen were available at the Edison National Historic Site in Menlo Park, New Jersey.

Finally, I drew heavily on some published periodicals of the late nineteenth and early twentieth centuries which are not strictly archival, but which are no longer widely available. For the railroads, specialized periodicals such as *Railway Age* and *American Railroad Journal* were quite useful. The *Transactions of the American Society of Mechanical Engineers* published early articles by proponents of systematic and scientific management philosophies. In the early twentieth century, several new managerial periodicals appeared to deal specifically with such issues. Of these, *System*, *Industrial Management*, and *Factory* were particularly valuable. Harvard Business School's Baker Library has a good collection of such old periodicals, as does the library system of the Massachusetts Institute of Technology.

Index

A. B. Dick Company, 53, 286n.86
Accounting, 2, 77, 78, 160, 161; appliances to aid, 64; in Du Pont, 206, 263, 264, 265; financial, 8; in Illinois Central, 109-10, 111, 124-25, 135-36, 140-42; in Scovill, 164-65, 168. *See also* Cost accounting
Ackerman, William K., 124, 126, 129
Adding machine, 64
Amberg File and Index Company, 34, 63. *See also* Cameron, Amberg, and Co.
American Railroad Journal, 7, 292n.85, 195n.33
American Railway Association, 153
Aniline dye: for mass duplicating, 50, 53; for press copying, 27-28, 34, 45-46
Annual reports. *See* Reports: annual
Armour & Company, 24-25
Arnold, Horace Lucian, 12, 18

Baldwin, Henry F., 230
Baltimore and Ohio Railroad, 59, 92-93
Barksdale, Hamilton, 225-27, 230, 233, 235-37, 244-46, 274
Bell, Alexander Graham, 21
Bell Telephone Company, 68
Beniger, James R., xvi, 274
Bernstein, Lester, 92, 93
Bishop, Philip W., 164, 171
Blouin, Francis X., Jr., 8
Bradshaw, Hamilton, 259, 260, 262
Brandes, Stuart D., 16
Brinton, Willard C., 85-91, 189
Brown, F. Donaldson, 264-66
Bruen, Mr., 138, 139
Bulletins, 66, 70, 72, 239, 240-41, 243, 244. *See also* Circular letters
Burlington Railroad, 6
Business communication. *See* Business English; *see also under* Communication
Business English, 65, 92, 94, 95, 97

Calhoun, John B., 102
Cameron, Amberg, and Co., 34. *See also* Amberg File and Index Company
Carbon copying, 22, 45, 46-50; cost of, 49; and downward communication, 70; in Du Pont, 214, 224, 250, 254, 256,

261, 320n.25; and filing, 56; in Illinois Central, 131, 156; in Scovill, 172, 180, 188; Taft Commission on, 48-49
Carney's Point, Du Pont plant at, 208, 216, 218, 269
Chandler, Alfred D., Jr., xv, xviii, 1, 7, 8, 24, 125, 231, 263, 264
Chicago, St. Louis, and New Orleans Railroad, 117, 127
Church, Alexander Hamilton, 11, 12, 13
Circular letters, 65, 66-71, 72; duplicating technology and, 52; in Du Pont, 224, 225, 239-41, 256, 262; in Illinois Central, 106, 107-8, 119-22, 150, 152; military models for, 68
Civil War, 114, 124
Clarke, J. C., 110, 111-12, 115, 123, 127-28, 129-30, 131, 135, 155
Committees: managerial, xvii, 18-19, 20, 98-100; shop, 17, 19, 20. *See also* Meetings, managerial
Communication
—with agents, 3; in Du Pont, 205-6, 207, 219-21; in Scovill, 160, 161-62
—downward, xvii, 2, 5-6, 50, 53-54, 66-77; in Du Pont, 215-16, 239-44, 256, 262; in Illinois Central, 104-8, 119-23, 153; in Scovill, 168, 172-76, 192-200
—internal: ad hoc, 23; centralized, 254-56; in factories, 3-4, 66, 68; formal, xvi-xvii, 5, 20, 45, 49, 61-63, 98, 201, 277n.2; lateral, xvii, 2, 15, 184-86; oral, 2-4, 65, 68, 100, 202-3, 209; vertical, 15; written, 12, 22, 25, 39, 61-63, 65, 68, 160-62, 164, 171-72. *See also* Genre; Reports
—upward, xvii, 2, 7, 13-14; in Du Pont, 205, 214; in Illinois Central, 123; in Scovill, 172, 176-80
Communication system, internal, 6, 10, 12, 271, 277n.2; in Du Pont, 201, 230, 235, 253, 254, 261-62, 269-70; in Illinois Central, 101, 116, 131; in Repauno, 227; in Scovill, 159, 163, 168, 180; technology and, 21-22, 25, 45
Communication technology, xviii, 21-64, 206, 270, 271-72; defined, 21; for written communication, 25, 39, 131. *See also* Carbon copying; Dictating

CONTROL THROUGH COMMUNICATION

Designed by Joyce Kachergis

Composed by BG Composition, Inc.
in Trump Mediaeval

Printed by Edwards Brothers
On 50-lb. Glatfelter Natural